Synchrotron Radiation and Free-Electron Lasers

Principles of Coherent X-Ray Generation

Learn about the latest advances in high-brightness X-ray physics and technology with this authoritative text. Drawing upon the most recent theoretical developments, pre-eminent leaders in the field guide you through the fundamental principles and techniques of high-brightness X-ray generation from both synchrotron and free-electron laser sources. A wide range of topics is covered, including high-brightness synchrotron radiation from undulators, self-amplified spontaneous emission, seeded high-gain amplifiers with harmonic generation, ultra-short pulses, tapering for higher power, free-electron laser oscillators, and X-ray oscillator and amplifier configuration. Novel mathematical approaches and numerous figures accompanied by intuitive explanations enable easy understanding of key concepts, while practical considerations of performance improving techniques and discussion of recent experimental results provide the tools and knowledge needed to address current research problems in the field.

This is a comprehensive resource for graduate students, researchers, and practitioners who design, manage, or use X-ray facilities.

Kwang-Je Kim is a Distinguished Fellow at the Argonne National Laboratory, Illinois, USA, and a Professor of Physics at the University of Chicago. He is also a Fellow of the American Physical Society.

Zhirong Huang is an Associate Professor at the SLAC National Accelerator Laboratory and in the Department of Applied Physics at Stanford University, and a Fellow of the American Physical Society.

Ryan Lindberg is a Staff Scientist at the Argonne National Laboratory, where he works on free-electron laser physics, and on analyzing and predicting collective effects in light-source storage rings.

Synchrotron Radiation and Free-Electron Lasers

Principles of Coherent X-Ray Generation

KWANG-JE KIM
Argonne National Laboratory, Illinois

ZHIRONG HUANG
SLAC National Accelerator Laboratory, California

RYAN LINDBERG
Argonne National Laboratory, Illinois

CAMBRIDGE
UNIVERSITY PRESS

CAMBRIDGE
UNIVERSITY PRESS

University Printing House, Cambridge CB2 8BS, United Kingdom

One Liberty Plaza, 20th Floor, New York, NY 10006, USA

477 Williamstown Road, Port Melbourne, VIC 3207, Australia

4843/24, 2nd Floor, Ansari Road, Daryaganj, Delhi – 110002, India

79 Anson Road, #06–04/06, Singapore 079906

Cambridge University Press is part of the University of Cambridge.

It furthers the University's mission by disseminating knowledge in the pursuit of education, learning, and research at the highest international levels of excellence.

www.cambridge.org
Information on this title: www.cambridge.org/9781107162617
DOI: 10.1017/9781316677377

First published 2017

Printed in the United Kingdom by Clays, St Ives plc

A catalogue record for this publication is available from the British Library.

ISBN 978-1-107-16261-7 Hardback

Contents

Preface

X-rays produced when highly relativistic electrons are accelerated along a curved trajectory, generally referred to as synchrotron radiation, have served as an important tool for studying the structure and dynamics of various atomic and molecular systems. The first dedicated synchrotron radiation facility was built in the 1970s using an electron storage ring, and since that time the demand for synchrotron radiation has steadily increased due to its high intensity, narrow angular opening, and broad spectral coverage.[1] Over the past few decades the effectiveness of synchrotron radiation has been further advanced by improvements in storage ring design that led to an increase in the electron beam phase space density, and by the use of magnetic devices such as undulators that dramatically increase the X-ray brightness over traditional bending magnets. These developments have widened and deepened the reach of "photon sciences" around the globe.

Another revolutionary advance in X-ray generation was made with the development of X-ray free-electron lasers (FELs). The radiation produced in an FEL acts back on the electron beam in a positive feedback loop, resulting in X-rays with dramatically improved intensity and coherence over those produced with storage-ring based sources. The X-ray FEL became feasible thanks to improvements in linear accelerator technology in general, and in particular to advances in the injector (electron source).

High-brightness, high-energy electron beams from a linear accelerator can now drive a high-gain X-ray FEL amplifier in a long undulator. The gain can be so high that the initially incoherent undulator radiation evolves to an intense, quasi-coherent field known as self-amplified spontaneous emission (SASE). The SASE pulse can be made ultrashort by using an ultrashort electron bunch. With X-ray FELs, experimental techniques developed for traditional synchrotron light sources can be made much more efficient, and new areas of material, chemistry, and biology research, such as ultrafast dynamics, have become accessible to study.

The advent of X-ray FELs, however, has by no means rendered other synchrotron radiation sources obsolete. For certain applications, including those that require high levels of stability or high levels of average flux, synchrotron radiation from storage rings can be more attractive than SASE. Future synchrotron radiation facilities can provide even brighter X-rays than those that are feasible with the current, "third"-generation

[1] The name "synchrotron radiation" came about because it was first observed in a synchrotron in 1947. However, the oft-used practice of referring to the storage ring as a "synchrotron source" should be avoided, since the source is a storage ring in which the electron beam is in a steady state, rather than a synchrotron in which electrons go through an acceleration cycle.

sources by improving the storage ring design; for example, "multi-bend achromats" can significantly improve the electron beam brightness and hence the radiation brightness. Also, advances in X-ray FEL capabilities are underway in several directions, including the production of fully-coherent soft X-rays via harmonic generation, improvements in hard X-ray coherence via the self-seeding technique, and the tailoring of X-ray pulse characteristics for each user including multi-color and/or multi-pulse X-ray production, among others. An X-ray FEL oscillator (XFELO) that provides full coherence and high spectral purity also appears to be feasible in the hard X-ray region by using an X-ray cavity in which Bragg crystals serve as the main mirrors. A grand X-ray facility could be envisaged in which the XFELO output serves as input for a high-gain amplifier, providing ultimate capabilities for future photon sciences.

The physics of X-ray production in synchrotron radiation sources and free-electron lasers is therefore of significant contemporary interest. We attempt to provide a unified, coherent account of X-ray generation in this book, even though there are several excellent references already devoted to these topics, some of which we list at the end of this book. While overlaps in exposition are unavoidable, we hope that the perspective and approach offered here can help readers develop a more complete understanding of coherent X-ray generation, and lay the foundation for potentially new innovations. We outline and highlight our philosophy and approach as follows.

First, our view is that synchrotron radiation and FELs should be treated as a unified subject, particularly in the X-ray spectral region. This is because the phenomenon of FEL feedback is always present in an undulator, although in many cases it is too weak to have an appreciable effect on the X-ray radiation properties.

Second, we have emphasized the importance of the phase space distribution for both particle and radiation beams. The phase space distribution of particle beams is familiar from accelerator physics, while that associated with an electromagnetic field can be defined following Wigner's construction. We are then able to identify the phase space density, or the brightness distribution, of synchrotron radiation in a logical manner. This places the electron and radiation beams on more equal footing, and can be used to answer some practically important questions including when are we allowed to neglect the effect of the electron beam phase space on the generated X-rays.

Third, we retain the discrete nature of the electrons by representing the electrons' phase space distribution as a sum of delta functions that indicates the position and momentum of each constituent electron. Such a distribution is known as the Klimontovich distribution function, and in general it contains all the classical information of interest. We then write the Klimontovich distribution as a sum of two parts: the first part describes the smooth, ensemble-averaged distribution, while the second contains both the fast oscillations due to the electromagnetic interaction as well as the fluctuations encapsulated in the discrete sum over delta functions. To render the problem soluble, we regard the second part as a small perturbation about the smooth equilibrium. If we neglect the transverse dimensions, a complete solution to the initial value problem can be found by employing the Laplace transform technique. To account for radiation diffraction and the transverse betatron oscillations of electrons requires a more sophisticated approach. In the low-gain case, the solution can be obtained by the method of integrating along the characteristics, while a formal solution to the high-gain case can be written in terms of the Van Kampen modes, and the resulting equations can be solved numerically once the smooth initial distribution is known.

The practical implementation of coherent X-ray production via FELs is a fast moving frontier. We have therefore not attempted to give an exhaustive account of the wide variety of proposed techniques. Nevertheless, we have provided discussions of several practical topics, including the effects of undulator errors and wakefields, along with several methods to produce coherent, soft X-ray output via nonlinear harmonic generation of an initially longer wavelength laser. We also present a short overview of currently existing and planned high-gain amplifier facilities. An FEL oscillator for hard X-rays making use of Bragg reflectors is discussed in some detail as a possible way to produce fully coherent, hard X-ray beams of high spectral purity.

This book began as a set of lecture notes for a course at the U.S. Particle Accelerator School, and has been gradually maturing as the course was repeated every two to three years over the last fifteen years. Feedback from students has played an important role in improving the notes. Suggestions and encouragements from our colleagues, too many to list here, were also essential for this book to become a reality. The construction of an X-ray FEL theory, and the application of that theory to the design and interpretation of subsequent FEL experiments and X-ray facilities, have been one of the most exciting and successful beam dynamics activities in recent years. It is our great pleasure to share some of these developments with students and with our colleagues, and to acknowledge and thank their contributions.

Kwang-Je Kim, Argonne, Illinois
Zhirong Huang, Stanford, California
Ryan Lindberg, Argonne, Illinois

Conventions and Notation

Throughout this book we use SI units, with the notable exception that we usually quote energies in eV. We use standard boldface to denote vectors (e.g., $\boldsymbol{x}$) and sans-serif fonts for matrices (M). Our definition of the Fourier transform follows the standard conventions in classical physics, so that the Fourier transform in space and time of the function $f(x,t)$ is defined as

$$f(\omega, x) = f_\omega(x) = \int_{-\infty}^{\infty} dt\; e^{i\omega t} f(x,t)$$

$$f(k,t) = \int_{-\infty}^{\infty} dx\; e^{-ikx} f(x,t).$$

Note that we usually write the frequency argument of the Fourier transform as a subscript. In addition, we typically do not include the limits of integration if they are over the entire line, so that the inverse transforms would appear as

$$f(x,t) = \frac{1}{2\pi}\int d\omega\; e^{-i\omega t} f_\omega(x)$$

$$f(x,t) = \frac{1}{2\pi}\int dk\; e^{ikx} f(k,t).$$

Finally, as is inevitable in a book like this, we introduce a lot of mathematical symbols to represent physical quantities. Rather than list every symbol introduced, here we include a table of only those symbols that appear across different sections and chapters of the book. We note that we also at times employ scaled versions of these variables that we denote with hats; for example, the scaled propagation distance along the undulator would be $\hat{z}$.

Symbol	Physical meaning/description
a_ν	Scaled electric field at dimensionless frequency ν
α	Fine structure constant $\alpha \equiv e^2/(4\pi\epsilon_0\hbar c) \approx 1/137$
$\alpha_{x,y}$	Twiss parameter for the correlation $\alpha_x = -\langle xx'\rangle/\varepsilon_x$
$\mathcal{B}$	Radiation brightness (or Wigner) function
B_0	Undulator peak magnetic field on axis
$\beta_{x,y}$	Transverse beta function $\beta_x = \langle x^2\rangle/\varepsilon_x$

$\bar{\beta}_x$	Average transverse beta function
β_n	Natural beta function determined by the undulator focusing
c	Speed of light in vacuum
$\Delta\nu$	Relative frequency detuning $\Delta\nu = \nu - 1 = (\omega - \omega_1)/\omega_1$
e	Magnitude of the electron charge
E_x	Transverse electric field
E	Slowly-varying transverse electric field amplitude
E_ν	Fourier component of the transverse electric field
$\mathcal{E}(\phi)$	Angular representation of the transverse electric field
η	Relative energy deviation from resonance $(\gamma - \gamma_r)/\gamma_r$
ϵ_0	Vacuum permittivity
ε or $\varepsilon_{x,y}$	Electron beam transverse emittance $\varepsilon_x = \langle x^2\rangle\langle x'^2\rangle - \langle xx'\rangle^2$
$\varepsilon_{x,n}$	Normalized transverse emittance of the electron beam ε/γ
ε_r	Radiation emittance $\varepsilon_r \geq \lambda/4\pi$
f or F	Electron phase space distribution function
f_ν	Fourier component of the distribution function
γ	Electron energy (in units of mc^2)
γ_r	Resonant electron energy (in units of mc^2)
γ_0	Initial/reference electron energy (in units of mc^2)
h	Odd harmonic order $h = 1, 3, 5, \ldots$
$\hbar$	Planck's constant over 2π
I	Peak electron bunch current
I_A	Alfvén current ≈ 17 kA
J_x	Transverse current
$\mathcal{J}_{x,y}$	Transverse particle action
J_n	Bessel function of order n $(n = 0, 1, 2, \ldots)$
[JJ]	Undulator Bessel function factor at the fundamental $[\mathrm{JJ}]_1 = [\mathrm{JJ}]$
$[\mathrm{JJ}]_h$	Undulator Bessel function factor at harmonic h
k_1	Fundamental radiation wavenumber ω_1/c
k_β	Average betatron focusing wavenumber $= 1/\bar{\beta}_x$
k_u	Undulator wavenumber
K	Undulator deflection parameter $K = eB_0/mck_u$
L_{G0}	1D FEL power gain length of a monoenergetic beam
L_G	3D FEL power gain length
L_u	Undulator length
λ_1	Fundamental FEL wavelength
λ_h	FEL wavelength at harmonic h
λ_u	Undulator period
m	Electron rest mass
M	Total number of independent modes in a radiation pulse
$M_{T,L}$	Number of transverse or longitudinal modes, respectively
μ	Scaled complex growth rate of the linear FEL
μ_3	Growth rate of the exponentially growing mode in 1D
$\mu_{\ell m}$	Growth rate of transverse mode with radial order ℓ and azimuthal order m

Symbol	Physical meaning/description
n_e	Electron volume density
N_e	Total number of electrons in a bunch
ω_1	Fundamental undulator radiation frequency
ν	Ratio of the radiation frequency ω to the fundamental ω_1
$\boldsymbol{p}$	Electron angle from the axis $\boldsymbol{p} = (x', y')$
P	Radiation power
P_{beam}	Electron beam power $(I/e)\gamma mc^2$
P_{sat}	FEL saturation power
$\boldsymbol{\phi}$	Radiation angle with respect to the optical axis
ρ	FEL Pierce parameter
σ_η	RMS relative energy spread of the electron beam
σ_r, $\sigma_{r'}$	RMS transverse size, divergence of the radiation
σ_ω	RMS radiation bandwidth
σ_x, $\sigma_{x'}$	RMS transverse size, divergence of the electron beam
$t_j(z)$	Electron's arrival time at the undulator location z
t_j	Shorthand for the electron's arrival time at $z = 0$, $t_j = t_j(0)$
$\bar{t}_j(z)$	Electron's arrival time averaged over an undulator period
T	Flat-top electron bunch duration
t_{coh}	Radiation temporal coherence time
θ	Electron's phase relative to the radiation wave
U	Total radiation energy
$\boldsymbol{v}$	Electron's transverse velocity
v_z	Electron's longitudinal velocity
$\bar{v}_z$	Electron's average longitudinal velocity in a planar undulator
$\boldsymbol{x}$	Electron's horizontal and vertical position (x, y)
$\boldsymbol{x}'$	Electron's horizontal and vertical angle (x', y')
z	Propagation distance from the undulator beginning
z_{sat}	FEL saturation distance
Z_R	Rayleigh length of the radiation

1 Preliminary Concepts

In this chapter we look at some physics of particle and radiation beams in the paraxial approximation. A *paraxial* beam is one that is well-collimated along its propagation direction, which we take to be the z axis. For a relativistic electron beam to be paraxial means that the angle between $\hat{z}$ and the velocity vector of a typical electron is small, so that $v_z \gg |\boldsymbol{v}_\perp|$ and $v_z \approx c$. In the case of electromagnetic radiation, a paraxial beam is one that can be characterized by a small angular divergence, which is equivalent to saying that the angle between the optical axis $\hat{z}$ and a typical ray is small. The similarity between particle trajectories and optical rays runs even deeper than is suggested here, and we will see several other similar characteristics between paraxial particle and radiation beams in what follows.

In the first section we cover certain essential points regarding paraxial particle beams. We start by introducing the relativistic electron phase space, and continue with a very brief description of electron beam transport and linear particle optics. This discussion will be self-contained but rather incomplete, covering only those beam properties and dynamics that are required for subsequent study of X-ray generation from relativistic beams; interested readers can consult any of the numerous texts on accelerator physics, some of which we list in the References. We conclude this section with a brief introduction to the particle distribution function on phase space, which will be essential to the treatment of FEL dynamics in Chapters 3, 5, and 6.

The second section introduces paraxial wave optics, starting with a treatment of diffraction and certain geometrical optics that will parallel the particle beam physics from the previous section. We then place these similarities on more formal grounds by introducing phase space methods for wave optics, in which we define a quasi-distribution function for paraxial radiation that is analogous to the particle distribution function, with subtle but important differences that arise due to the wave nature of light. We conclude this section by discussing temporal coherence and its attendant intensity enhancement that is one of the distinctive features of laser light.

1.1 Particle (Electron) Beams

Our primary goal is to study radiation from highly relativistic electrons, by which we mean electrons whose speed $|\boldsymbol{v}|$ approaches the speed of light in vacuum. To get a handle on the approximate orders of magnitude involved, we first consider the electron energy $U_e = \gamma mc^2$, where γ is the relativistic Lorentz factor, m is the electron mass, and c

is the speed of light in vacuum. Thus, the Lorentz factor γ is the ratio of the electron energy to its rest mass energy which, in terms of commonly used units, is given by

$$\gamma = \frac{U_e}{mc^2} = \frac{U_e[\text{GeV}]}{0.511 \times 10^{-3}} = 1957\, U_e[\text{GeV}]. \tag{1.1}$$

Typical FEL and synchrotron radiation sources use electron beams with energies from one to tens of GeV, so that $\gamma \sim 10^3$ to 10^4. Additionally, if we define $\boldsymbol{\beta} \equiv \boldsymbol{v}/c$ to be the electron velocity scaled by the velocity of light c, the scaled speed $\beta \equiv |\boldsymbol{v}|/c$ is related to the Lorentz factor by

$$\beta = \sqrt{1 - \frac{1}{\gamma^2}} \approx 1 - \frac{1}{2\gamma^2}, \tag{1.2}$$

and

$$1 - \beta \approx 5 \times 10^{-8} \text{ for } \gamma mc^2 = 1.5 \text{ GeV}. \tag{1.3}$$

Thus, the electron speed approaches that of light to within a factor less than or of order 10^{-7}. While it might therefore be tempting to approximately set $\beta \to 1$ from the beginning, we will see that certain essential radiation physics result from the fact that the electron speed is less than the speed of light c; mathematically, these effects typically appear as terms $\sim (1-\beta)^{-1}$. Therefore, we will have to take care to set $\beta \to 1$ only once we are sure that it is safe to do so.

1.1.1 Electron Beam Phase Space

The accelerating phase of modern radio-frequency (rf) accelerators typically accommodates tens to thousands of pico-Coulombs of charge, or 10^7 to 10^{10} individual electrons. We will refer to this collection of electrons as an electron beam or bunch, so that an electron bunch is composed of $N_e \gg 1$ electrons whose relativistic velocity is primarily directed along $\hat{z}$. To describe this beam, it is convenient to adopt z, the propagation distance along a reference trajectory, as the independent variable or evolution parameter. This convention matches that of usual accelerator physics in the straight sections we study. We denote derivatives with respect to z with a prime; for example,

$$x' \equiv \frac{dx}{dz} = \frac{dx/dt}{dz/dt} = \frac{1}{v_z}\frac{dx}{dt}. \tag{1.4}$$

Additionally, we will find that the physics of X-ray production are such that the transverse motion is largely independent of the longitudinal degrees of freedom. Hence, we employ the notation

$$\boldsymbol{x} \equiv (x, y) \qquad\qquad \boldsymbol{x}' \equiv (x', y')$$

for the transverse coordinates. For the relativistic paraxial beams that we consider, x', $y' \ll 1$, and

$$|\boldsymbol{x}'| = \sqrt{x'^2 + y'^2} \approx \frac{1}{c}\sqrt{v_x^2 + v_y^2} \ll 1 \tag{1.5}$$

is the angle between $\hat{z}$ and the electron's velocity vector.

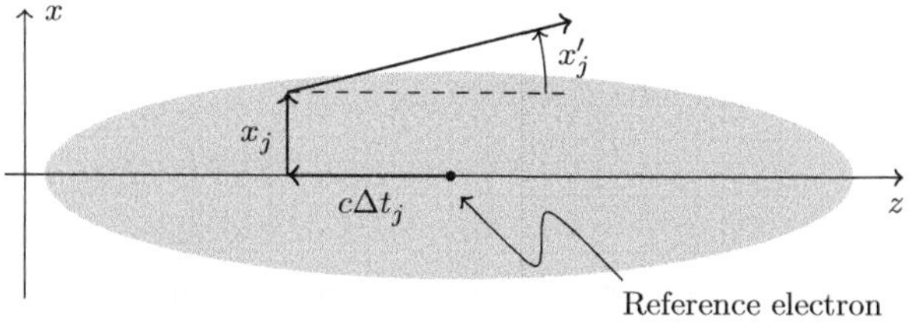

Figure 1.1 Schematic representation of the transverse and longitudinal phase space variables of electrons in a bunch.

To fully describe the electron dynamics requires six independent phase space coordinates for every electron (three "positions" and three "momenta"), each of which is a function of the independent variable z. We further divide this phase space into transverse and longitudinal (or temporal) degrees of freedom. The transverse phase space variables are $(\boldsymbol{x}_j, \boldsymbol{x}'_j)$, where $\boldsymbol{x}_j = (x_j, y_j)$ are the transverse coordinates of the jth electron relative to the reference trajectory, while $\boldsymbol{x}'_j = (x'_j, y'_j)$ denote the transverse angles or divergence with respect to the z axis, and $j = 1, 2, \ldots, N_e$. The longitudinal/temporal phase space variables are given by $(\Delta t_j, \Delta\gamma_j)$, where Δt_j is the time that the electron arrives at the transverse plane located at z relative to the nominal electron bunch time, and the energy deviation $\Delta\gamma_j \equiv \gamma_j - \gamma_0$ is defined with respect to the central beam energy $\gamma_0 mc^2$. We show the meaning of these coordinates in Figure 1.1. Note in particular that an electron with $\Delta t < 0$ arrives at a particular z before the reference electron; in many cases it is more natural to use a longitudinal coordinate proportional to $-\Delta t$, since $(x, y, -\Delta t)$ form a right-hand coordinate system moving with the beam.[1]

1.1.2 Beam Transport and Linear Optics

The transport and control of particle beams is the purview of accelerator physics, and here we will only take a cursory look at certain transformations of transverse phase space for paraxial beams; more complete discussions of the physics of charged particle transport can be found in, e.g., [1, 2]. To simplify matters, we use the fact that the coordinates $\boldsymbol{x}$ and $\boldsymbol{x}'$ are small, and consider only those forces that are linear in $(\boldsymbol{x}, \boldsymbol{x}')$. In this case, the coordinates at the output plane $(\boldsymbol{x}, \boldsymbol{x}')_{\text{out}}$ can be written as linear combinations of the initial coordinates $(\boldsymbol{x}, \boldsymbol{x}')_{\text{in}}$ at the input plane. Thus, there exists a transformation matrix M that relates the two; for example, if the motion along x is decoupled from that along y, we have

$$\begin{bmatrix} x \\ x' \end{bmatrix}_{\text{out}} = \mathsf{M}_{z_{\text{in}} \to z_{\text{out}}} \begin{bmatrix} x \\ x' \end{bmatrix}_{\text{in}}. \tag{1.6}$$

Furthermore, we will be specifically interested in straight transport sections that are composed of drift spaces where the beam propagates essentially in vacuum, interrupted by quadrupole magnets that provide transverse focusing. For free space propagation over a length ℓ, we have the transformation

[1] Furthermore, in the Hamiltonian formulation $(-ct, mc\gamma)$ are canonical position–momentum conjugates; see Appendix A.

$$(x,x') \to (x+\ell x',x'), \tag{1.7}$$

which in matrix form is written as

$$\begin{bmatrix} x \\ x' \end{bmatrix}_{\text{out}} = \begin{bmatrix} 1 & \ell \\ 0 & 1 \end{bmatrix} \begin{bmatrix} x \\ x' \end{bmatrix}_{\text{in}} \equiv \mathsf{M}_\ell \begin{bmatrix} x \\ x' \end{bmatrix}_{\text{in}}. \tag{1.8}$$

A magnetic quadrupole acts like a focusing element which, under the thin-lens approximation, gives an electron a sudden change in angle (or "kick") that is proportional to its displacement from the axis. In the two planes we have the transformation rules

$$(x,x') \to (x,x'-x/f) \qquad\qquad (y,y') \to (y,y'+y/f). \tag{1.9}$$

We see that if the quadrupole focuses the beam in the x direction, it is defocusing along y. The 4×4 matrix representation of (1.9) is

$$\begin{bmatrix} x \\ x' \\ y \\ y' \end{bmatrix}_{\text{out}} = \mathsf{M}_f \begin{bmatrix} x \\ x' \\ y \\ y' \end{bmatrix}_{\text{in}} = \begin{bmatrix} 1 & 0 & 0 & 0 \\ -1/f & 1 & 0 & 0 \\ 0 & 0 & 1 & 0 \\ 0 & 0 & 1/f & 1 \end{bmatrix} \begin{bmatrix} x \\ x' \\ y \\ y' \end{bmatrix}_{\text{in}}, \tag{1.10}$$

where we note that this matrix can be decomposed into two independent 2×2 blocks, one for each transverse direction.

In general, the total transport matrix of a beam line is obtained via the matrix multiplications of all the individual elements in the appropriate order. For N elements numbered sequentially $n = 1, 2, \ldots, N$ (so that the beam first passes through element 1, than 2, etc.), the total transformation is given by

$$\mathsf{M} = \mathsf{M}_N \, \mathsf{M}_{N-1} \, \ldots \, \mathsf{M}_2 \, \mathsf{M}_1. \tag{1.11}$$

To conclude this section, we note that an equivalent and complementary description of linear beam transport can be formulated in terms of solutions to the linear differential equation

$$x'' + K(z)x = 0, \tag{1.12}$$

where $K(z)$ is an arbitrary function of z whose form depends upon the linear element under consideration. For example,

$$K(z) = \begin{cases} 0 & \text{for free space propagation} \\ +\dfrac{\delta(z)}{f} & \text{for a focusing thin lens.} \end{cases} \tag{1.13}$$

It is easy to see that this formulation is equivalent to (1.7) and (1.9).

1.1.3 Beam Emittance and Envelope Functions

Tracking the millions to trillions of individual particle orbits in an electron beam requires significant computational resources. Often we are not even interested in the individual particle trajectories, but rather wish to know the overall "beam" properties that we can easily measure. We express these beam properties as averages (or moments) of products of the transverse coordinates over the beam. If we orient our coordinate axes such that the average motion is along $\hat{z}$, the first-order moments vanish:

$$\langle x \rangle \equiv \frac{1}{N_e} \sum_j x_j = 0 = \langle x' \rangle \equiv \frac{1}{N_e} \sum_j x'_j. \tag{1.14}$$

In this case, the simplest beam properties are the second-order moments. Restricting ourselves to one transverse dimension for simplicity, the second-order beam moments have the following physical interpretation:

The squared RMS beam size: $$\sigma_x^2(z) = \left\langle x^2 \right\rangle = \frac{1}{N_e} \sum_j x_j^2. \tag{1.15}$$

The squared RMS beam angular divergence: $$\sigma_{x'}^2(z) = \left\langle x'^2 \right\rangle = \frac{1}{N_e} \sum_j x_j'^2. \tag{1.16}$$

The RMS beam correlation: $$\left\langle x x' \right\rangle = \frac{1}{N_e} \sum_j x_j x'_j. \tag{1.17}$$

We further illustrate some example beam distributions using ellipses in phase space in Figure 1.2, showing the RMS size, divergence, and correlation. These ellipses are particularly convenient because they remain ellipses under linear transport, with phase space area proportional to the *geometric emittance* (or simply the emittance) defined by

$$\varepsilon_x \equiv \sqrt{\langle x^2 \rangle \langle x'^2 \rangle - \langle x x' \rangle^2}. \tag{1.18}$$

Since phase space area is conserved, the emittance (1.18) is an invariant of the linear transport lattice.[2]

The geometric emittance quantifies the phase space area occupied by an electron beam, with (1.18) specifying how much one must increase the RMS size to decrease the divergence and vice versa. As such, ε_x gives an invariant measure of the beam quality

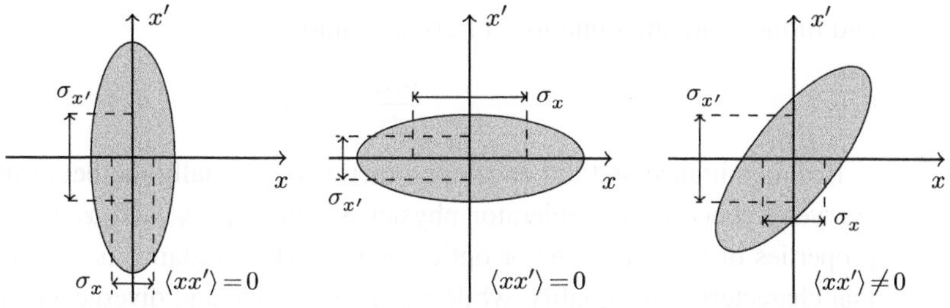

Figure 1.2 Representative phase space ellipses, each of which have the same emittance ε_x. Note that either the beam size or divergence is minimized when the correlation $\langle xx' \rangle$ vanishes.

[2] While every Hamiltonian system conserves phase space area, nonlinear forces do not generically map initial ellipses to final ellipses in phase space. For this reason, ε_x as defined in (1.18) is only invariant for linear forces. Additionally, generalizations to (1.18) are required if the linear forces couple motion in the x direction to that along y (or t).

which serves as an important metric for how the electron beam couples to the radiation. Since the FEL interaction typically puts a minimum requirement on the beam emittance, we might also like to know what beam quality we require from the injector at the beginning of the accelerator. However, acceleration will change the emittance ε_x even if the transverse forces are linear. This is because our transverse coordinates $(\boldsymbol{x}, \boldsymbol{x}')$ are not canonically conjugate variables when γ varies, so that the usual area conservation theorems of Hamiltonian mechanics do not apply in the space spanned by $(\boldsymbol{x}, \boldsymbol{x}')$. If we instead adopt the Hamiltonian position and momentum pair $(\boldsymbol{x}, \boldsymbol{p} = mc\beta_z\gamma\boldsymbol{x}')$, we can construct the *normalized emittance* $\varepsilon_{x,n}$ in a manner analogous to (1.18), with the property that $\varepsilon_{x,n}$ ideally remains invariant from the injector through the accelerating stages to the final insertion device. The value of the normalized emittance is related to the usual geometric emittance via

$$\varepsilon_{x,n} = \beta_z\gamma\varepsilon_x \approx \gamma\varepsilon_x. \tag{1.19}$$

The normalized emittance $\varepsilon_{x,n}$ is conserved through a linear system including acceleration, and is therefore often used as an important criterion for the beam produced by an electron gun.

Returning to the particle beam phase space with its transverse moments and geometric emittance ε_x, we introduce certain electron beam envelope functions that conveniently describe the beam properties. These envelope functions, which are referred to as the Courant–Snyder or Twiss parameters in the accelerator community, are usually defined as

$$\beta_x = \frac{\langle x^2 \rangle}{\varepsilon_x} \qquad \gamma_x = \frac{\langle x'^2 \rangle}{\varepsilon_x} \qquad \alpha_x = -\frac{\langle xx' \rangle}{\varepsilon_x}. \tag{1.20}$$

Although we typically reserve β for the normalized electron velocity and γ for its relativistic Lorentz factor, the notation (1.20) has been universally accepted and so we adopt it too; we will try to be clear when we are discussing the envelope functions. In view of the definition of the emittance (1.18), we have

$$\beta_x\gamma_x - \alpha_x^2 = 1, \tag{1.21}$$

and furthermore the Equations (1.20) also imply that

$$\frac{d\beta_x}{dz} = -2\alpha_x. \tag{1.22}$$

Beam emittance and the envelope functions, especially the beta function β_x, are the everyday concern of accelerator physicists, with β_x, γ_x, and α_x typically regarded as properties of the beam line or optical lattice. The emittance is a property of the beam that characterizes its quality, while the physical size and divergence are determined by both the beam and the lattice properties.

1.1.4 Beam Properties under Simple Transport

Now, we investigate the beam envelope functions and the RMS moments in two simple but physically important situations. First, we consider electron transport through free space, after which we study beam evolution in a focusing lattice comprised of periodically spaced quadrupoles.

Free Space Transport

We calculate the free space transport of a beam over a distance z using the matrix (1.8) with $\ell = z$, writing

$$\begin{bmatrix} x \\ x' \end{bmatrix}_{\text{out}} = \begin{bmatrix} 1 & z \\ 0 & 1 \end{bmatrix} \begin{bmatrix} x \\ x' \end{bmatrix}_{\text{in}}. \tag{1.23}$$

Thus, the evolution of the RMS beam size is given by

$$\left\langle x_{\text{out}}^2 \right\rangle = \left\langle \left(x_{\text{in}} + z x'_{\text{in}}\right)^2 \right\rangle = \left\langle x_{\text{in}}^2 \right\rangle + 2z \left\langle x_{\text{in}} x'_{\text{in}} \right\rangle + z^2 \left\langle x'^2_{\text{in}} \right\rangle, \tag{1.24}$$

while the divergence is constant, $\sigma_{x'}(z) = \sigma_{x'}(0)$. If we assume that the initial correlation is zero (i.e., that $\langle x_{\text{in}} x'_{\text{in}} \rangle = 0$) and apply the definitions (1.20), we obtain

$$\beta_x(z) = \beta_x(0) + z^2 \gamma_x(0). \tag{1.25}$$

Since we have assumed that there is initially no correlation, $\alpha_x(0) = 0$ and the identity (1.21) implies that $\gamma_x(0) = 1/\beta_x(0)$. Thus, we have

$$\beta_x(z) = Z_\beta + \frac{z^2}{Z_\beta}, \tag{1.26}$$

where the focusing parameter[3] $Z_\beta = \beta_x(0)$ is determined by the focusing conditions outside the drift space. The RMS beam size is given by

$$\sigma_x(z) = \sqrt{\varepsilon_x \beta_x(z)} = \sqrt{\varepsilon_x \left(Z_\beta + \frac{z^2}{Z_\beta} \right)}. \tag{1.27}$$

The RMS size is a minimum at $z = 0$ when the correlation vanishes; this location is known as the beam waist. At small longitudinal distances the beam expands quadratically with z, while the beam size increases linearly if $z \gg Z_\beta$. This beam spreading occurs because different particles travel at different angles even though each individual particle trajectory is a straight line, as shown in Figure 1.3.

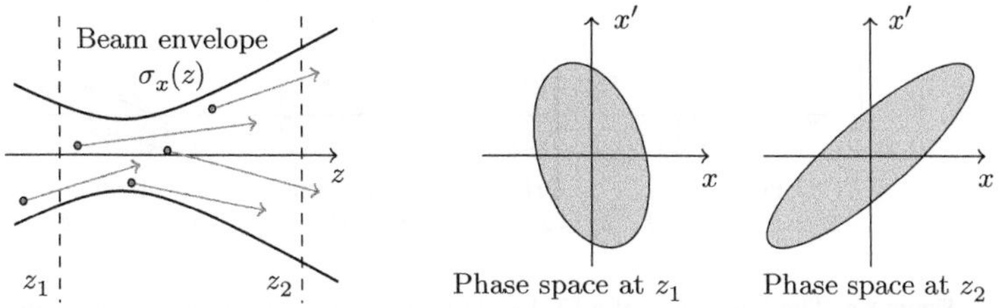

Figure 1.3 Illustration of straight electron trajectories giving rise to a diverging beam envelope. On the left are several electron trajectories and the free-space beam size in the x-z plane. On the right we draw the phase space ellipses in the (x, x') plane at position z_1 and z_2.

[3] Particle collider physicists typically use β_x^* to denote our Z_β.

Alternating Gradient Focusing Lattice

We now consider a simple focusing lattice consisting of a periodic series of quadrupoles and drift spaces. Since a quadrupole only focuses in one plane at a time (i.e., either x or y), one typically achieves focusing in both planes by using a lattice whose elementary cell consists of two quadrupoles separated by drift sections as shown in Figure 1.4. The first quad focuses the beam in the x direction while it defocuses in y, and the second quad is defocusing in x while focusing in y. The net effect of the focusing-drift–defocusing-drift (FODO) lattice is to confine the beam in both directions, which can be seen from the phase space diagrams in Figure 1.4.

For high-gain FELs, one typically wants small variations in the beam size (or β_x-function) while minimizing the angular divergence (the γ_x-function). This can be done by using a particular FODO lattice whose phase space transformations are shown pictorially in Figure 1.4. Note how the projection on the x axis has small variations, while the projection on the x' axis is small at all times. Physically, this focusing lattice uses lenses whose focal lengths are much longer than the drift space distance, with $|f| \gg \ell$.

The FODO lattice can be simply analyzed if we consider the input and output planes to be "halfway" through the first focusing quad shown in Figure 1.4. In this case, the correlation $\langle xx' \rangle$ is zero at the input and output planes as indicated in the figure. Since the focal length of one-half a lens is twice the focal length of the full lens (as can be seen through a matrix multiplication), the matrix transformation of one cell is given by

$$\mathsf{M}_{\mathrm{FODO}} = \begin{bmatrix} 1 & 0 \\ -1/2f & 1 \end{bmatrix} \begin{bmatrix} 1 & \ell \\ 0 & 1 \end{bmatrix} \begin{bmatrix} 1 & 0 \\ 1/f & 1 \end{bmatrix} \begin{bmatrix} 1 & \ell \\ 0 & 1 \end{bmatrix} \begin{bmatrix} 1 & 0 \\ -1/2f & 1 \end{bmatrix}$$
$$= \begin{bmatrix} 1 - \frac{\ell^2}{2f^2} & 2\ell\left(1 + \frac{\ell}{2f}\right) \\ -\frac{\ell}{2f^2}\left(1 - \frac{\ell}{2f}\right) & 1 - \frac{\ell^2}{2f^2} \end{bmatrix}. \tag{1.28}$$

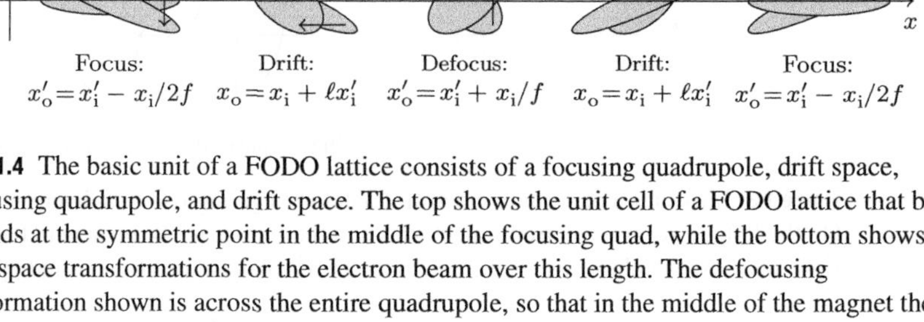

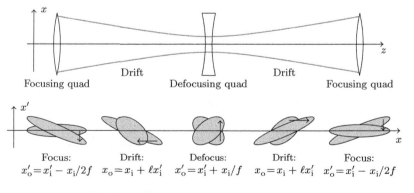

Figure 1.4 The basic unit of a FODO lattice consists of a focusing quadrupole, drift space, defocusing quadrupole, and drift space. The top shows the unit cell of a FODO lattice that begins and ends at the symmetric point in the middle of the focusing quad, while the bottom shows the phase space transformations for the electron beam over this length. The defocusing transformation shown is across the entire quadrupole, so that in the middle of the magnet the beam is uncorrelated in xx'.

For periodic motion we have $\beta_x(0) = \beta_x(2\ell)$ and $\gamma_x(0) = \gamma_x(2\ell)$, while vanishing correlation α_x at the two planes implies that $\beta_x(0) = 1/\gamma_x(0)$ from (1.21). We compute the beam size in a manner similar to that done for free space propagation in (1.24); solving the resulting quadratic equation for $\beta_x(0)$, we find that

$$\beta_x(0) = 2\sqrt{\frac{2f^3 + f^2\ell}{2f - \ell}} \approx 2\,|f|\left(1 + \frac{\ell}{2f}\right) \tag{1.29}$$

for $f \gg \ell$. If we had set the input and output planes at consecutive defocusing quadrupoles the calculation would have been identical but with $f \to -f$, so that

$$\beta_x(\ell) \approx 2\,|f|\left(1 - \frac{\ell}{2f}\right). \tag{1.30}$$

Between the quadrupoles the beta-function is an approximately linear function of z, so that from Equation (1.22) we have $d\beta_x/dz = -2\alpha_x \approx -2$ in the first half-period, while $d\beta_x/dz = -2\alpha_x \approx +2$ in the second half-period. Nevertheless, in the limit $|f| \gg \ell$ the envelope functions β_x, γ_x, and α_x are approximately constant throughout the lattice. We could calculate the average γ_x-function in a manner similar to what we did for β_x, but it is simpler to apply the identity

$$\gamma_x = \frac{1 + \alpha_x^2}{\beta_x} \approx \frac{1}{|f|}, \tag{1.31}$$

where the final result follows from $\alpha_x \approx \pm 1$ and the formulas for β_x above.

Now, we can write down the average values of the lattice functions for the FODO lattice in the limit $|f| \gg \ell$. Including the corresponding RMS beam size, divergence, and correlation, we have

$$\beta_x(z) \approx \bar{\beta}_x = 2f \quad \rightarrow \quad \langle x^2 \rangle \approx 2\varepsilon_x f \tag{1.32}$$

$$\gamma_x(z) \approx \frac{2}{\bar{\beta}_x} = \frac{1}{f} \quad \rightarrow \quad \langle x'^2 \rangle \approx \frac{\varepsilon_x}{f} \tag{1.33}$$

$$\alpha_x^2(z) \approx \bar{\beta}_x \bar{\gamma}_x - 1 = 1 \quad \rightarrow \quad \langle xx' \rangle \approx \pm\varepsilon_x. \tag{1.34}$$

Note how the envelope functions depend on only the lattice parameters ℓ and f (in this approximation, they depend only on the focal strength f), while the physical size, divergence, and correlation also depend on the beam quality through the emittance ε_x.

1.1.5 Electron Distribution Function on Phase Space

We close our discussion of electron beams with a brief introduction to the electron distribution function in phase space. In general, we take the electron distribution function to be a nonnegative function F of the phase space coordinates, whose value is proportional to the number of electrons per unit phase space volume. For example, in the previous section the ellipses of Figures 1.2 and 1.3 could be thought of as representing the distribution function in 4D transverse phase space, with the shaded areas proportional to the local density of electrons. In this simple idealization, the transverse distribution function is uniform inside the ellipse and zero outside.

A complete, classical description of the N_e electrons in the bunch can be obtained by using the Klimontovich distribution function, which represents each point-like electron by a delta-function centered about its trajectory in 6D phase space:

$$F(\Delta t, \Delta\gamma, \boldsymbol{x}, \boldsymbol{x}'; z) = \frac{1}{N_e}\sum_{j=1}^{N_e} \delta[\Delta t - \Delta t_j(z)]\delta[\Delta\gamma - \Delta\gamma_j(z)] \\ \times \delta[\boldsymbol{x} - \boldsymbol{x}_j(z)]\delta[\boldsymbol{x}' - \boldsymbol{x}_j'(z)]. \tag{1.35}$$

We choose the normalization so that integrating F over all the coordinates is unity. Furthermore,

$$N_e F(\Delta t, \Delta\gamma, \boldsymbol{x}, \boldsymbol{x}'; z)\, d\boldsymbol{x} d\boldsymbol{x}' d(\Delta t) d(\Delta\gamma) \tag{1.36}$$

is the number of particles per unit phase space volume whose coordinates are in the 6-cube defined by the sides $(\boldsymbol{x}, \boldsymbol{x} + d\boldsymbol{x})$, $(\boldsymbol{x}', \boldsymbol{x}' + d\boldsymbol{x}')$, etc.

Conservative (Hamiltonian) dynamics transports the value of the distribution function along single particle trajectories. In this case F behaves as an incompressible fluid in phase space that satisfies the Liouville equation

$$\frac{d}{dz}F = \left[\frac{\partial}{\partial z} + (\Delta t)'\frac{\partial}{\partial \Delta t} + (\Delta\gamma)'\frac{\partial}{\partial \Delta\gamma} + \boldsymbol{x}' \cdot \frac{\partial}{\partial \boldsymbol{x}} + \boldsymbol{x}'' \cdot \frac{\partial}{\partial \boldsymbol{x}'}\right] F = 0. \tag{1.37}$$

Although the Klimontovich distribution function is a complete description and will be an important tool for analysis in subsequent chapters, it frequently contains more information than is physically required. Often, we can approximate the distribution as a smooth phase space fluid whose value is proportional to the local density. As we mentioned in the beginning of the section, the constant-valued density contained in the phase space ellipse is one such simplification (these types of constant distributions are often referred to as "waterbag" distributions). Another convenient distribution function that can be handled analytically is a Gaussian function. In fact, if the only things we know about F are the first- and second-order moments (the former being zero, the latter being given by the envelope functions and ε_x), than a Gaussian F is the distribution of maximum entropy, and hence is the best we can do with the given information.[4]

As mentioned previously, for X-ray generation the transverse electron dynamics typically does not depend on the longitudinal physics, so that we can consider the transverse distribution function independently of the longitudinal distribution. Subsequent chapters mainly focus on the longitudinal dynamics relevant to radiation generation and the FEL interaction, while here we introduce some general concepts that, in keeping with our prior discussion, we apply to the transverse part of F. We introduce the Gaussian phase space distribution in one transverse dimension for simplicity, as extension to 2D is trivial. Assuming a Gaussian transverse distribution function whose second-order moments are consistent with the envelope functions (1.20) implies that

$$F(x, x'; z) = \frac{1}{2\pi\varepsilon_x}\exp\left\{-\frac{1}{2\varepsilon_x}\left[\gamma_x(z)x^2 + \beta_x(z)x'^2 + 2\alpha_x(z)xx'\right]\right\} \tag{1.38}$$

[4] If higher-order moments of F are known, than an appropriate distribution consistent with this data should be chosen.

as can be seen by direct computation. For free space, we have $\beta_x(z) = Z_\beta + z^2/Z_\beta$, $\gamma_x(z) = 1/Z_\beta$, and $\alpha_x(z) = -z/Z_\beta$, so that

$$F(x,x';z) = \frac{1}{2\pi\varepsilon_x}\exp\left[-\frac{(x-x'z)^2}{2Z_\beta\varepsilon_x} - \frac{x'^2}{2\varepsilon_x/Z_\beta}\right]. \tag{1.39}$$

This expression can be derived in another manner that is independent of the functional form of F, relying only on the fact that F obeys the Liouville equation (1.37); because of its generality, this method will prove to be useful in later computations, and we review it here. The Liouville equation implies that the value of F is transported along particle trajectories, which are the characteristic curves of the partial differential equation (1.37). This means that if at some initial position s the distribution function is given by $F(\boldsymbol{x},\boldsymbol{x}';s)$, at position z along the beam axis we have

$$F(\boldsymbol{x},\boldsymbol{x}';z) = F[\boldsymbol{x}_0(\boldsymbol{x},\boldsymbol{x}',z;s),\boldsymbol{x}_0'(\boldsymbol{x},\boldsymbol{x}',z;s);s], \tag{1.40}$$

where $(\boldsymbol{x},\boldsymbol{x}')$ are the new coordinates at z that have evolved from the initial coordinates $(\boldsymbol{x}_0,\boldsymbol{x}_0')$ at $z = s$. We show pictorially the meaning of Equation (1.40) in Figure 1.5. Panel (a) shows how the distribution function is transported along the single particle trajectory from the phase space plane located at s to that at z. Our notation $\boldsymbol{x}_0(\boldsymbol{x},\boldsymbol{x}',z;s)$ in Equation (1.40) conveys the fact that the initial coordinate $\boldsymbol{x}_0$ is considered to be a function of the final coordinates $(\boldsymbol{x},\boldsymbol{x}')$, the final position z, and the starting position s, with $\boldsymbol{x}_0(\boldsymbol{x},\boldsymbol{x}',s;s) = \boldsymbol{x}$.

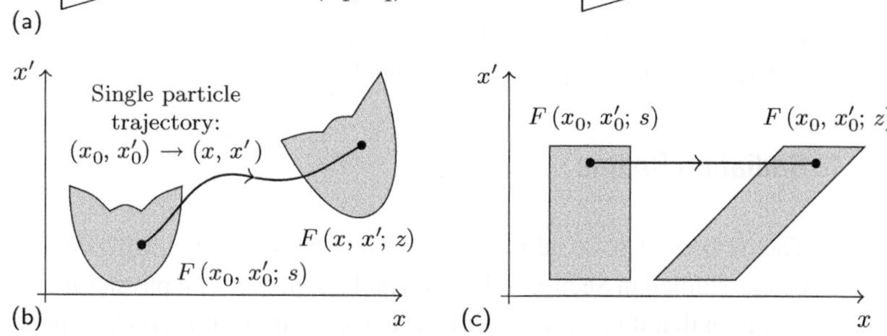

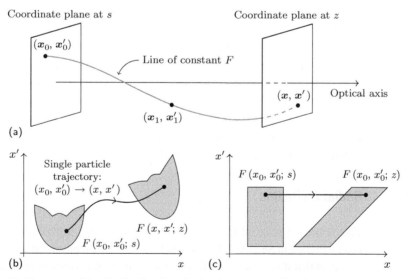

Figure 1.5 Transport of the distribution function in phase space. The value of F is transported from the plane s to that at located at z along the single particle trajectories as shown in (a). The initial coordinates $(\boldsymbol{x}_0,\boldsymbol{x}_0')$ can be considered as functions of the final coordinates $(\boldsymbol{x},\boldsymbol{x}')$ and the elapsed distance $z-s$. Panel (b) shows the same basic physics in the (x,x') plane, with the value of F being transported such that the phase space volume of the enclosed shaded region is conserved by the flow. (c) shows the evolution for free space propagation, for which the enclosed region is simply sheared along x'.

Figure 1.5(b) shows the same physics in the phase space plane, where the two shaded regions are enclosed by phase space contours at locations s and z. The Hamiltonian flow transports F along the single particle trajectory such that the shaded area of the figure at s equals that at z. For the simple case of free-space propagation in one transverse dimension, we have

$$x = x_0 + x_0'(z - s) \qquad x' = x_0', \tag{1.41}$$

which we depict in Figure 1.5(c). This clearly shows that free-space propagation results in a shearing of the phase space along x by an amount proportional to x'. We apply this to the Gaussian beam by considering it to come to a waist at $z = 0$,

$$F(x, x'; z = 0) = \frac{1}{2\pi\varepsilon_x}\exp\left(-\frac{x_0^2}{2\sigma_x^2} - \frac{x_0'^2}{2\sigma_{x'}^2}\right) \tag{1.42}$$

with $\sigma_x^2 = \varepsilon_x\beta_x(0) = \varepsilon_x Z_\beta = Z_\beta^2\sigma_{x'}^2$. Solving (1.41) for the initial coordinates (x_0, x_0'), we obtain

$$x_0(x, x', z; s) = x - x'(z - s) \qquad x_0'(x, x', z; s) = x'. \tag{1.43}$$

Setting $s = 0$ and inserting these expressions for the initial coordinates into the initial distribution function (1.42) yields $F(x, x'; z)$ for arbitrary z; the result is identical to (1.39). Finally, we note that the distribution in physical space can be obtained by marginalizing F over the angles. Integrating F with respect to x' yields

$$\int dx'\, F(x, x'; z) = \frac{\exp\left[-\frac{x^2}{2\sigma_x^2\left(1+z^2/Z_\beta^2\right)}\right]}{\sqrt{2\pi}\,\sigma_x\sqrt{1 + z^2/Z_\beta^2}}. \tag{1.44}$$

It is evident from (1.44) that the transverse width $\sigma_x(z) = \sqrt{\varepsilon_x(Z_\beta + z^2/Z_\beta)}$, in accordance with the previously derived (1.27).

1.2 Radiation Beams

The physics of paraxial radiation shares many qualities with that of relativistic particle beams studied in Section 1.1. Nevertheless, in classical physics the electromagnetic field differs in that it is a wave. For this reason, at times it is convenient to describe the longitudinal profile in terms of its Fourier transform or spectrum. An important characteristic of laser-like light is its central energy, frequency, or wavelength, which are related by $U_\gamma = \hbar\omega = 2\pi\hbar c/\lambda$, with $\hbar$ Planck's constant divided by 2π. In terms of practical units, we have

$$U_\gamma[\text{keV}] = \frac{12.4}{\lambda[\text{Å}]}. \tag{1.45}$$

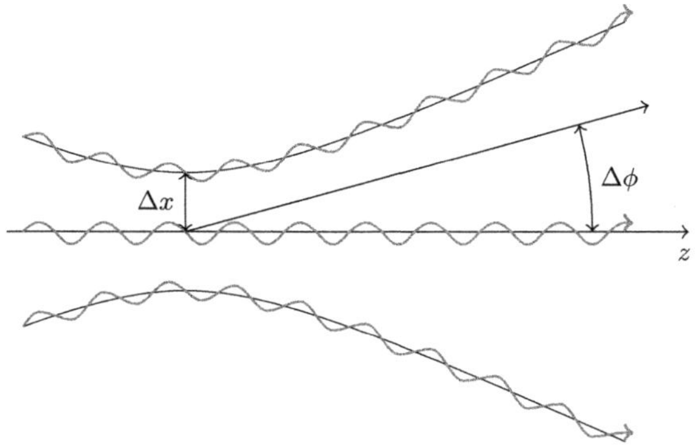

Figure 1.6 Diffraction of a transversely confined radiation beam.

Visible light, for example, spans the wavelength range between 3800 and 7600 Å, while ultraviolet (UV) and infrared (IR) lie just below and above this range in λ, respectively. We will be most concerned with X-ray generation, where X-rays are typically defined to have characteristic energies $\gtrsim$ 100 eV, meaning that $\lambda \lesssim$ 100 Å. X-rays are also further subdivided into soft X-rays that roughly cover the range 100Å $\lesssim \lambda \lesssim$ 5Å, and hard X-rays that have wavelengths shorter than 5 Å.

1.2.1 Diffraction of Paraxial Beams

If coherent electromagnetic radiation of wavelength λ is focused to a transverse spot of size Δx, the beam will be diffracted with an angle $\Delta\phi \sim \lambda/\Delta x$, as shown in Figure 1.6. This can be understood as a consequence of the Fourier "uncertainty relations" $\Delta x \Delta k \sim \Delta x \Delta\phi/\lambda \sim 1$. Since in the paraxial approximation we assume that the angular spread $\Delta\phi \ll 1$, our subsequent developments strictly hold only if the transverse size is much larger than the wavelength, $\Delta x \gg \lambda$. Nevertheless, experience has shown that for most purposes the paraxial approximation gives reliable results if $\Delta x > \lambda$.

To investigate paraxial wave propagation and diffraction, we first write down Maxwell's wave equation for the electric field in vacuum:

$$\left[\frac{\partial^2}{\partial t^2} - c^2\frac{\partial^2}{\partial z^2} - c^2\frac{\partial^2}{\partial \boldsymbol{x}^2}\right] E(\boldsymbol{x}, z, t) = 0. \tag{1.46}$$

The paraxial solutions to (1.46) are easy to identify in Fourier space with respect to the time t and space $\boldsymbol{x}$. We begin with the former, and consider a time-harmonic component of the electric field in the form $E_\omega(\boldsymbol{x}; z)e^{-i\omega t}$, where ω is the angular frequency, and we assume that the transverse plane $\boldsymbol{x}$ is perpendicular to the propagation direction (optical axis) z. The wave equation then becomes

$$\left[\frac{\partial^2}{\partial z^2} + \frac{\partial^2}{\partial \boldsymbol{x}^2} + k^2\right] E_\omega(\boldsymbol{x}; z) = 0, \quad k = \frac{\omega}{c} = \frac{2\pi}{\lambda}. \tag{1.47}$$

Now, we introduce the angular representation of the electric field $\mathcal{E}_\omega(\boldsymbol{\phi}; z)$ which is defined in terms of the transverse Fourier transform,

$$\mathcal{E}_\omega(\boldsymbol{\phi}; z) = \frac{1}{\lambda^2} \int d\boldsymbol{x}\, e^{-ik\boldsymbol{\phi}\cdot\boldsymbol{x}} E_\omega(\boldsymbol{x}; z) \tag{1.48}$$

$$E_\omega(\boldsymbol{x}; z) = \int d\boldsymbol{\phi}\, e^{ik\boldsymbol{\phi}\cdot\boldsymbol{x}} \mathcal{E}_\omega(\boldsymbol{\phi}; z). \tag{1.49}$$

A general solution for E_ω can then be represented by

$$E_\omega(\boldsymbol{x}; z) = \int d\boldsymbol{\phi}\, \exp\left[ik(\boldsymbol{\phi}\cdot\boldsymbol{x} \pm z\sqrt{1-\boldsymbol{\phi}^2})\right] \mathcal{E}_\omega(\boldsymbol{\phi}; 0), \tag{1.50}$$

where $\mathcal{E}_\omega(\boldsymbol{\phi}; 0)$ is the angular representation at the plane $z = 0$. We make the paraxial approximation by assuming that the wave is concentrated in the forward direction, so that the field $\mathcal{E}_\omega(\boldsymbol{\phi}; 0)$ is non-negligible only in the region where $\phi^2 \ll 1$.[5] Hence, we Taylor-expand the square root as

$$\sqrt{1-\boldsymbol{\phi}^2} \approx 1 - \frac{\boldsymbol{\phi}^2}{2}. \tag{1.51}$$

Additionally, we assume that the waves travel exclusively along $+\hat{z}$, taking the positive sign of the expanded square root in (1.50). Thus, we have

$$E_\omega(\boldsymbol{x}; z) = \int d\boldsymbol{\phi}\, e^{ik\boldsymbol{\phi}\cdot\boldsymbol{x}} e^{ik(1-\boldsymbol{\phi}^2/2)z} \mathcal{E}_\omega(\boldsymbol{\phi}; 0) \tag{1.52}$$

$$= \frac{ke^{ikz}}{2\pi iz} \int d\boldsymbol{y}\, e^{ik(\boldsymbol{x}-\boldsymbol{y})^2/2z} E_\omega(\boldsymbol{y}; 0), \tag{1.53}$$

where the final line results from writing $\mathcal{E}_\omega(\boldsymbol{\phi}; 0)$ in the $\boldsymbol{x}$-representation and doing the Gaussian integral over $\boldsymbol{\phi}$. Equation (1.53) is identical to the Fresnel diffraction integral, appropriate for free space evolution provided $|\boldsymbol{x}| \ll z$. Furthermore, Fourier transforming (1.52) shows that vacuum propagation in the angular representation is obtained by simple phase multiplication:

$$\mathcal{E}_\omega(\boldsymbol{\phi}; z) = e^{ik(1-\boldsymbol{\phi}^2/2)z} \mathcal{E}_\omega(\boldsymbol{\phi}; 0). \tag{1.54}$$

To understand the diffractive characteristics of Equations (1.53) and (1.54), we need to be able to compute the physical beam size and divergence in a manner similar to what we did for the particle beams in Section 1.1.3. The radiation intensity profile $|E_\omega(\boldsymbol{x}; z)|^2$ serves as a natural choice for the positive measure, and we define the square of the RMS beam size by

$$\sigma_r^2(z) = \frac{1}{2}\left\langle \boldsymbol{x}^2 \right\rangle = \frac{\int d\boldsymbol{x}\, \boldsymbol{x}^2\, |E_\omega(\boldsymbol{x}; z)|^2}{\int d\boldsymbol{x}\, |E_\omega(\boldsymbol{x}; z)|^2}, \tag{1.55}$$

while the RMS divergence squared is more compactly computed using the angular representation for the intensity:

$$\sigma_{r'}^2(z) = \frac{1}{2}\left\langle \boldsymbol{\phi}^2 \right\rangle = \frac{\int d\boldsymbol{\phi}\, \boldsymbol{\phi}^2\, |\mathcal{E}_\omega(\boldsymbol{\phi}; z)|^2}{\int d\boldsymbol{\phi}\, |\mathcal{E}_\omega(\boldsymbol{\phi}; z)|^2}. \tag{1.56}$$

[5] Note that for $\phi \ll 1$, we are justified in referring to $\phi = \boldsymbol{x} \cdot \boldsymbol{k}/k$ as the angle that the wavevector makes with the optical axis z.

The formula for the correlation term $\langle \boldsymbol{x} \cdot \boldsymbol{\phi} \rangle$, on the other hand, cannot be easily justified at this point, but we will return to its definition after we introduce the radiation brightness function.

Returning to diffraction, the first thing that we find is that the angular divergence is constant for free space propagation. This is because the evolution in angular space (1.54) is given by multiplying by the phase $e^{-ik\phi^2 z/2}$, which leaves (1.56) invariant. We will illustrate the essential features of diffraction in the spatial representation by considering a single frequency Gaussian profile in one transverse dimension. Thus, we consider the simple coherent beam

$$E_\omega(x;0) = E_0 \exp\left(-\frac{x^2}{4\sigma_r^2}\right), \tag{1.57}$$

whose corresponding angular representation is

$$\mathcal{E}_\omega(\phi;0) = \mathcal{E}_0 \exp\left(-\frac{\phi^2}{4\sigma_{r'}^2}\right). \tag{1.58}$$

Since (1.57) and (1.58) are related by the Fourier transform, it is easy to show that

$$\sigma_r \sigma_{r'} = \frac{\lambda}{4\pi} \equiv \varepsilon_r. \tag{1.59}$$

The quantity ε_r closely resembles the emittance of the particle beam that we introduced in Section 1.1.3, so that the RMS emittance of a Gaussian light beam is $\lambda/4\pi$. Additionally, because ε_r is proportional to the Fourier uncertainty product that is minimized by a Gaussian profile, $\lambda/4\pi$ is the minimum emittance for a radiation beam.

Inserting (1.57) into the free space diffraction formula (1.53), we find the general expression for the Gaussian beam

$$\begin{aligned} E_\omega(x;z) &= \frac{E_0 e^{ikz}}{\sqrt{1 + i\sigma_{r'} z/\sigma_r}} \exp\left[-\frac{x^2}{4\sigma_r^2(1 + i\sigma_{r'} z/\sigma_r)}\right] \\ &= \frac{E_0 e^{ikz}}{\left(1 + z^2/Z_R^2\right)^{1/4}} \exp\left[-\frac{x^2(1 - iz/Z_R)}{4\sigma_r^2(1 + z^2/Z_R^2)} - \frac{i}{2}\tan^{-1}\left(\frac{z}{Z_R}\right)\right], \end{aligned} \tag{1.60}$$

where $Z_R \equiv \sigma_r/\sigma_{r'} = 2k\sigma_r^2$ is known as the Rayleigh length. We already know that vacuum propagation preserves the RMS divergence, $\sigma_{r'}(z) = \sigma_{r'}$, while the RMS beam size can be calculated using $|E(x;z)|^2$ from (1.60):

$$\sigma_r(z) = \sqrt{\frac{\lambda}{4\pi}\left(Z_R + \frac{z^2}{Z_R}\right)}. \tag{1.61}$$

This equation is analogous to Equation (1.27) derived for drift space propagation in particle beam optics. The diffraction law of a coherent radiation beam is therefore formally equivalent to the free-space propagation of particle beams, provided that one makes the following identification

$$\varepsilon_x \leftrightarrow \frac{\lambda}{4\pi} \qquad\qquad Z_\beta \leftrightarrow Z_R. \tag{1.62}$$

Additionally, by drawing the analogy (1.62) we see that the intensity of the Gaussian beam obtained from (1.60) corresponds in a similar manner to the distribution of the Gaussian electron beam in physical space, namely $\int dx'\, F(x,x';z)$ given by (1.44). After a brief discussion of the paraxial wave equation and its transport of electromagnetic energy in the following section, we will investigate the similarities between paraxial particle and radiation beams more completely in Section 1.2.3.

1.2.2 The Paraxial Wave Equation and Energy Transport

In this section we write down a wave equation associated with the paraxial solutions (1.53) and (1.54), and then find expressions for the electromagnetic energy and power in this approximation. First, we show how the paraxial wave equation arises. To do this, we define the slowly varying amplitude, which is obtained from the electric field by factoring out the carrier oscillations along $\hat{z}$ at the wavelength $\lambda = 2\pi/k$:

$$\mathcal{E}_\omega(\boldsymbol{\phi}; z) \equiv e^{ikz}\tilde{\mathcal{E}}_\omega(\boldsymbol{\phi}; z) \tag{1.63}$$

$$E_\omega(\boldsymbol{x}; z) = e^{ikz}\tilde{E}_\omega(\boldsymbol{x}; z). \tag{1.64}$$

Inserting (1.63) into the free-space propagation formula (1.54), it is straightforward to see that the amplitude $\tilde{\mathcal{E}}_\omega$ is a solution to the equation

$$\left[\frac{\partial}{\partial z} + \frac{ik}{2}\boldsymbol{\phi}^2\right]\tilde{\mathcal{E}}_\omega(\boldsymbol{\phi}, z) = 0. \tag{1.65}$$

Taking the Fourier transform of (1.65), we find the usual form of the paraxial wave equation

$$\left[\frac{\partial}{\partial z} - \frac{i}{2k}\frac{\partial^2}{\partial \boldsymbol{x}^2}\right]\tilde{E}_\omega(\boldsymbol{x}; z) = 0 \tag{1.66}$$

in terms of the physical space amplitude. The paraxial wave equation (1.66) is formally identical to the Schrödinger equation in two dimensions, and its Green function solution is precisely the Fresnel diffraction integral (1.53).

There is another way to obtain the paraxial wave equation that will be particularly useful when we include source currents relevant to X-ray generation: we insert the definition of the slowly varying amplitude (1.64) into the full wave equation (1.47) and then neglect the second-order longitudinal derivative. In other words, we can reduce the second-order Maxwell wave equation to the first-order (in z) paraxial wave equation (1.66) by assuming that $\left|\frac{\partial}{\partial z}\tilde{E}_\omega\right| \ll k\left|\tilde{E}_\omega\right|$. This implies that the longitudinal variations in the wave amplitude $\tilde{E}_\omega(\boldsymbol{x}; z)$ occur over a much longer spatial scale than that set by the wavelength λ, so that $\tilde{E}_\omega(\boldsymbol{x}; z)$ serves as an envelope function within which there are the fast oscillations at the carrier frequency. We illustrate this in Figure 1.7, where the full field is drawn as a gray solid line, while the slowly varying envelope function is denoted by the black dotted line.

Having defined the slowly varying amplitude, we are now in a position to write down expressions for the electromagnetic energy, power density, and flux relevant to paraxial beams. We consider a general field in time, obtained by superposing the frequency components using the Fourier integral

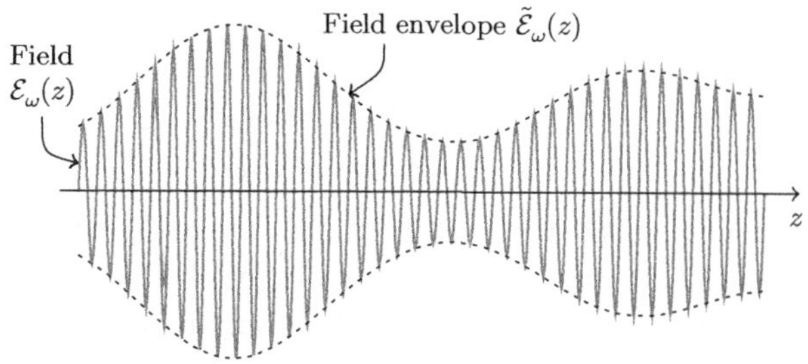

Figure 1.7 An illustration of the envelope function $\tilde{\mathcal{E}}_\omega(z)$ (black dotted line) for the field $\mathcal{E}_\omega(z)$ (gray solid line). The envelope is assumed to vary slowly on the scale-length of the typical wavelength λ as shown.

$$E(\boldsymbol{x}, t; z) = \int d\omega\, e^{-i\omega t} E_\omega(\boldsymbol{x}; z) = \int d\omega d\boldsymbol{\phi}\, e^{-i(\omega t - k\boldsymbol{\phi}\cdot\boldsymbol{x})} \mathcal{E}_\omega(\boldsymbol{\phi}; z), \tag{1.67}$$

where $E_\omega = E^*_{-\omega}$, to ensure that E is real. The power passing through any transverse plane is obtained by integrating the projection of the Poynting vector along the unit normal $\hat{z}$ over the plane $\boldsymbol{x}$:

$$P(t; z) = \int d\boldsymbol{x}\, (\hat{z}\cdot \boldsymbol{S}) = \epsilon_0 c^2 \int d\boldsymbol{x}\, [\hat{z}\cdot(\boldsymbol{E}\times\boldsymbol{B})]. \tag{1.68}$$

We rewrite the Poynting flux in terms of the electric field using the Maxwell equation $\partial \boldsymbol{B}/\partial t = -\nabla \times \boldsymbol{E}$ and the Fourier decomposition (1.67). From the vector identity $\hat{z}\cdot[\boldsymbol{A}\times(\nabla\times\boldsymbol{B})] = \boldsymbol{A}\cdot(\partial\boldsymbol{B}/\partial z) - \boldsymbol{A}\cdot(\nabla B_z)$ we find

$$\begin{aligned}
\hat{z}\cdot(\boldsymbol{E}\times\boldsymbol{B}) &= \int d\omega d\omega'\, \frac{e^{-i(\omega+\omega')t}}{i\omega'} \boldsymbol{E}_\omega(\boldsymbol{x}; z)\cdot\frac{\partial}{\partial z}\boldsymbol{E}_{\omega'}(\boldsymbol{x}; z) \\
&\approx \frac{1}{c}\int d\omega d\omega'\, e^{-i(\omega+\omega')(t-z/c)} \tilde{\boldsymbol{E}}_\omega(\boldsymbol{x}; z)\cdot\tilde{\boldsymbol{E}}_{\omega'}(\boldsymbol{x}; z),
\end{aligned} \tag{1.69}$$

where we have assumed that $\boldsymbol{E}$ is transverse and the last line follows by expressing $\boldsymbol{E}$ in terms of the slowly varying envelope and neglecting $\partial\tilde{E}/\partial z$ with respect to $k\tilde{E}$. From Equation (1.68), the power is given by summing the Poynting flux over the transverse plane. Integrating the power in time yields the total energy U, which we find to be

$$\begin{aligned}
U(z) &= 2\pi\epsilon_0 c \int d\boldsymbol{x} \int d\omega\, \left|\tilde{\boldsymbol{E}}_\omega(\boldsymbol{x}; z)\right|^2 \\
&= 4\pi\epsilon_0 c \int d\boldsymbol{x} \int_0^\infty d\omega\, \left|\tilde{\boldsymbol{E}}_\omega(\boldsymbol{x}; z)\right|^2.
\end{aligned} \tag{1.70}$$

Thus, we can interpret the quantity $4\pi\epsilon_0 c\left|\tilde{\boldsymbol{E}}_\omega(\boldsymbol{x}; z)\right|^2$ as the spectral energy density. In a similar manner, expressing the electric field in the angular representation shows that $(4\pi\epsilon_0 c\lambda^2)\left|\tilde{\boldsymbol{\mathcal{E}}}_\omega(\boldsymbol{\phi}; z)\right|^2$ is the angular spectral energy density. From here, we write expressions for the power density in a somewhat heuristic manner. If we measure the electromagnetic energy over a time of duration T, the average power $P = U/T$. Therefore, we define the spatial power spectral density and the angular power spectral density respectively as follows:

$$\frac{dP}{d\omega d\boldsymbol{x}} = \frac{4\pi\epsilon_0 c}{T}\left|\tilde{\boldsymbol{E}}_\omega(\boldsymbol{x};z)\right|^2 \tag{1.71}$$

$$\frac{dP}{d\omega d\boldsymbol{\phi}} = \frac{4\pi\epsilon_0 c\lambda^2}{T}\left|\tilde{\mathcal{E}}_\omega(\boldsymbol{\phi};z)\right|^2. \tag{1.72}$$

This definition is the most useful if the wave has many oscillations during the observation time ($T \gg 2\pi/\omega$), and if T can also be chosen to be less than the time over which the field envelope $\tilde{E}$ varies. Finally, the spectral photon flux can be found by dividing the power density by the characteristic photon energy $\hbar\omega$. For example, the number of photons per unit time per unit frequency is given by

$$\frac{d\mathcal{F}}{d\omega} = \frac{1}{\hbar\omega}\frac{4\pi\epsilon_0 c\lambda^2}{T}\int d\boldsymbol{\phi}\,\left|\tilde{\mathcal{E}}_\omega(\boldsymbol{\phi};z)\right|^2. \tag{1.73}$$

A similar expression can also be written in physical space, which can be considered to be a consequence of Parseval's theorem. In the next section, we extend the idea of a photon density to the phase space $(\boldsymbol{x}, \boldsymbol{\phi})$ by introducing the wave brightness function. This will make explicit the similarities between paraxial particle beams and paraxial optics, as alluded to in the previous section.

1.2.3 Phase Space Methods in Wave Optics

Brightness is defined as the photon flux per unit area and the photon flux per unit solid angle at the source. Brightness is a conserved quantity in perfect optical systems as illustrated in Figure 1.8(a), and is therefore a very useful figure of merit in designing synchrotron radiation and FEL facilities. Typical experiments use a monochromator to select a very narrow spectrum as in Figure 1.8(b), and it is conventional to define the spectral brightness as the density of photons in 6D phase space:

$$\mathcal{B} = \frac{d\mathcal{N}_{\text{ph}}}{\Delta A \Delta\Omega d\omega dt}. \tag{1.74}$$

Here $\mathcal{N}_{\text{ph}}$ is the number of photons, A is the area, and Ω is the solid angle.

The brightness of a single pulse obtained via Equation (1.74) is also called the peak spectral brightness to differentiate it from the average photon output. Although technically not a source brightness, applying Equation (1.74) with the average photon flux over a long time (and many pulses) yields what is referred to as the average spectral brightness.

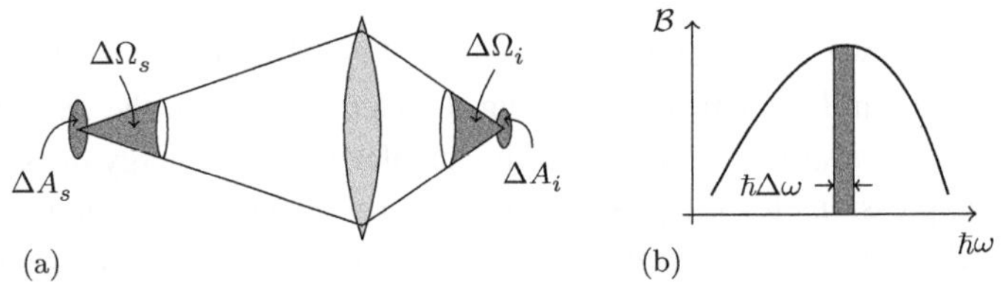

Figure 1.8 (a) A perfect optical system with 100 percent efficiency transmits all collected photons such that $\Delta A_s \Delta\Omega_s = \Delta A_i \Delta\Omega_i$. (b) Illustration of the brightness as a function of radiation energy, where a typical experiment may select those photons in the shaded region.

A more formal definition of the wave optics brightness function [3, 4] also makes the analogy between paraxial particle and radiation beams more precise. We define the brightness $\mathcal{B}(\boldsymbol{x}, \boldsymbol{\phi}; z)$ on the position–angle phase space $(\boldsymbol{x}, \boldsymbol{\phi})$ to be given by the Wigner transform [5] of the electric field:

$$\mathcal{B}(\boldsymbol{x}, \boldsymbol{\phi}; z) = \frac{1}{\hbar\omega}\frac{4\pi\epsilon_0 c}{T}\int d\boldsymbol{\xi}\,\left\langle \mathcal{E}^*\left(\boldsymbol{\phi} + \tfrac{1}{2}\boldsymbol{\xi}; z\right)\mathcal{E}\left(\boldsymbol{\phi} - \tfrac{1}{2}\boldsymbol{\xi}; z\right)\right\rangle e^{-ik\boldsymbol{\xi}\cdot\boldsymbol{x}} \tag{1.75}$$

$$= \frac{1}{\hbar\omega}\frac{4\pi\epsilon_0 c}{\lambda^2 T}\int d\boldsymbol{y}\,\left\langle E^*\left(\boldsymbol{x} + \tfrac{1}{2}\boldsymbol{y}; z\right)E\left(\boldsymbol{x} - \tfrac{1}{2}\boldsymbol{y}; z\right)\right\rangle e^{ik\boldsymbol{y}\cdot\boldsymbol{\phi}}. \tag{1.76}$$

In (1.75) and (1.76) the angular brackets $\langle\cdot\rangle$ denote an ensemble average (which is important for partially coherent or chaotic fields), and the prefactor is chosen so that integrating over positions and angles yields the total photon flux, in agreement with (1.73). Thus, the quantity $T\mathcal{B}(\boldsymbol{x}, \boldsymbol{\phi})$ can be roughly interpreted as the total number of photons per unit phase space area per unit frequency $d\omega$. We say roughly because $\mathcal{B}$ can in fact take on negative values, so that $\mathcal{B}(\boldsymbol{x}, \boldsymbol{\phi}; z)$ is not a true photon number density. This is due to the wave nature of light, which prevents an exact correspondence between the electron distribution function F and the radiation brightness $\mathcal{B}$. On the other hand the integral of the brightness over a phase space area greater than $2\pi\varepsilon_r = \lambda/2$ is a positive definite quantity that can be identified with a photon number.[6]

Nevertheless, the brightness defined above behaves very much like a distribution function. For example, integrating $\mathcal{B}$ over the transverse coordinates yields the angular flux density, while integrating with respect to $\boldsymbol{\phi}$ gives the intensity in physical space; both of these are measurable physical quantities. Furthermore, this helps show that the brightness plays the role of a distribution function when calculating radiation beam moments. The RMS beam size and divergence can be computed via

$$\sigma_r^2(z) = \frac{1}{2}\left\langle \boldsymbol{x}^2\right\rangle = \frac{1}{2}\frac{\int d\boldsymbol{x}d\boldsymbol{\phi}\;\boldsymbol{x}^2\mathcal{B}(\boldsymbol{x}, \boldsymbol{\phi}; z)}{\int d\boldsymbol{x}d\boldsymbol{\phi}\;\mathcal{B}(\boldsymbol{x}, \boldsymbol{\phi}; z)} \tag{1.77}$$

$$\sigma_{r'}^2(z) = \frac{1}{2}\left\langle \boldsymbol{\phi}^2\right\rangle = \frac{1}{2}\frac{\int d\boldsymbol{x}d\boldsymbol{\phi}\;\boldsymbol{\phi}^2\mathcal{B}(\boldsymbol{x}, \boldsymbol{\phi}; z)}{\int d\boldsymbol{x}d\boldsymbol{\phi}\;\mathcal{B}(\boldsymbol{x}, \boldsymbol{\phi}; z)}, \tag{1.78}$$

while the second-order correlation of the radiation field is given by

$$\langle \boldsymbol{x}\cdot\boldsymbol{\phi}\rangle = \frac{\int d\boldsymbol{x}d\boldsymbol{\phi}\;(\boldsymbol{x}\cdot\boldsymbol{\phi})\mathcal{B}(\boldsymbol{x}, \boldsymbol{\phi}; z)}{\int d\boldsymbol{x}d\boldsymbol{\phi}\;\mathcal{B}(\boldsymbol{x}, \boldsymbol{\phi}; z)}. \tag{1.79}$$

In addition to providing the measure for radiation moments, the brightness also obeys a conservation equation in phase space. Multiplying the paraxial wave equation (1.66) by $E^*(\boldsymbol{x} + \boldsymbol{y}; z)$, Fourier transforming the result, and adding the complex conjugate equation shows that in vacuum the brightness function satisfies the Liouville-type equation

$$\left[\frac{\partial}{\partial z} + \boldsymbol{\phi}\cdot\frac{\partial}{\partial \boldsymbol{x}}\right]\mathcal{B}(\boldsymbol{x}, \boldsymbol{\phi}; z) = 0. \tag{1.80}$$

[6] Technically, the convolution of $\mathcal{B}$ with a Gaussian function whose phase space area is greater than $\lambda/2$ is positive [6]; alternatively, the convolution of any two brightness (Wigner) functions is everywhere positive [7].

This equation is formally equivalent to the force-free ($\boldsymbol{x}'' = 0$) Liouville equation for the particle beam distribution given in Equation (1.37). Thus, by analogy we can conclude that the free-space transformation of the brightness over the propagation distance ℓ is given by

$$\mathcal{B}(\boldsymbol{x}, \boldsymbol{\phi}; z + \ell) = \mathcal{B}(\boldsymbol{x} - \ell\boldsymbol{\phi}, \boldsymbol{\phi}; z). \tag{1.81}$$

This result can also be obtained directly from the Fresnel diffraction formula (1.53) and the definition of $\mathcal{B}$. Using the free-space transformation above, we find that in vacuum the RMS beam size behaves as

$$\sigma_r^2(\ell) = \sigma_r^2(0) + \ell^2\sigma_{r'}^2(0) \tag{1.82}$$

if we define the plane $z = 0$ to be at the beam waist where the correlation vanishes, i.e., $\int d\boldsymbol{x}d\boldsymbol{\phi}\ (\boldsymbol{x} \cdot \boldsymbol{\phi})\mathcal{B}(\boldsymbol{x}, \boldsymbol{\phi}; 0) = 0$. Again, this compares directly with the particle beam formula (1.24). For completeness, we also include how the brightness transforms due to a thin lens. An ideal thin lens with focal length f acts by multiplying the electric field by a phase that increases quadratically with $\boldsymbol{x}$:

$$E(\boldsymbol{x}; z)_{\text{out}} = e^{ikx^2/2f} E(\boldsymbol{x}; z)_{\text{in}}. \tag{1.83}$$

Inserting (1.83) into the definition of the brightness function yields the thin lens transformation

$$\mathcal{B}(\boldsymbol{x}, \boldsymbol{\phi}; z)_{\text{out}} = \mathcal{B}(\boldsymbol{x}, \boldsymbol{\phi} + \boldsymbol{x}/f; z)_{\text{in}}. \tag{1.84}$$

Again, this result should be compared to the particle beam optics equation (1.9). Alternatively, we could have anticipated this result by comparing free space propagation in the angular representation with the lens transformation in physical space. Both multiply the field by a quadratically increasing phase, the former by $e^{-ik\ell\phi^2/2}$, the latter by $e^{ikx^2/2f}$. Thus, the transformation properties of free space and a thin lens are analogous, with the roles of $\boldsymbol{\phi}$ and $\boldsymbol{x}$ interchanged, and $\ell \leftrightarrow -1/f$.

The free-space and thin lens transformations that we have considered here transport the brightness along the ray trajectories. Furthermore, these optical elements preserve the brightness at the phase space origin $\boldsymbol{x} = \boldsymbol{\phi} = \boldsymbol{0}$, so that

$$\mathcal{B}(\boldsymbol{0}, \boldsymbol{0}; z_1) = \mathcal{B}(\boldsymbol{0}, \boldsymbol{0}; z_2) \equiv \mathcal{B}(\boldsymbol{0}, \boldsymbol{0}) \tag{1.85}$$

is an invariant measure of the strength of a radiation source.

As a simple example of the radiation brightness in the two-dimensional phase space (x, ϕ), we again consider a Gaussian profile. Inserting the field expression (1.60) into the definition for $\mathcal{B}$ (1.77), we find that the brightness for a Gaussian field is

$$\mathcal{B}(x, \phi; z) = \mathcal{B}(\boldsymbol{0}, \boldsymbol{0}) \exp\left[-\frac{(x - z\phi)^2}{2\sigma_r^2} - \frac{\phi^2}{2\sigma_{r'}^2} \right]. \tag{1.86}$$

At $z = 0$, the brightness is a Gaussian in both position and angle. Free-space propagation along z shears $\mathcal{B}$ in phase space at fixed angle ϕ. Thus, the brightness of a Gaussian laser beam is directly analogous to the Gaussian particle beam distribution function written in (1.39).

While we mentioned in the beginning of this section that the radiation brightness is not a true probability function on phase space because it can be negative, we have yet to show any explicit examples. Lest we give the mistaken impression that this point is a subtlety that one can typically ignore, we conclude this section with a simple example illustrating the potential complexities in interpreting $\mathcal{B}$. This example will also serve as an instructive prequel to our subsequent discussion of coherence. We consider two Gaussian beams of equal magnitudes but with different transverse positions; each is focused to a waist at the plane $z = 0$, but the waist location is displaced by $\pm x_0$ from the optical axis, so that

$$E(x;0) = E_0 \exp\left[-\frac{(x-x_0)^2}{4\sigma_r^2} + i\psi_1\right] + E_0 \exp\left[-\frac{(x+x_0)^2}{4\sigma_r^2} + i\psi_2\right], \tag{1.87}$$

where $\psi_{1,2}$ are two independent phases. The brightness of the two displaced Gaussians is

$$\begin{aligned}\mathcal{B}(x,\phi;0) = \mathcal{B}_G \Bigg\{ &\exp\left[-\frac{(x-x_0)^2}{2\sigma_r^2} - \frac{\phi^2}{2\sigma_{r'}^2}\right] \\ &+ \exp\left[-\frac{(x+x_0)^2}{2\sigma_r^2} - \frac{\phi^2}{2\sigma_{r'}^2}\right] \\ &+ 2\exp\left[-\frac{x^2}{2\sigma_r^2} - \frac{\phi^2}{2\sigma_{r'}^2}\right] \cos[2kx_0\phi - (\psi_1 - \psi_2)] \Bigg\},\end{aligned} \tag{1.88}$$

where $\mathcal{B}_G$ is the maximum brightness for a single Gaussian beam. The first two terms in (1.88) are the brightness functions of the individual beams, now centered at $x = \pm x_0$. The third term arises from the product between the two Gaussian pulses; this interference term oscillates in ϕ with a frequency proportional to the displacement x_0, while its peak amplitude is governed by the phase difference $\psi_1 - \psi_2$. When we consider many such beams the statistical properties of the latter will quantify the degree to which the field is temporally coherent: an approximately constant phase difference will mean that the field is largely coherent, while if the phase is a random variable the resulting statistical mixture will wash out these cross terms. Furthermore, if $x_0 \lesssim \sigma_r$ the individual beams act as a single source in the transverse plane (transverse coherence), while for larger separations the field can be transversely coherent only if it is also temporally coherent. We discuss transverse coherence in Section 1.2.4 and temporal coherence and its effects in Sections 1.2.5 and 1.2.6. Here, we merely wish to investigate (1.88).

For the present purpose, we take the fields to be real, setting $\psi_1 = \psi_2 = 0$. If the magnitude of the displacement x_0 is much less than the Gaussian width σ_r, than the oscillations in ϕ occur where $\mathcal{B}$ is comparatively small, and the total brightness closely resembles a single Gaussian whose magnitude is nearly four times that of a single Gaussian. As we will discuss further in the Section 1.2.6, if each field was produced by a distinct source this is indicative of a coherent addition of the radiation amplitude. On the other hand, if $x_0 \gg \sigma_r$ we have two well-separated Gaussian functions localized about $x = \pm x_0$, and another Gaussian function near the origin that oscillates rapidly along the angular direction ϕ.

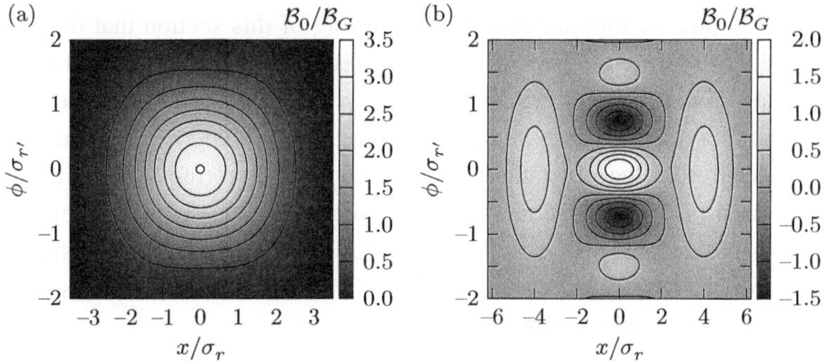

Figure 1.9 Radiation brightness corresponding to two displaced Gaussian beams. Panel (a) shows $\mathcal{B}$ for two closely spaced Gaussians, with the displacement $x_0 = \sigma_r$. In this case the radiation adds coherently leading to a single peak with $\mathcal{B}(\mathbf{0}, \mathbf{0}) = 3.5\mathcal{B}_G$. In panel (b) we increase the displacement to $x_0 = 4\sigma_r$, and the two individual Gaussian beams centered at $(\pm x_0, 0)$ become visible. Additionally, there is another Gaussian peak about the origin that oscillates significantly in ϕ. The peak at the origin is $2\mathcal{B}_G$, while the intensity $|E(x = 0)|^2 = \int d\phi\ \mathcal{B}(x = 0, \phi)$ is very small.

We show two cases of $\mathcal{B}$ for the displaced Gaussians in Figure 1.9. In panel (a), we choose $x_0 = \sigma_r$, in which case the brightness is a single peaked function whose magnitude at the origin is $\sim$3.5 times the brightness of a single Gaussian $\mathcal{B}_G$. This is close to the coherent increase that one would get if $x_0 = 0$, in which case the amplitude doubles while the power increases by four. In panel (b), we set $x_0 = 4\sigma_r$, so that the two individual beam brightnesses are well separated and centered at $x = \pm 4\sigma_r$. Additionally, $\mathcal{B}$ in Figure 1.9(b) has clear oscillations along $x = 0$ due to the interference term. The brightness at the origin $\mathcal{B}(\mathbf{0}, \mathbf{0})$ is now twice that of a single Gaussian, indicating that the source strength is now a linear sum of the two. Furthermore, although $\mathcal{B}(\mathbf{0}, \mathbf{0})$ is large, integrating along ϕ to get the intensity at $x = 0$ results in significant cancellation and correspondingly small intensity.

In this simple example, we started with a purely real and positive electric field profile, and found that the corresponding radiation brightness can include highly oscillatory regions. This behavior is caused by the interference of separate coherent regions in transverse phase space, which we investigate further in the next section by discussing transverse coherence.

1.2.4 Transverse Coherence

Coherence is a unique feature of a wave that quantifies the degree of phase correlation at two distinct points. The level of coherence can be ascertained by measuring the sharpness of the radiation interference pattern. Two types of coherence can be distinguished: transverse coherence refers to the degree that a wave is correlated at two points in the transverse plane, while temporal coherence quantifies the phase correlation at two points separated in time. We discuss transverse coherence in this section, and defer temporal coherence to the next section.

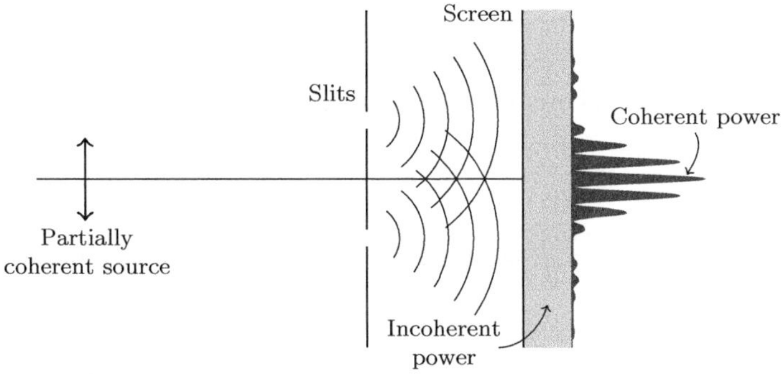

Figure 1.10 Schematic of an interference experiment to determine the transverse coherence. A partially coherent source gives rise to an intensity pattern composed of an interference pattern due to the coherent fraction on top of the constant, incoherent background.

As alluded to in the previous paragraph, transverse coherence can be measured via the interference pattern in Young's two slit experiment. For this reason, we consider the simple thought experiment shown in Figure 1.10: radiation from a source passes through a pair of pinholes located symmetrically about the optical axis at the transverse plane $z = 0$ to form an interference pattern at an image plane far from the pinholes. Let $\boldsymbol{x}$ and $-\boldsymbol{x}$ be the positions of the pinholes, and $d\boldsymbol{x} = dxdy$ be the area of each individual pinhole. In general, the resulting interference pattern on the screen consists of a smooth background and an oscillatory part having maxima and minima, and one can show that the flux on the screen is given by

$$\begin{aligned} d\mathcal{F} = \frac{d\omega}{\hbar\omega}\frac{4\pi\epsilon_0 c}{T}d\boldsymbol{x}\Big\{&|\langle E(\boldsymbol{x};z)\rangle|^2 + |\langle E(-\boldsymbol{x};z)\rangle|^2 \\ &+ 2\left|\langle E(\boldsymbol{x};z)E^*(-\boldsymbol{x};z)\rangle\right|\cos[\psi(\boldsymbol{x},-\boldsymbol{x})]\Big\}. \end{aligned} \tag{1.89}$$

The second line in (1.89) represents the interference term, with the angle ψ depending upon both the relative phase of the field at the pinholes and on the imaging screen position. The correlation function $|\langle E(\boldsymbol{x};z)E^*(-\boldsymbol{x};z)\rangle|$ plays a central role in traditional coherence theory where it is typically referred to as the mutual coherence function (see, e.g., [8, 9]). We take a slightly different perspective, following [10] to use phase space methods to characterize the coherence. While not as general or rigorous, the phase space approach yields valuable physical insight and is more in line with our subsequent developments. Returning to the thought experiment in Figure 1.10, the differential flux contained in the interference portion of the image is given by

$$\frac{d\mathcal{F}_{\text{int}}}{d\omega} = 2\frac{1}{\hbar\omega}\frac{4\pi\epsilon_0 c}{T}\left|\langle E^*(\boldsymbol{x};z)E(-\boldsymbol{x};z)\rangle\right| d\boldsymbol{x}, \tag{1.90}$$

while the flux in the smooth background $d\mathcal{F}_{\text{BG}}$ is the remainder:

$$d\mathcal{F}_{\text{BG}} = d\mathcal{F} - d\mathcal{F}_{\text{int}}. \tag{1.91}$$

We now sum the differential contributions to the coherent flux over the entire beam; integrating (1.90) over different pinhole locations and using the fact that this integral is half the integral over the entire transverse plane, we obtains

$$\begin{aligned}\frac{d\mathcal{F}_{\text{int}}}{d\omega} &= \frac{1}{\hbar\omega}\frac{4\pi\epsilon_0 c}{T}\int d\boldsymbol{x}\;\left|\langle E^*(\boldsymbol{x};z)E(-\boldsymbol{x};z)\rangle\right| \\ &= \frac{1}{4}\int d\boldsymbol{x}\;\left|\int d\boldsymbol{\phi}\; e^{-ik\boldsymbol{\phi}\cdot\boldsymbol{x}}\mathcal{B}(\boldsymbol{0},\boldsymbol{\phi};z)\right| \geq \frac{1}{4}\left|\int d\boldsymbol{x}d\boldsymbol{\phi}\; e^{-ik\boldsymbol{\phi}\cdot\boldsymbol{x}}\mathcal{B}(\boldsymbol{0},\boldsymbol{\phi};z)\right|.\end{aligned}$$

Integrating over $\boldsymbol{x}$ yields $\lambda^2\delta(\boldsymbol{\phi})$, so that the flux comprising the interference pattern satisfies $d\mathcal{F}_{\text{int}}/d\omega \geq (\lambda/2)^2\,|\mathcal{B}(\boldsymbol{0},\boldsymbol{0})|$. Since the total flux $d\mathcal{F} \geq d\mathcal{F}_{\text{int}}$, it follows that

$$\frac{d\mathcal{F}/d\omega}{|\mathcal{B}(\boldsymbol{0},\boldsymbol{0})|} \geq (\lambda/2)^2. \tag{1.92}$$

The ratio $\mathcal{F}/\,|\mathcal{B}(\boldsymbol{0},\boldsymbol{0})|$ above may be roughly associated with the phase space area of the radiation beam, and the inequality (1.92) states that the minimum possible phase space area in wave optics is $(\lambda/2)^2$. It also follows from the above argument that if the phase space area associated with a radiation beam is determined to be the minimum possible value $(\lambda/2)^2$, then it will exhibit a sharp interference pattern without a smooth background for any pinhole pair symmetrically placed about the optical axis. We are therefore justified in identifying the transversely coherent flux $\mathcal{F}_{\text{coh}}$ associated with a radiation beam as

$$\frac{d\mathcal{F}_{\text{coh}}}{d\omega} = (\lambda/2)^2\,|\mathcal{B}(\boldsymbol{0},\boldsymbol{0})|\,. \tag{1.93}$$

Note that this is an invariant quantity (for ideal optical elements and no apertures) since the peak brightness is constant.

We illustrate the phase space treatment developed above with a simple example that explains the well-known fact that incoherent radiation can be made transversely coherent by introducing an aperture. To simplify the discussion, we consider one transverse dimension only. We imagine a source of radius R that emits uniformly over all angles ($\Delta\phi \sim 1$). The radiation brightness occupies the area $\Delta x\Delta\phi \sim 2R$ in phase space as shown on the left of Figure 1.11, and the source is incoherent if $R \gg \lambda$. At a distance D from the source, the phase space distribution is tilted in phase space as indicated by the brightness transformation formula (1.81). This sheared distribution is shown on the right-hand side of Figure 1.11, where the angular width at any transverse position is $2R/D$. If one introduces a slit of width $2a$, represented in the figure by the vertical lines, the phase space area becomes $4Ra/D$. According to the phase space area criterion developed above, the radiation will be transversely coherent if $4Ra/D \lesssim \lambda/2$, in rough agreement with the well-known result in optics.

We also wish to understand the coherence of radiation that originates from many separate sources. To investigate this, we consider an extension of the two displaced Gaussian beams from the previous section, in which the radiation is composed of many transverse Gaussian wave packets located at positions x_j. The straightforward generalization of the brightness (1.88) to N_e displaced (electron) sources is

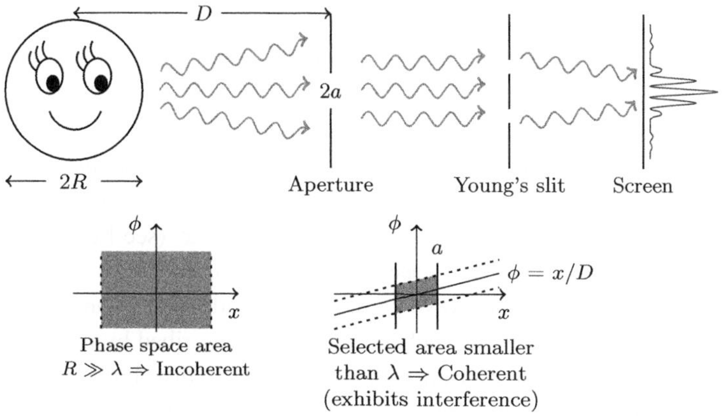

Figure 1.11 Phase space area criteria for the first-order coherence. On the left is the incoherent source that fills a large area of phase space. By choosing a sufficiently small slit labeled by the lines at $x = \pm a$, one can select a region of phase space whose area is less than $\lambda/2$ which is therefore coherent. Adapted from Ref. [3].

$$\begin{aligned}\mathcal{B}(x,\phi;0) &= \mathcal{B}_G\left\langle\sum_{j=1}^{N_e}\exp\left[-\frac{(x-x_j)^2}{2\sigma_r^2}-\frac{\phi^2}{2\sigma_{r'}^2}\right]\right\rangle\\ &+\mathcal{B}_G\left\langle\sum_{j\neq k}^{N_e}\exp\left\{-\frac{[x-(x_j+x_k)/2]^2}{2\sigma_r^2}-\frac{\phi^2}{2\sigma_{r'}^2}\right\}\right.\\ &\left.\times\cos[k(x_j-x_k)\phi-(\psi_j-\psi_k)]\right\rangle,\end{aligned} \tag{1.94}$$

where the angular brackets imply an ensemble average over the electron displacements x_j and phases ψ_j. The second sum has $N_e(N_e-1)$ terms involving the electron temporal phase difference $(\psi_j-\psi_k)$. We will discuss how this sum relates to temporal coherence in the next section, but here we assume that the electron phases are uncorrelated, so that these $N_e(N_e-1)$ terms largely cancel. In this case, only the first sum containing the N_e diagonal terms in (1.94) contributes to the brightness. If we further assume that the sources are distributed with a Gaussian probability distribution

$$f(x_j) = \frac{1}{\sqrt{2\pi}\,\sigma_x}e^{-x_j^2/2\sigma_x^2}, \tag{1.95}$$

the ensemble average can be computed in closed form. Integrating the first term in (1.94) over x_j yields

$$\mathcal{B}(x,\phi;0) = N_e\mathcal{B}_G\frac{\sigma_r}{\sqrt{\sigma_x^2+\sigma_r^2}}\exp\left[-\frac{x^2}{2(\sigma_x^2+\sigma_r^2)}-\frac{\phi^2}{2\sigma_{r'}^2}\right]. \tag{1.96}$$

The total power $\int d\boldsymbol{x}d\boldsymbol{\phi}\ \mathcal{B}(\boldsymbol{x},\boldsymbol{\phi})$ scales linearly with the number of sources, while the beam size is given by the convolution of the beam and radiation areas. In addition, the peak brightness is proportional to the transversely coherent fraction $\mathcal{F}_{\mathrm{coh}}/\mathcal{F} = \sigma_r/\sqrt{\sigma_x^2+\sigma_r^2}$. If the total electron beam size is much larger than that of the radiation, the coherent fraction and peak brightness scale as $\sigma_r/\sigma_x \ll 1$. On the other hand, the peak brightness is maximized and the coherent fraction approaches unity in the opposite limit, namely, when the characteristic radiation size is larger than that of the electron beam.

The previous derivations were obtained for a source that was distributed in the transverse position, but similar considerations apply in the more general situation where the source varies in both the position $\boldsymbol{x}$ and angle $\boldsymbol{x}'$ (see Equation (2.97) and accompanying discussion). In this case, the radiation beam can be modeled as a large number of individual Gaussian wave packets with random positions and orientations. Much like the integration above, such a beam can be described by the convolution of the coherent Gaussian radiation beam with the probability distribution of electrons in phase space. The resulting beam size and angular divergence are

$$\Sigma_x = \sqrt{\sigma_x^2 + \sigma_r^2} \qquad \Sigma_{x'} = \sqrt{\sigma_{x'}^2 + \sigma_{r'}^2}. \tag{1.97}$$

If the electron beam moments σ_x^2 and $\sigma_{x'}^2$ can be neglected as compared to the moments of the coherent radiation mode σ_r^2 and $\sigma_{r'}^2$, then we have

$$\Sigma_x \Sigma_{x'} \approx \sigma_r \sigma_{r'} = \frac{\lambda}{4\pi} \tag{1.98}$$

and the resulting radiation is transversely coherent, exhibiting Young's interference phenomenon. In the limit that the electron beam moments dominate the coherent mode moments, the phase space area

$$\Sigma_x \Sigma_{x'} \gg \frac{\lambda}{4\pi} \tag{1.99}$$

and the radiation is incoherent and shows no interference. The intermediate case is known as partial coherence, and the ratio

$$M_T = \frac{\Sigma_x \Sigma_{x'}}{\lambda/4\pi} \approx \frac{\varepsilon_x}{\varepsilon_r} \tag{1.100}$$

gives the number of coherent transverse "modes" of the radiation beam.

1.2.5 Temporal Coherence

We have seen that spatial coherence measures the correlation of the field at two separate spatial locations. In a similar manner, temporal coherence specifies the extent to which the radiation maintains a definite phase relationship at two different times. Temporal coherence is characterized by the coherence time, which can be experimentally determined by measuring the path length difference over which fringes can be observed in a Michelson interferometer. A simple representation of a coherent wave in time is given by

$$E_0(t) = e_0 \exp\left(-\frac{t^2}{4\sigma_\tau^2} - i\omega_1 t\right). \tag{1.101}$$

Here σ_τ is the RMS temporal width of the intensity profile $|E_0(t)^2|$. The coherence time t_{coh} can be defined as

$$t_{\text{coh}} \equiv \int d\tau \; |\mathcal{C}(\tau)|^2 , \tag{1.102}$$

where $\mathcal{C}(\tau)$ is the normalized, first-order correlation function (or complex degree of temporal coherence) given by

$$\mathcal{C}(\tau) \equiv \frac{\left\langle \int dt \; E(t)E^*(t+\tau)\right\rangle}{\left\langle \int dt \; |E(t)|^2\right\rangle}, \tag{1.103}$$

and the brackets denote ensemble averaging. In the simple Gaussian model of Equation (1.101), the coherence time $t_{\text{coh}} = 2\sqrt{\pi}\sigma_\tau$.

In the frequency domain, we have

$$E_\omega^0 = \int dt \; e^{i\omega t} E_0(t) = \frac{e_0\sqrt{\pi}}{\sigma_\omega} \exp\left[-\frac{(\omega-\omega_1)^2}{4\sigma_\omega^2}\right], \tag{1.104}$$

where $\sigma_\omega = (2\sigma_\tau)^{-1}$ is the RMS width of the frequency profile $|E_\omega|^2$. Let us introduce the temporal (longitudinal) phase space variables ct and $(\omega - \omega_1)/\omega_1 = \Delta\omega/\omega_1$. The Gaussian wave packet then satisfies

$$c\sigma_\tau \cdot \frac{\sigma_\omega}{\omega_1} = \frac{\lambda_1}{4\pi}, \tag{1.105}$$

which is the same phase space area relationship as (1.59) obtained for a transversely coherent Gaussian beam.

Most radiation observed in nature, however, is temporally incoherent. Sunlight, fluorescent light bulbs, black-body radiation, and undulator radiation (which we study in the next chapter) are all temporally incoherent, and are often referred to as chaotic light or as a partially coherent wave. As a mathematical model of such chaotic light, we consider a collection of coherent Gaussian pulses that are displaced randomly in time with respect to each other:

$$E(t) = \sum_{j=1}^{N_e} E_0(t-t_j) = e_0 \sum_{j=1}^{N_e} \exp\left[-\frac{(t-t_j)^2}{4\sigma_\tau^2} - i\omega_1(t-t_j)\right]. \tag{1.106}$$

In Equation (1.106), t_j is a random number, and the sum extends to N_e to suggest that these wave packets have been created by electrons. We illustrate this partially coherent wave (chaotic light) in Figure 1.12, which we obtained by using $N_e = 100$ wave packets with $\lambda_1 = 2\pi/\omega_1 = 1$ and $\sigma_\tau = 2$ ($\sigma_\omega = 0.25$), assuming that the t_js are randomly distributed with equal probability over the bunch length duration $T = 100$. Panel (a) shows 10 randomly chosen such wave packets; plotting many more than this results in a jumbled disarray. Figure 1.12(b) shows the $E(t)$ that results by summing over all 100 waves. The remarkable feature of this plot is that the resultant wave is a relatively regular oscillation that is interrupted only a few times, much fewer than one might have naively guessed based on the fact that it is a random superposition of 100 wave packets. In fact, the duration of each regular region is independent of the number of wave packets, and is instead governed by the time over which the wave maintains a definite phase

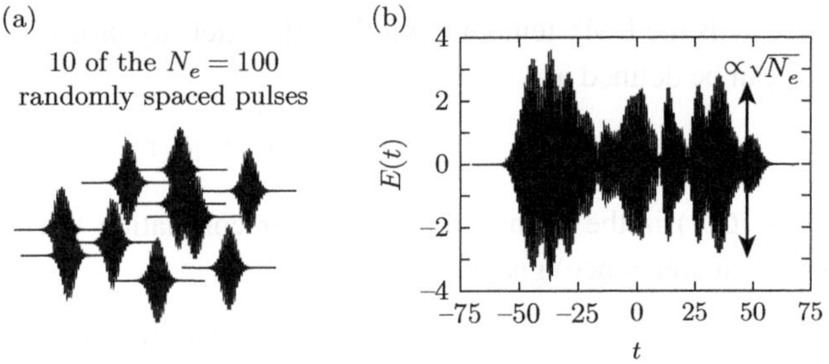

Figure 1.12 (a) Representation of the randomly phased wave packets that chooses 10 out of the 100 total waves. The individual waves are shown transversely displaced for illustration purpose only. (b) Total electric field, given by the incoherent sum of the 100 wave packets. The field consists of order $T/4\sigma_\tau \approx 10$ regular regions (i.e., $M_L \approx 10$ longitudinal modes).

relationship, namely, the coherence time. Note that the coherence time of a random collection of Gaussian waves (1.106) equals that of the single mode (1.101). Thus, each regular region can be identified with a coherent mode whose temporal width is of order the coherence time t_{coh}. The number of regular regions equals the number of coherent longitudinal modes M_L, which is roughly the ratio of the bunch length to the coherence length. Approximately, we have

$$M_L \approx \frac{T}{t_{\text{coh}}} = \frac{T}{2\sqrt{\pi}\sigma_\tau} \approx \frac{T}{4\sigma_\tau}. \tag{1.107}$$

The average field intensity scales linearly with the number of sources, while the instantaneous intensity fluctuates as a function of time. Associated with this intensity variation will be a fluctuation in the observed number of photons $\mathcal{N}_{\text{ph}}$ over a given time. Denoting the average photon number by $\langle\mathcal{N}_{\text{ph}}\rangle$, the RMS squared fluctuation in the number of photons observed is

$$\sigma^2_{\mathcal{N}_{\text{ph}}} = \frac{\langle\mathcal{N}_{\text{ph}}\rangle^2}{M_L}, \tag{1.108}$$

where M_L is the number of longitudinal modes in the observation time T.

The formula (1.108) for the photon number variation can be generalized in two respects. First, the mode counting must include the number of transverse modes M_T in both the x and y directions, so that the total number of modes

$$M = M_L M_T^2. \tag{1.109}$$

Second, there are inherent intensity fluctuations arising from quantum mechanical uncertainty in the form of photon shot noise. This number uncertainty is attributable to the discrete quantum nature of electromagnetic radiation, and, like any shot noise, it adds a contribution to $\sigma^2_{\mathcal{N}_{\text{ph}}}$ equal to the average number $\langle\mathcal{N}_{\text{ph}}\rangle$. Thus, the RMS squared photon number fluctuation is

$$\sigma^2_{\mathcal{N}_{\text{ph}}} = \frac{\langle\mathcal{N}_{\text{ph}}\rangle^2}{M} + \langle\mathcal{N}_{\text{ph}}\rangle = \frac{\langle\mathcal{N}_{\text{ph}}\rangle^2}{M}\left(1 + \frac{1}{\delta_{\text{degen}}}\right). \tag{1.110}$$

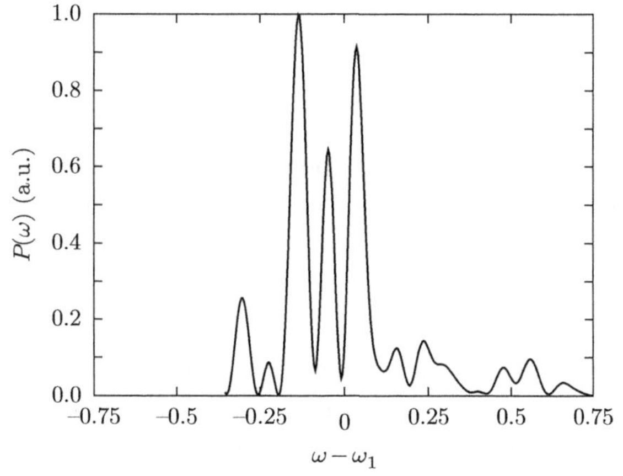

Figure 1.13 Intensity spectrum of Equation (1.111) using identical parameters as Figure 1.12(b). The spectrum consists of $M \sim 10$ sharp frequency spikes of approximate width $2/T \approx 0.02$, which are distributed within a Gaussian envelope of RMS width $\sigma_\omega \sim 0.25$. The height and placement of the spectral peaks fluctuate by 100 percent for different sets of random numbers.

The second term in parentheses is the inverse of the number of photons per mode, which is also known as the degeneracy parameter. In the classical devices that we consider there are many photons per mode, $\langle \mathcal{N}_{\text{ph}} \rangle / M \equiv \delta_{\text{degen}} \gg 1$, and the fluctuations due to quantum uncertainty are negligible. In this classical limit the length of the radiation pulse can be determined by measuring its intensity fluctuations, from which the source electron beam length may be deduced [11].

It is interesting to note that the mode counting we performed in the time domain can also be done in the frequency domain. Figure 1.13 shows the intensity spectrum $P(\omega) \propto |E_\omega|^2$, where

$$E_\omega = \frac{e_0\sqrt{\pi}}{\sigma_\omega} \sum_{j=1}^{N_e} \exp\left[-\frac{(\omega - \omega_1)^2}{4\sigma_\omega^2} + i\omega t_j \right] \tag{1.111}$$

using the same wave parameters as in Figure 1.12. The spectrum consists of sharp peaks of width $\Delta\omega \sim 2/T$ that are randomly distributed within the radiation bandwidth $\sigma_\omega = (2\sigma_\tau)^{-1}$. In other words, the frequency bandwidth $\Delta\omega$ of each mode is set by the duration of the entire radiation pulse T, while the frequency range over which the modes exist is given by the inverse coherence time. Thus, the total number of spectral peaks is the same as the number of the coherent modes in the time domain.

1.2.6 Bunching and Intensity Enhancement

Let us calculate the average electric field intensity generated by many electrons as expressed in Equation (1.111). Defining $|E_\omega^0|^2$ to be the intensity due to a single electron, we have

$$\left\langle \left| E(\omega) \right|^2 \right\rangle = \left| E^0_\omega \right|^2 \left\langle \left| \sum_{j=1}^{N_e} e^{i\omega t_j} \right|^2 \right\rangle, \tag{1.112}$$

where the angular bracket denotes an ensemble average over many instances of the initial particle distribution. Dividing the double sum into the piece where the particles are identical (the phase factor being unity) and the remaining terms, we obtain

$$\left\langle \left| \sum_{j=1}^{N_e} e^{i\omega t_j} \right|^2 \right\rangle = N_e + \left\langle \sum_{j \neq k}^{N_e} e^{i\omega(t_j - t_k)} \right\rangle. \tag{1.113}$$

We assume that the electrons are uncorrelated, so that the probability of finding any electron at position t_j is independent of the positions of all the other electrons. Thus, the temporal statistics are completely specified by the single particle probability distribution function $f(t)$, and the sum in Equation (1.113) can be expressed as $N_e(N_e - 1)$ identical integrals over f:

$$\left\langle \sum_{j \neq k}^{N_e} e^{i\omega(t_j - t_k)} \right\rangle = N_e(N_e - 1) \left| \int dt\, f(t) e^{i\omega t} \right|^2 \tag{1.114}$$

$$= N_e(N_e - 1) \left| f(\omega) \right|^2. \tag{1.115}$$

This expression is general for an arbitrary collection of electrons that are independently distributed in time according to $f(t)$. To get a physical understanding of (1.114), we consider a Gaussian distributed electron bunch,

$$f(t) = \frac{1}{\sqrt{2\pi}\, \sigma_e} \exp\left(-\frac{t^2}{2\sigma_e^2} \right),$$

where σ_e is the bunch length. Carrying out the Gaussian integral, we find

$$\left\langle \left| E(\omega) \right|^2 \right\rangle = N_e \left| E^0_\omega \right|^2 \left[1 + (N_e - 1) e^{-\omega^2 \sigma_e^2} \right]. \tag{1.116}$$

Typically, we have

$$(N_e - 1) e^{-\omega^2 \sigma_e^2} \ll 1$$

for the frequency range we are interested in. As an example, one nano-Coulomb of charge has $N_e \sim 10^{10}$, in which case $N_e e^{-\omega^2 \sigma_e^2} \sim 1$ when $c\sigma_e \sim \lambda$. Therefore, at wavelengths much shorter than the electron bunch length the second term in (1.116) is negligible, and the average radiation intensity due to N_e electrons is simply N_e times the intensity calculated for a single electron. This follows the usual rule of intensity addition for incoherent radiation arising from unbunched electron beams.

If, however, the bunch length becomes comparable to the radiation wavelength, we may have

$$(N_e - 1) e^{-\omega^2 \sigma_e^2} \geq 1,$$

resulting in a significant intensity enhancement over the incoherent case. In the extreme case where $c\sigma_e \ll \lambda$, i.e., in the limit of a vanishing bunch length, the intensity equals $N_e^2 \left| E^0_\omega \right|^2$, leading to an enhancement over the incoherent case by a factor of the number of electrons N_e. Two examples of situations in which such an intensity enhancement is often observed include coherent transition radiation, produced when an electron bunch traverses an interface between two media with differing indices of refraction, and coherent synchrotron radiation, generated when the beam trajectory is bent in a magnetic field. However, the coherent radiation produced through such processes is limited to the optical region of the spectrum $\lambda \gtrsim 300$ nm even for very high-charge, single femto-second electron beams.

For these reasons, typical synchrotron radiation sources generate temporally incoherent X-rays whose average intensity scales as the number of electrons in the bunch. We will discuss the properties of such incoherent radiation in the next chapter, both from simple bending magnets and from undulators, which are a periodic series of magnets with alternating polarities. There is, however, another mechanism to produce coherent radiation at wavelengths much shorter than the electron bunch. If we again consider either expression (1.114) or (1.115), we find that the coherent term scales as the absolute square of the Fourier transform of the distribution function. Thus, one can observe a coherent intensity enhancement of $|E^0_\omega|^2$ if the distribution function $f(t)$ has structure (microbunching) at the frequency ω:

$$\left\langle |E(\omega)|^2 \right\rangle = N_e |E^0_\omega|^2 \left(1 + (N_e - 1)\,|f(\omega)|^2\right). \tag{1.117}$$

Free-electron lasers (FELs) are devices in which the electron beam distribution develops a periodic density modulation on the scale of the radiation wavelength, resulting in a coherent enhancement of the intensity. The density modulation arises from the resonant interaction of an electron beam with the X-rays in a periodic undulator; we will see that if the undulator is sufficiently long, the bunch current is sufficiently high, and the e-beam phase space is of sufficient quality (small emittance and energy spread), than the radiation acts on the particles to generate a periodic density modulation whose length scale is near the resonant X-ray wavelength. This leads to a significant intensity enhancement as compared to the incoherent undulator radiation, even though the electron bunch length is much longer than the radiation wavelength.

References

[1] A. W. Chao, *Physics of Collective Beam Instabilities in High Energy Accelerators*. New York: Wiley, 1993.

[2] H. Wiedemann, *Particle Accelerator Physics I and II*, 2nd ed. Berlin: Springer-Verlag, 1999.

[3] K.-J. Kim, "Brightness, coherence, and propagation characteristics of synchrotron radiation," *Nucl. Instrum. Methods Phys. Res., Sect. A*, vol. 246, p. 71, 1986.

[4] R. Coisson and R. P. Walker, "Phase space distribution of brilliance of undulator sources," in *Proc. of SPIE*, I. E. Lindau and R. O. Tatchyn, Eds., no. 582. United States: SPIE, p. 24, 1986.

[5] E. Wigner, "On the quantum correction for thermodynamic equilibrium," *Phys. Rev.*, vol. 40, p. 749, 1932.

[6] N. D. Cartwright, "A non-negative Wigner-type distribution," *Physica*, vol. 83, p. 210, 1976.

[7] R. Jagannathan, R. Simon, E. C. G. Sudarshan, and R. Vasudevan, "Dynamical maps and nonnegative phase-space distribution functions in quantum mechanics," *Phys. Lett. A*, vol. 120, p. 161, 1987.

[8] M. Born and E. Wolf, *Principles of Optics*, 6th ed. New York: Pergamon Press, New York, 6th ed., 1980.

[9] L. Mandel and E. Wolf, "Coherence properties of optical fields," *Rev. Mod. Phys.*, vol. 37, p. 231, 1965.

[10] K.-J. Kim, "Characteristics of synchrotron radiation," in *Proc. US Particle Accelerator School*, ser. AIP Conference Proceedings, M. Month and M. Dienes, Eds., no. 184. New York: AIP, p. 565, 1989.

[11] P. Catravas, W. P. Leemans, J. S. Wurtele, M. S. Zolotorev, M. Babzien, I. Ben-Zvi, Z. Segalov, X.-J. Wang, and V. Yakimenko, "Measurement of electron-beam bunch length and emittance using shot-noise-driven fluctuations in incoherent radiation," *Phys. Rev. Lett*, vol. 82, p. 5261, 1999.

2 Synchrotron Radiation

Synchrotron radiation – the radiation produced by a charged particle that moves with relativistic speed along a curved trajectory – has become an important experimental tool in many areas of basic and applied science. Modern synchrotron facilities, often referred to as light sources, use high-quality electron beams to provide a high-brightness flux of photons in the soft to hard X-ray region of the spectrum that is not accessible with other sources.

There are three basic kinds of synchrotron radiation sources: bending magnets, wigglers, and undulators. In bending magnets, electrons move in a circular trajectory and produce a smooth X-ray spectrum. Wigglers can be regarded as a sequence of bending magnets of alternate polarities that bend the particle back and forth about a nominally straight trajectory. Thus, the radiation characteristics are similar to those of bending magnets, apart from an intensity increase proportional to the number of magnetic poles; for a wiggler comprised of $2N_u$ magnets we have a $2N_u$-fold intensity enhancement. We will not further discuss wiggler radiation in this book, although the emission from wigglers can also be considered as a limiting case of undulator radiation. An undulator is an N_u-period magnetic structure that produces a gentle, periodic orbit. Because of the interference of radiation from different parts of the trajectory, undulator radiation is squeezed into a discrete spectrum characterized by a narrower emission angle. The result is an N_u^2-fold enhancement of the angular flux density in the forward direction. The electron trajectories and spectral characteristics of radiation from bending magnets, wigglers and undulators are schematically illustrated and compared in Figure 2.1.

For a given experiment, a wide range of source properties may be desired including, but not limited to, peak power, average power, total number of photons, degree of temporal and/or transverse coherence, pulse duration, stability, etc. Nevertheless, probably the single most useful intrinsic metric by which synchrotron radiation sources can be judged is the brightness $\mathcal{B}$, namely, the photon flux per unit bandwidth per unit phase space area.[1] High brightness means that the radiation can travel a long distance without significant spread and can be focused to a small spot, while subsequent manipulations of the radiation typically cannot improve brightness. For example, introducing an aperture to improve the transverse coherence will at best preserve the brightness while decreasing the total flux in proportion to the decrease in phase space area; similar considerations apply when using a monochromator to select a narrow radiation bandwidth.

[1] Here we consider the brightness to be a photon spectral flux density in the transverse phase space $(\boldsymbol{x}, \boldsymbol{\phi})$, but it can also be thought of as the photon number density in the six-dimensional phase space $(\boldsymbol{x}, t, \boldsymbol{\phi}, \omega)$; the former is more in line with previous notation, while the latter makes apparent that changing the integrated $\mathcal{B}$ requires the radiation or absorption of photons.

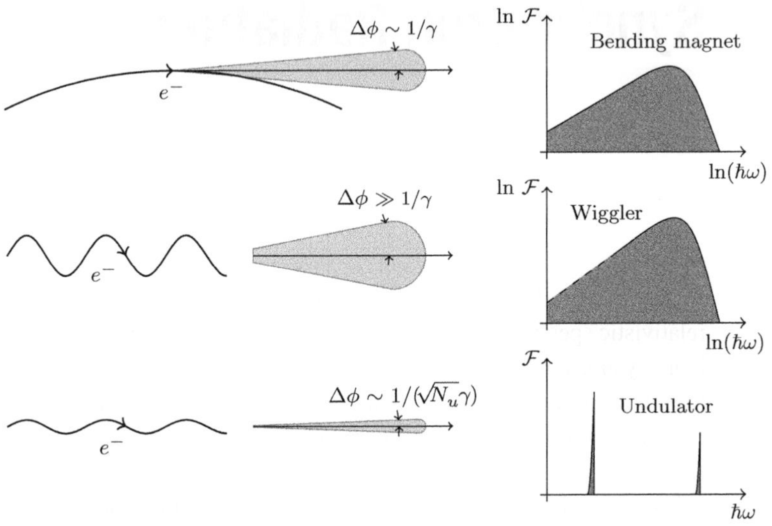

Figure 2.1 An illustration of the angular distribution and spectral flux characteristics of radiation from bending magnets, wigglers, and undulators. Adapted from Ref. [1].

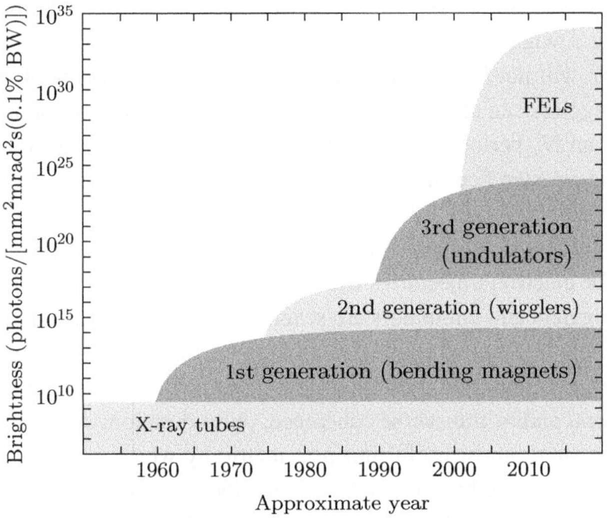

Figure 2.2 Illustration of the increase in the peak X-ray source brightness over the years from various sources.

The brightness of synchrotron radiation has steadily increased over the years thanks to advances in the accelerator technology that is used to store high-current, tightly focused, and stable electron beams, and to the development of strong and precise magnets that control the electron orbit for optimal production of radiation. In Figure 2.2 we approximately trace the increase in X-ray brightness over the years as light sources have improved. We can see that synchrotron radiation is orders of magnitude brighter than traditional X-ray tubes: the increase in $\mathcal{B}$ is at least two orders of magnitude for bending magnets, over ten orders of magnitude for undulators in modern facilities, and even more for X-ray FELs.

The goal of this chapter is to introduce the basic physics of synchrotron radiation, with a particular emphasis on those concepts relevant to the study of free-electron lasers. For this reason, our discussion of bending magnet radiation is mostly conceptual in nature, while we treat undulator radiation more carefully, but by no means exhaustively. The interested reader may consult, e.g., the review article [1] or the books [2, 3]. We begin this chapter with a quick overview of the physics of radiation from relativistic electrons in Section 2.1, and proceed with the discussion of the paraxial equation in 2.2 and its application to bending magnet radiation in Section 2.3. Section 2.4 treats undulator radiation in some detail, including its basic physics, analytic expressions of the flux, and a treatment of the effects of the electron beam distribution on the brightness. Finally, we conclude by surveying how the peak brightness of undulator radiation can be increased by approximately ten orders of magnitude with free-electron lasers.

2.1 Radiation by Relativistic Electrons

Synchrotron radiation was first characterized by Schott [4] and later by Schwinger [5]. Its unique properties come from the fact that the motion of an electron moving close to the speed of light toward a stationary observer appears to occur over a much shorter time scale than that of the real motion [6]. In this section we discuss the general physics of radiation from a relativistic electron.

Consider a relativistic electron moving on an arbitrary trajectory and a stationary observer as shown in Figure 2.3(a). We orient our coordinate system such that the observer is situated at the origin, and let $\boldsymbol{r}(t')$ be the position of the electron at time t'. For this geometry, an electromagnetic signal emitted by the electron at time t' and traveling on a straight path will arrive at the observer at the time

$$t = t' + \left|\boldsymbol{r}(t')\right| / c. \tag{2.1}$$

Equation (2.1) is a relation between the two times t and t'. We designate the time t to be the *observer time*. The time t' is typically referred to by the somewhat mysterious "retarded time," but here we will call it the *emitter time* in order to more clearly contrast it with the observer time. As the definition suggests, the electron motion $\boldsymbol{r}(t')$ is a specified function of the emitter time. However, the motion appears different to the stationary observer situated at the origin: the apparent motion in terms of the observer time is obtained by stretching or squeezing the time axis according to Equation (2.1):

$$\boldsymbol{r}_{\text{obs}}(t) = \boldsymbol{r}(t'(t)), \tag{2.2}$$

where $t'(t)$ is a solution of (2.1). To see how the time scale changes, we consider a small emitter time interval $\Delta t'$ and the corresponding observer time interval Δt. We have

$$\Delta t \approx \frac{dt}{dt'}\Delta t' = \left[1 - \beta(t')\cos\phi(t')\right]\Delta t', \tag{2.3}$$

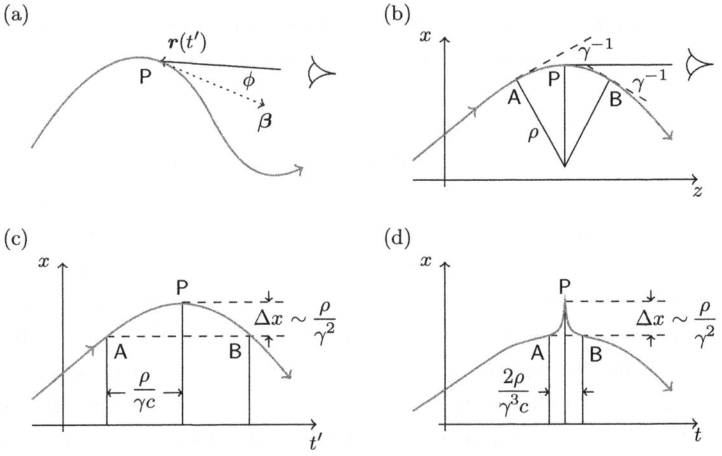

Figure 2.3 (a) The electron trajectory and a stationary observer. (b) Trajectory with local radius of curvature ρ as a function of z. Tangent lines at points A and B make an angle γ^{-1} from the observation direction at point P. (c) Plot of the same trajectory and points as a function of the emitter time t'. (d) Shows the motion as a function of the observer time t, where the distance between points A and B are squeezed by the factor $1/\gamma^2$. Adapted from Ref. [1].

where $\beta(t')$ is the instantaneous particle speed scaled by c and $\cos\phi(t')$ is the angle that the particle's velocity vector makes with its position vector as shown in Figure 2.3(a). The function $\left[1-\beta(t')\cos\phi(t')\right]$ is the t'-dependent scale-change factor that yields the time squeezing or stretching relevant to synchrotron radiation. Note that while this scale change factor may be reminiscent of the Lorentz transformation of special relativity, it does not originate from a change in inertial frame. Nevertheless, the basic physics of synchrotron radiation relies on the special relativistic tenet that the speed of light is independent of the motion of the emitter.

For extremely relativistic electrons the time squeezing depends sensitively on the angle ϕ. If we replace β by the Lorentz factor γ using

$$\beta = \sqrt{1-\frac{1}{\gamma^2}} \approx 1-\frac{1}{2\gamma^2} \tag{2.4}$$

for $\gamma \gg 1$, the time squeezing factor (2.3) becomes

$$1-\beta\cos\phi \approx \begin{cases} O(1) & \text{for } \phi \sim 1 \\ \dfrac{1}{2}\left(\dfrac{1}{\gamma^2}+\phi^2\right) & \text{for } \phi \ll 1 \end{cases}. \tag{2.5}$$

From Equation (2.5), we see that the scale difference in the emitter time and the observer time is greatest when $\phi \lesssim \gamma^{-1}$:

$$\Delta t \sim \Delta t'/\gamma^2 \quad \text{if} \quad \phi \lesssim \gamma^{-1}\,. \tag{2.6}$$

In other words, when the electron velocity is pointing toward the observer within an angle γ^{-1}, the electron motion over a time interval $\Delta t'$ appears to the observer to be squeezed into the much shorter interval given by (2.6). We show this effective squeezing of the motion in Figure 2.3(c)–(d). At larger angles, this time-squeezing effect is not as significant.

The time-squeezing effect in (2.6) is essentially due to the fact that a highly relativistic electron can closely follow the photons that it emitted at earlier times. Therefore, any photons that the trailing electron emits at subsequent times will be measured over a comparatively short time interval by the stationary observer. As an extreme example, consider a hypothetical electron traveling at exactly the speed of light c toward the observer. In this case, signals emitted at two different points along the trajectory will reach the observer at the same time ($\Delta t = 0$!), and the time squeezing effect will be infinite.

The electric field strength detected by an observer is proportional to the apparent acceleration of the motion as seen by the observer, while the apparent acceleration becomes large when the time squeezing effect is important. Thus, a significant amount of radiation will be observed when the electron motion and the observation direction maximizes the time-squeezing effect. This is the basic explanation of synchrotron radiation.

A complete mathematical formulation of this basic physics can be based on the Liénard–Wiechert potentials, as explained in [7, 8]. These expressions have been used by several authors (see, e.g., [1, 9, 10]) to make a thorough quantitative analysis of synchrotron radiation in the far-field. Our approach is based on the paraxial wave equation driven by an appropriate electron source, which is more comparable to the paraxial study of Ref. [11]. Our formulation is relatively simple, and will be directly relevant to our subsequent study of FEL theory.

2.2 The Driven Paraxial Wave Equation

To calculate the radiation field due to relativistically moving charges, we first write out the full wave equation for E_ω in the frequency domain:

$$\left[\frac{\partial^2}{\partial z^2} + \frac{\partial^2}{\partial \boldsymbol{x}^2} + k^2\right]\boldsymbol{E}_\omega(\boldsymbol{x}; z) = \frac{1}{\epsilon_0 c^2}\frac{1}{2\pi}\int dt\, e^{i\omega t}\left[\frac{\partial \boldsymbol{J}}{\partial t} + c^2\frac{\partial \rho_e}{\partial \boldsymbol{x}}\right]. \tag{2.7}$$

Our derivation of the paraxial equation begins with the introduction of the slowly varying amplitude in the angular representation $\tilde{\mathcal{E}}_\omega$ as

$$E_\omega(\boldsymbol{x}; z) = e^{ikz}\tilde{E}_\omega(\boldsymbol{x}; z) = e^{ikz}\int d\boldsymbol{\phi}\; e^{ik\boldsymbol{\phi}\cdot\boldsymbol{x}}\tilde{\mathcal{E}}_\omega(\boldsymbol{\phi}; z). \tag{2.8}$$

The envelope $\tilde{\mathcal{E}}_\omega$ has factored out the fast oscillations along z with wavelength $\lambda = 2\pi/k = 2\pi c/\omega$ in an identical manner as when we first derived the paraxial wave equation in Chapter 1.2.2. To arrive at the first-order (in z) paraxial wave equation, we again assume that the variation of the envelope $\tilde{\mathcal{E}}_\omega$ occurs over a much longer scale length than $1/k$, in which case we drop the second-order derivatives with respect to z from the wave equation. The precise requirements and range of validity of this assumption will depend upon the radiation source considered, but in any case the resulting angular representation of the driven paraxial wave equation is given by

$$\left[\frac{\partial}{\partial z} + \frac{ik}{2}\boldsymbol{\phi}^2\right]\tilde{\mathcal{E}}_\omega(\boldsymbol{\phi};z) = -\frac{1}{4\pi\epsilon_0 c\lambda^2}\int dt\, d\boldsymbol{x}\; e^{ik(ct-z-\boldsymbol{\phi}\cdot\boldsymbol{x})} \times \left[\boldsymbol{J}(\boldsymbol{x},t;z) - c\rho_e(\boldsymbol{x},t;z)\boldsymbol{\phi}\right], \tag{2.9}$$

where we have integrated both the charge and current density by parts. The charge density due to electron j is $\rho_e = -e\delta[z - z_j(t)]\delta[\boldsymbol{x} - \boldsymbol{x}_j(t)]$, while the current density is $\boldsymbol{J} = \rho_e \boldsymbol{v}_j(t)$; both of these can be expressed in terms of the particle time coordinate $t_j(z)$ by using the identity $\delta[z - z_j(t)] = \delta[t - t_j(z)]/|dz/dt_j|$. Thus, the paraxial wave equation relevant to a single electron source is

$$\left[\frac{\partial}{\partial z} + \frac{ik}{2}\boldsymbol{\phi}^2\right]\tilde{\mathcal{E}}_\omega(\boldsymbol{\phi};z) = \frac{e[\boldsymbol{\beta}_j(z) - \boldsymbol{\phi}]}{4\pi\epsilon_0\lambda^2|dz/dt_j|}e^{ik[ct_j(z)-z-\boldsymbol{\phi}\cdot\boldsymbol{x}_j(z)]}. \tag{2.10}$$

The expression (2.10) is general for all paraxial synchrotron radiation, with the terms proportional to $\boldsymbol{\beta}_j$ and $\boldsymbol{\phi}$ arising from the time derivative of the particle current density and the gradient of the charge density, respectively. The former is given by the transverse velocity, whose order of magnitude $|\boldsymbol{\beta}_j| \sim 1/\gamma$, while the latter is appreciable if $\phi \approx 1/\gamma$ and is negligible both for $\phi \to 0$ and for $\phi \gg 1/\gamma$.

The paraxial equation in the angular representation (2.10) is that of a driven oscillator with "frequency" $k\boldsymbol{\phi}^2/2$, and the solution can be obtained using the Green function $e^{-ik(z-z')\boldsymbol{\phi}^2/2}\Theta(z - z')$. Alternatively, we can multiply both sides of (2.10) by $e^{ik\boldsymbol{\phi}^2 z/2}$ and write the left-hand side as

$$e^{ik\boldsymbol{\phi}^2 z/2}\left[\frac{\partial}{\partial z} + \frac{ik}{2}\boldsymbol{\phi}^2\right]\tilde{\mathcal{E}}_\omega(\boldsymbol{\phi};z) = \frac{\partial}{\partial z}\left[e^{ik\boldsymbol{\phi}^2 z/2}\tilde{\mathcal{E}}_\omega(\boldsymbol{\phi};z)\right]. \tag{2.11}$$

Now the integration is trivial, and we find that

$$\tilde{\mathcal{E}}_\omega(\boldsymbol{\phi};z) = \int_{-\infty}^{z} dz'\; \frac{e[\boldsymbol{\beta}_j(z') - \boldsymbol{\phi}]}{4\pi\epsilon_0\lambda^2 c^2}e^{ik[ct_j(z')-z'-\boldsymbol{\phi}\cdot\boldsymbol{x}_{j}z'+\boldsymbol{\phi}^2(z'-z)/2]} \tag{2.12}$$

$$\equiv \frac{e\omega^2}{16\pi^3\epsilon_0 c^2}\int_{-\infty}^{z/c} d\zeta\; [\boldsymbol{\beta}_j(\zeta) - \boldsymbol{\phi}]e^{i\omega\tau(\zeta,\boldsymbol{\phi})}, \tag{2.13}$$

where we have set $|dz_j/dt| \approx c$ for a relativistic particle directed along the optical axis. The second line equation (2.13) replaces the integration variable z' by ζ/c and defines the variable

$$\tau(\zeta,\boldsymbol{\phi}) \equiv t_j(\zeta) - \zeta - \boldsymbol{\phi}\cdot\boldsymbol{x}_j(\zeta)/c + \boldsymbol{\phi}^2(\zeta - z/c)/2], \tag{2.14}$$

which is closely related to the emitter time introduced in the previous section. In fact, it turns out that the time-squeezing factor defined in (2.3) equals the derivative $\frac{\partial}{\partial\zeta}\tau(\zeta,\boldsymbol{\phi}) \equiv \dot{\tau}(\zeta,\boldsymbol{\phi})$.

The formula (2.12) gives the electric field in the spectral-angular representation once the electron trajectory is known. In addition to the field, it is also useful to have expressions for the power and power density. We start by inserting the field (2.13) into Equation (1.72)

$$\frac{dP}{d\omega d\boldsymbol{\phi}} = \frac{e^2\omega^2}{16\pi^3\epsilon_0 cT}\int d\zeta d\zeta'\, [\boldsymbol{\beta}_j(\zeta) - \boldsymbol{\phi}]\cdot[\boldsymbol{\beta}_j(\zeta') - \boldsymbol{\phi}]e^{i\omega[\tau(\zeta,\boldsymbol{\phi})-\tau(\zeta',\boldsymbol{\phi})]}. \tag{2.15}$$

The angular power density obtains by integrating over frequency; since this integral is not straightforward, we write it out in steps. First, we rewrite the factor ω^2 using derivatives with respect to ζ as follows:

$$\begin{aligned}\frac{dP}{d\boldsymbol{\phi}} &= \frac{e^2}{16\pi^3\epsilon_0 cT}\int d\zeta\, \frac{[\boldsymbol{\beta}_j(\zeta) - \boldsymbol{\phi}]}{\dot{\tau}(\zeta,\boldsymbol{\phi})}\cdot\frac{\partial}{\partial\zeta}\left\{\frac{-1}{\dot{\tau}(\zeta,\boldsymbol{\phi})}\right.\\ &\quad\times\left.\frac{\partial}{\partial\zeta}\int d\zeta'\,[\boldsymbol{\beta}_j(\zeta') - \boldsymbol{\phi}]\int d\omega\; e^{i\omega[\tau(\zeta,\boldsymbol{\phi})-\tau(\zeta',\boldsymbol{\phi})]}\right\}.\end{aligned} \tag{2.16}$$

Note that the integration is only over positive ω. Nevertheless, since P is real, the integral over frequency can be performed in the following way:

$$\begin{aligned}\int_0^\infty d\omega\; e^{i\omega[\tau(\zeta,\boldsymbol{\phi})-\tau(\zeta',\boldsymbol{\phi})]} &= \frac{1}{2}\int_{-\infty}^{\infty} d\omega\; e^{i\omega[\tau(\zeta,\boldsymbol{\phi})-\tau(\zeta',\boldsymbol{\phi})]}\\ &= \pi\delta[\tau(\zeta,\boldsymbol{\phi}) - \tau(\zeta',\boldsymbol{\phi})] = \frac{\pi\delta(\zeta-\zeta')}{|\dot{\tau}(\zeta,\boldsymbol{\phi})|}.\end{aligned} \tag{2.17}$$

This in turn makes the integral over ζ' in (2.16) trivial, so that the angular power density is

$$\begin{aligned}\frac{dP}{d\boldsymbol{\phi}} &= -\frac{e^2}{16\pi^2\epsilon_0 cT}\int d\zeta\, \frac{[\boldsymbol{\beta}_j(\zeta) - \boldsymbol{\phi}]}{\dot{\tau}(\zeta,\boldsymbol{\phi})}\cdot\frac{\partial}{\partial\zeta}\left\{\frac{1}{\dot{\tau}(\zeta,\boldsymbol{\phi})}\frac{\partial}{\partial\zeta}\frac{[\boldsymbol{\beta}_j(\zeta) - \boldsymbol{\phi}]}{\dot{\tau}(\zeta,\boldsymbol{\phi})}\right\}\\ &= \frac{e^2}{16\pi^2\epsilon_0 cT}\int d\zeta\, \frac{\{\dot{\boldsymbol{\beta}}_j(\zeta)\dot{\tau}(\zeta,\boldsymbol{\phi}) - [\boldsymbol{\beta}_j(\zeta) - \boldsymbol{\phi}]\ddot{\tau}(\zeta,\boldsymbol{\phi})\}^2}{\dot{\tau}^5(\zeta,\boldsymbol{\phi})}.\end{aligned} \tag{2.18}$$

The final result (2.18) integrates by parts and notes that the boundary terms are proportional to the acceleration $\dot{\beta}(\zeta)$, which vanishes if the observation point is beyond the magnetic structure.

Now, we further simplify the frequency integrated power density by inserting the time derivative $dt_j/d\zeta = 1/\beta_z \approx 1 + \boldsymbol{\beta}_\perp^2/2 + 1/2\gamma^2$ into the phase derivatives $\dot{\tau}$ and $\ddot{\tau}$. Then, we have

$$\dot{\tau}(\zeta,\boldsymbol{\phi}) = \frac{1}{2\gamma^2} + \frac{1}{2}(\boldsymbol{\beta}_j - \boldsymbol{\phi})^2 \qquad \ddot{\tau}(\zeta,\boldsymbol{\phi}) = \dot{\boldsymbol{\beta}}_j\cdot(\boldsymbol{\beta}_j - \boldsymbol{\phi}), \tag{2.19}$$

and the numerator of (2.18) becomes

$$\begin{aligned}\dot{\boldsymbol{\beta}}_j\dot{\tau} - (\boldsymbol{\beta}_j - \boldsymbol{\phi})\ddot{\tau} &= \frac{\dot{\beta}_x}{2\gamma^2}[1 - \gamma^2(\beta_x - \phi_x)^2 + \gamma^2\phi_y^2]\hat{x}\\ &\quad + \frac{\dot{\beta}_x}{2\gamma^2}2\gamma(\beta_x - \phi_x)\gamma\phi_y\hat{y}.\end{aligned} \tag{2.20}$$

Here, our final result has assumed that the motion is in the x-z plane, so that $\boldsymbol{\beta}_j$ is directed solely along $\hat{x}$. When inserted into (2.18), the first line in (2.20) will give the power contribution in the component polarized along x, while the second line contains the power in the y-polarization. To avoid potentially confusing x, y subscripts on the power, we will denote the power in the x component as P_σ, and the power along y as P_π; this is in keeping with traditional notation when the subsequent reflecting optic is oriented vertically. Collecting all these results, we find the paraxial power density

$$\frac{dP_\sigma}{d\boldsymbol{\phi}} = \frac{e^2\gamma^6}{2\pi^2\epsilon_0 cT} \int d\zeta \, \frac{\dot{\beta}_x^2\left[1 - \gamma^2(\beta_x - \phi_x)^2 + \gamma^2\phi_y^2\right]^2}{\left[1 + \gamma^2(\beta_x - \phi_x)^2 + \gamma^2\phi_x^2\right]^5}, \tag{2.21}$$

$$\frac{dP_\pi}{d\boldsymbol{\phi}} = \frac{e^2\gamma^6}{2\pi^2\epsilon_0 cT} \int d\zeta \, \frac{4\dot{\beta}_x^2\gamma^2(\beta_x - \phi_x)^2\gamma^2\phi_y^2}{\left[1 + \gamma^2(\beta_x - \phi_x)^2 + \gamma^2\phi_x^2\right]^5}. \tag{2.22}$$

2.3 Bending Magnet Radiation

As a quick illustration of the analysis and properties of the paraxial equation (2.10) and its power expressions (2.21)–(2.22), in this section we consider the synchrotron radiation produced by a bending magnet. In this case, we approximate the motion by a circular trajectory in the x-z plane with radius ρ at a constant speed v:

$$x_j(t) = \rho - \rho\cos(vt/\rho) \qquad y_j(t) = 0 \qquad z_j(t) = \rho\sin(vt/\rho). \tag{2.23}$$

To obtain $t_j(z)$ for a bending magnet, we invert the final equation in (2.23). The radiation along the optical axis is predominantly generated when the particle makes an angle $\lesssim 1/\gamma$ with respect to $\hat{z}$, i.e., when $vt/\rho \lesssim 1/\gamma \ll 1$. Thus, for the relevant part of the trajectory we can solve for the particle time $t_j(z)$ by expanding $z_j(t)$ in vt/ρ and inverting order-by-order. At third order we have

$$ct_j(z) \approx \frac{z}{\beta} + \frac{\rho}{6\beta}\frac{z^3}{\rho^3} \approx \left(1 + \frac{1}{2\gamma^2}\right)z + \frac{\rho}{6}\frac{z^3}{\rho^3}, \tag{2.24}$$

where we have expanded the scaled speed $\beta = (1 - 1/\gamma^2)^{1/2}$ assuming $\gamma \gg 1$. Plugging (2.24) into the paraxial wave equation (2.10), we find that the first term in the expansion for $t_j(z)$ cancels the like term equal to kz in the phase of current. The remaining phase $\sim kz/\gamma^2$ is a hallmark of the time-squeezing effect in the paraxial approximation: because the particle speed along the axis approaches that of light, the characteristic frequency of the radiation is reduced by a factor $\sim 1/\gamma^2$.

In the forward direction $\boldsymbol{\phi} = \mathbf{0}$, Equation (2.10) can be directly integrated in z using the transverse velocity $\beta_j(z) \approx (v^2t/c\rho)\hat{x} \approx (z/\rho)\hat{x}$. Introducing the critical frequency for bending magnet radiation,

$$\omega_c \equiv \frac{3\gamma^3 c}{2\rho}, \tag{2.25}$$

the on-axis field is given polarized along x and is given by

$$\tilde{\mathcal{E}}_\omega(\mathbf{0}) = \frac{e}{4\pi\epsilon_0\lambda^2}\frac{3\gamma}{2\omega_c}\int d\zeta \; \zeta \exp\left[\frac{3i\omega}{4\omega_c}(\zeta + \zeta^3/3)\right], \tag{2.26}$$

where the dimensionless variable $\zeta \equiv \gamma z/\rho$. The integral (2.26) can be evaluated in terms of modified Bessel functions, but we will merely discuss its qualitative behavior. For frequencies less than or of order ω_c, the integral is a number $O(1)$, while when $\omega \gg \omega_c$ the integrand oscillates rapidly and the integral largely cancels to zero. Thus, the radiation is broadband, with $4\omega_c$ customarily considered the maximum frequency of useful photon flux. The total number of radiated photons associated with the field (2.26) can be approximated by integrating the total flux $T\mathcal{F}$ from (1.73) over frequency:

$$\begin{aligned}\mathcal{N}_{\text{ph}} \equiv \int d\omega\, T\frac{d\mathcal{F}}{d\omega} &= \int d\omega\, \frac{4\pi\epsilon_0 c\lambda^2}{\hbar\omega} \int d\boldsymbol{\phi}\, \left|\tilde{\mathcal{E}}_\omega(\boldsymbol{\phi}; z)\right|^2 \\ &\sim \Delta\omega\Delta\boldsymbol{\phi}\frac{4\pi\epsilon_0 c\lambda_c^2}{\hbar\omega_c}\left|\tilde{\mathcal{E}}_{\omega_c}(\mathbf{0}; z)\right|^2, \end{aligned} \tag{2.27}$$

where $\lambda_c = 2\pi c/\omega_c$ is the critical wavelength of bending magnet radiation. The radiation has a broad bandwidth, implying that $\Delta\omega \sim 2\omega_c$, while the angular width is $\sim 1/\gamma$ in each transverse angle, so that $\Delta\boldsymbol{\phi} \sim 2\pi/\gamma^2$. Using (2.26), the on-axis intensity is

$$\left|\tilde{\mathcal{E}}_{\omega_c}(\mathbf{0}; z)\right|^2 \sim \left(\frac{e}{4\pi\epsilon_0\lambda_c^2}\frac{3\gamma}{2\omega_c}\right)^2 = \frac{\alpha\hbar}{4\pi\epsilon_0 c\lambda_c^2}\frac{9\gamma^2}{16\pi^2}, \tag{2.28}$$

where the fine structure constant $\alpha \equiv e^2/(4\pi\epsilon_0\hbar c) \approx 1/137$. Combining (2.28) and (2.27) with $\Delta\omega\Delta\boldsymbol{\phi} \sim 4\pi\omega_c/\gamma^2$, we find that

$$\mathcal{N}_{\text{ph}} \approx \frac{9}{4\pi}\alpha \sim \alpha, \tag{2.29}$$

reproducing the well-known result that on average electrons radiate approximately α photons when bent through an angle $1/\gamma$.

More detailed analysis of the paraxial equation (2.10) with the circular trajectory (2.26) can be done to solve for the full angular flux density of bending magnet radiation. Here, we present a simple physical explanation of the characteristic source size and divergence of bending magnet radiation, and then close with more detailed analysis of the frequency integrated power distribution. In view of our discussion of the apparent motion shown in Figure 2.3, we expect that the acceleration over an arc of length $\ell \sim \rho/\gamma$ contributes to the field. The effective source size in the horizontal and vertical directions is therefore $\Delta x \sim \ell\Delta\phi_x$ and $\Delta y \sim \ell\Delta\phi_y$. Additionally, the opening angle in the horizontal direction $\Delta\phi_x \sim 1/\gamma$ for frequencies much lower than the critical frequency $\omega_c = 3c\gamma^3/2\rho$, while being smaller than this in the high frequency limit. Hence, we have

$$\Delta x \sim \frac{\rho}{\gamma}\Delta\phi_x \sim \rho\begin{cases}(\Delta\phi_x)^2 & \text{if } \omega \ll \omega_c \\ \Delta\phi_x/\gamma & \text{if } \omega \gg \omega_c\end{cases}, \qquad \Delta y \sim \Delta x\frac{\Delta\phi_y}{\Delta\phi_x}. \tag{2.30}$$

Since a spatially confined radiation field will expand due to diffraction, we also have the conditions $\Delta x \sim \lambda/\Delta\phi_x$ and $\Delta y \sim \lambda/\Delta\phi_y$. Combining the diffractive limits with the bending magnet sizes (2.30), we find that

$$\Delta\phi_x \sim \Delta\phi_y \sim \begin{cases} \left(\dfrac{\lambda}{\rho}\right)^{1/3} \sim \dfrac{1}{\gamma}\left(\dfrac{\omega_c}{\omega}\right)^{1/3} & \text{if } \omega \ll \omega_c \\ \left(\dfrac{\gamma\lambda}{\rho}\right)^{1/2} \sim \dfrac{1}{\gamma}\left(\dfrac{\omega_c}{\omega}\right)^{1/2} & \text{if } \omega \gg \omega_c, \end{cases} \tag{2.31}$$

while

$$\Delta x \sim \Delta y \sim \begin{cases} \rho\left(\dfrac{\lambda}{\rho}\right)^{2/3} & \text{if } \omega \ll \omega_c \\ \left(\dfrac{\rho\lambda}{\gamma}\right)^{1/2} & \text{if } \omega \gg \omega_c. \end{cases} \tag{2.32}$$

The source size introduced above is the size of the image when focused by 1:1 optics. The idea of source size and divergence will be quite useful in the next section when we discuss undulator radiation.

The previous discussion showed the frequency dependence of the source size and divergence. If we integrate over frequency, we will find the associated power distribution. For this analysis, we use the results from the end of Section 2.2 given by Equations (2.21)–(2.22). Again, in these formulas the power in the x component is denoted as P_σ, and the power in the y polarization as P_π. Using these expressions with the bending magnet velocity $\beta_x \approx z/\rho = c\zeta/\rho$ and acceleration $\dot{\beta}_x = c/\rho$, we find that the power

$$\frac{dP_\sigma}{d\boldsymbol{\phi}} = \frac{e^2\gamma^5}{\pi\epsilon_0\rho T}\int d\xi\, \frac{[1-\xi^2+\gamma^2\phi_y^2]^2}{[1+\xi^2+\gamma^2\phi_y^2]^5} \tag{2.33}$$

$$\frac{dP_\pi}{d\boldsymbol{\phi}} = \frac{e^2\gamma^5}{\pi\epsilon_0\rho T}\int d\xi\, \frac{4\xi^2\gamma^2\phi_y^2}{[1+\xi^2+\gamma^2\phi_y^2]^5}, \tag{2.34}$$

where $\xi = \gamma(c\zeta/\rho - \phi_x)$. The integrations can be performed in the complex plane by closing the contour at infinity and using the residue theorem. We find that

$$\frac{dP_\sigma}{d\boldsymbol{\phi}} = \frac{7}{64\pi}\frac{e^2\gamma^4}{\epsilon_0 mc}\frac{eB}{mc}\frac{I}{e}\frac{1}{[1+\gamma^2\phi_y^2]^{5/2}} \tag{2.35}$$

$$\frac{dP_\pi}{d\boldsymbol{\phi}} = \frac{7}{64\pi}\frac{e^2\gamma^4}{\epsilon_0 mc}\frac{eB}{mc}\frac{I}{e}\frac{5\gamma^2\phi_y^2}{[1+\gamma^2\phi_y^2]^{7/2}}. \tag{2.36}$$

Along the axis the X-rays are polarized purely along x, and the power density in the forward direction

$$\left.\frac{dP_\sigma}{d\boldsymbol{\phi}}\right|_{\boldsymbol{\phi}=\mathbf{0}} = \frac{7}{64\pi}\frac{e^2\gamma^4}{\epsilon_0 mc}\frac{eB}{mc}\frac{I}{e} = 5.43B[\mathrm{T}](\gamma mc^2)^4[\mathrm{GeV}]I[\mathrm{A}]\ \mathrm{W/mrad}^2. \tag{2.37}$$

If we integrate both (2.35) and (2.36) over the vertical angle ϕ_y, we find that the total power in the σ/x polarization is seven times that in the π/y polarization; integrating the sum of these two results over the horizontal acceptance angle $\Delta\phi_x$ gives

$$P = \frac{e^2\gamma^2}{6\pi\epsilon_0}\frac{I}{e}\left(\frac{eB}{mc}\right)^2(\rho\Delta\phi_x), \tag{2.38}$$

$$\approx 1.27\,(\gamma mc^2)[\text{GeV}]B^2[\text{T}]I[\text{A}](\rho\Delta\phi_x)[\text{m}]\ \text{kW}. \tag{2.39}$$

2.4 Undulator Radiation

Thus far, we have introduced some of the basic physics of synchrotron radiation, derived the driven paraxial wave equation, and illustrated certain characteristics of the generated X-ray flux and power from bending magnets. In this section we analyze undulator radiation in some detail, both because undulators radiation is important in its own right as a bright source of X-rays, and because undulator radiation can act back on the electrons to produce FEL gain. An undulator is a periodic magnetic structure built of magnets with opposite polarities. Most undulators in use today are constructed from permanent magnets following the pioneering work by Halbach [12, 13]. A typical modern insertion device based on a permanent-magnet/steel hybrid design is illustrated in Figure 2.4. With optimal design, the peak magnetic field strength is given by the Halbach formula

$$B_0[T] \approx 3.44\ \exp\left[-\frac{g}{\lambda_u}\left(5.08 - 1.54\frac{g}{\lambda_u}\right)\right] \qquad \text{[neodymium-iron]},$$

$$B_0[T] \approx 3.33\ \exp\left[-\frac{g}{\lambda_u}\left(5.47 - 1.8\frac{g}{\lambda_u}\right)\right] \qquad \text{[samarium-cobalt]}, \tag{2.40}$$

where g is the full gap between the magnets through which the electron beam propagates and $g < \lambda_u$.

This book will primarily consider planar undulators for which the magnetic field $\boldsymbol{B}$ is directed predominantly along the vertical axis (the y direction) and varies periodically

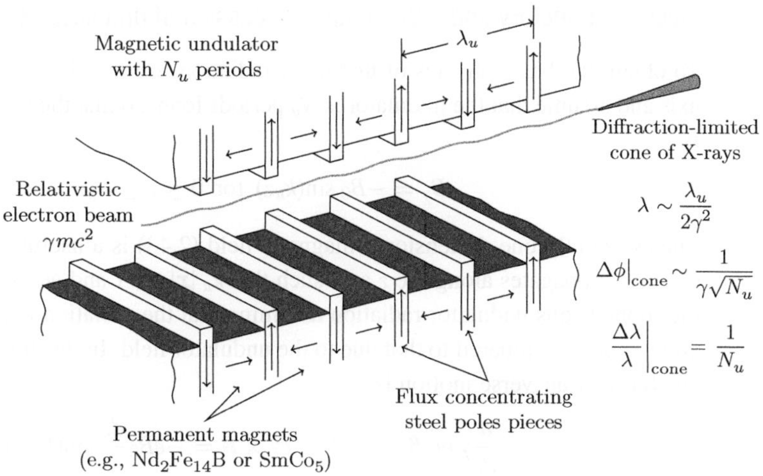

Figure 2.4 Schematic of a periodic magnetic device (an undulator) of period λ_u and with a number of periods, N_u. The structure is based on permanent magnets. Adapted from Ref. [1].

along the electron propagation direction z. Occasionally, we may mention helical undulators, whose principal magnetic direction makes a helix along z. To determine a model for the $\boldsymbol{B}$-field in a planar undulator, we start with the fact that in free space the magnetic field can be derived from a scalar potential Φ_B that satisfies Laplace's equation $\nabla^2\Phi_B = 0$. The simplest such potential describing a B_y field that oscillates along z and is nonzero at $x = y = 0$ is

$$\Phi_B(x, y, z) = -\frac{B_0}{k_u}\sinh(k_u y)\sin(k_u z). \tag{2.41}$$

Taking the gradient of (2.41), we find that a reasonably good approximation to the magnetic field in a planar undulator is

$$\boldsymbol{B}(x, y, z) = -B_0\cosh(k_u y)\sin(k_u z)\hat{y} - B_0\sinh(k_u y)\cos(k_u z)\hat{x}. \tag{2.42}$$

Here, B_0 is the peak on-axis field and k_u is the wavevector of the undulator, which is related to its period λ_u via $k_u = 2\pi/\lambda_u$. The magnetic field equation (2.42) is exact in the limit that the undulator is infinitely long in z and wide in x, and furthermore satisfies Maxwell's equations in vacuum (namely, has vanishing divergence and curl) since it comes from an appropriate magnetic potential.

For a fixed gap between poles, Equation (2.40) indicates that the peak on-axis magnetic field decreases exponentially as one decreases the undulator period. Typical parameters of permanent magnetic devices have $B_0 \sim 1$ Tesla and $\lambda_u \sim 2$ to 3 cm. To further reduce the undulator period while maintaining sufficient magnetic field strength, alternative technologies are under active development. Among them, cryogenic undulators, RF undulators, and superconducting undulators are leading the way to the future. We will briefly return to superconducting undulators when we explore some future directions of synchrotron radiation sources at the end of this chapter.

2.4.1 Electron Trajectory and a Qualitative Discussion of Undulator Radiation

To obtain the basic physics of undulator radiation, we use the field profile along the z axis and assume that the undulator is N_u periods long, so that the relevant magnetic field is

$$B_y = -B_0\sin(k_u z) \text{ for } 0 \le z \le N_u\lambda_u. \tag{2.43}$$

The essentially one-dimensional magnetic field (2.43) is a useful approximation if the electron propagates along the z axis such that $k_u|\boldsymbol{x}| \ll 1$ and $N_u \ll \gamma/K$. We calculate the spontaneous undulator radiation assuming that the radiation itself produces a negligible force as compared to that due to the undulator field. In this limit, the Lorentz force law for the transverse motion is

$$\begin{aligned}\frac{d}{dt}\gamma mc\boldsymbol{\beta} = -e\boldsymbol{E} - e\boldsymbol{v}\times\boldsymbol{B} &= -eB_0\frac{dz}{dt}\sin(k_u z)\hat{x}\\ &= \frac{eB_0}{k_u}\frac{d}{dt}\cos(k_u z)\hat{x}.\end{aligned} \tag{2.44}$$

Assuming that the electron is initially directed along the optical axis, the velocity along y vanishes, while in the x direction integrating (2.44) yields

$$\beta_x = \frac{eB_0}{\gamma mck_u}\cos(k_u z) \equiv \frac{K}{\gamma}\cos(k_u z), \tag{2.45}$$

where we have introduced the dimensionless undulator deflection parameter K, and we see that the maximum slope of the electron trajectory is K/γ. Undulators are typically defined to be those magnetic structures for which the maximum slope is less than the radiation opening angle $1/\gamma$, i.e., for $K \lesssim 1$. When K is large, the device is usually called a wiggler. A practical engineering formula for the deflection parameter is

$$K = \frac{eB_0}{mck_u} = 0.934\,\lambda_u[\text{cm}]B_0[\text{T}]. \tag{2.46}$$

From the Hamiltonian perspective discussed in Appendix A, the result (2.45) follows from the conservation of canonical momentum $\boldsymbol{p} = \gamma\boldsymbol{\beta} - \boldsymbol{a} = \gamma\boldsymbol{\beta} - K\cos(k_u z)\hat{x}$.[2]

Moving forward with our calculation of the particle trajectory, the relationship between the velocity and the (constant) energy can be used to obtain the longitudinal velocity via

$$\begin{aligned}\beta_z = \sqrt{1 - \frac{1}{\gamma^2} - \beta_x^2} &\approx 1 - \frac{1}{2\gamma^2} - \frac{K^2}{2\gamma^2}\cos^2(k_u z) \\ &= 1 - \frac{1 + K^2/2}{2\gamma^2} - \frac{K^2}{4\gamma^2}\cos(2k_u z).\end{aligned} \tag{2.47}$$

The average longitudinal velocity $\bar{\beta}_z = 1 - (1 + K^2/2)/2\gamma^2$ is decreased from its maximal value $1 - 1/2\gamma^2$ due to the fact that the electron executes oscillations along $\hat{x}$. These oscillations result in a sinusoidal (with z) variation in the longitudinal velocity, whose period is twice that given by the undulator. In an inertial frame moving at the average beam velocity, this motion traces out a figure-eight pattern in the x-z plane.

We deduce the particle time coordinate from the longitudinal velocity using the definition $dt/dz = 1/c\beta_z$, where we use (2.47) to expand $1/\beta_z$ through $O(1/\gamma^2)$ and then integrate. We find that the particle crosses the plane z at a time $t_j(z)$ given by

$$ct_j(z) = \left[1 + \frac{1 + K^2/2}{2\gamma^2}\right]z + \frac{K^2}{8\gamma^2 k_u}\sin(2k_u z) + ct_j(0). \tag{2.48}$$

Note that this is identical to what we would have found by integrating Hamilton's equation of motion $ct' = -\tau' = -(\partial\mathcal{H}/\partial\gamma)$.

To understand the properties of the emitted X-rays, imagine that the electrons emit radiation at the beginning of each undulator period. An observer far away in the ϕ direction would see a different period $\lambda_1(\phi)$ as shown in Figure 2.5. Since $\overline{AA''}$ is the distance light travels while the electron moves along the arc length $\widetilde{AB}$,

$$\frac{\lambda_1(\phi)}{c} = \frac{\widetilde{AB}}{v} - \frac{\overline{AA'}}{c}. \tag{2.49}$$

[2] The associated Hamiltonian $\mathcal{H} = [p_x + K\cos(k_u z)]^2/2\gamma + p_y^2/2\gamma - (\gamma - 1/2\gamma)$ is independent of $\boldsymbol{x}$, so that p_x and p_y and conserved.

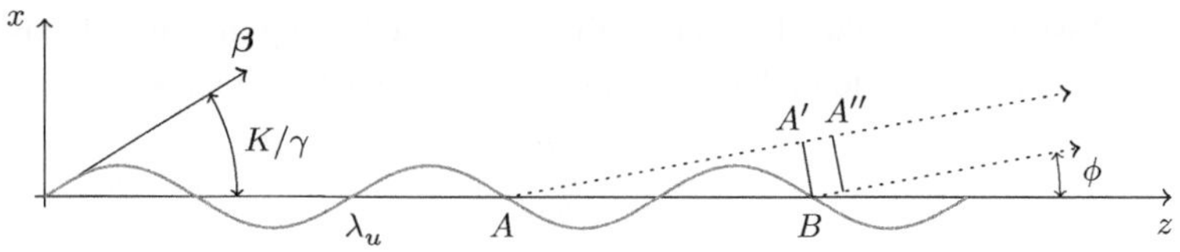

Figure 2.5 Illustration of the resonance condition in an undulator.

The arc length $\widetilde{AB}$ is given by integrating along the electron trajectory

$$\widetilde{AB} = \int_0^{\lambda_u} dz\,\sqrt{1+(x')^2} \approx \int_0^{\lambda_u} dz\,\left(1+\frac{1}{2}x'^2\right) = \lambda_u\left(1+\frac{K^2}{4\gamma^2}\right), \tag{2.50}$$

while for small angles the length

$$\overline{AA'} = \lambda_u \cos\phi \approx \lambda_u\left(1-\frac{\phi^2}{2}\right). \tag{2.51}$$

Inserting the length expressions (2.50) and (2.51) into Equation (2.49), we obtain

$$\begin{aligned}\frac{\lambda_1(\phi)}{c} &= \frac{\lambda_u}{c}\left[\frac{1+K^2/(4\gamma^2)}{\beta} - \left(1-\frac{\phi^2}{2}\right)\right] \\ &\approx \frac{\lambda_u}{c}\frac{1+K^2/2+\gamma^2\phi^2}{2\gamma^2},\end{aligned} \tag{2.52}$$

where the final line uses $1/\beta \approx 1 + 1/2\gamma^2$. Thus, for $\phi \lesssim 1/\gamma$ the radiation wavelength seen by the observer is reduced from λ_u by a factor $\sim 1/\gamma^2$, which is precisely the time-squeezing effect. We have used the subscript 1 to indicate that λ_1 is the fundamental wavelength; in the next section we will derive the general properties of harmonic emission as well. The fundamental radiation frequency is

$$\omega_1(\phi) = \frac{2\pi c}{\lambda_1(\phi)} = ck_u\frac{2\gamma^2}{1+K^2/2+\gamma^2\phi^2}, \tag{2.53}$$

and the photon energy is $\hbar\omega_1(\phi)$.

Since any electron makes N_u oscillations in an undulator composed of N_u periods, the resulting wave train has N_u cycles. Thus, the spectrum of the undulator radiation at observation angle ϕ is peaked around $\omega_1(\phi)$ with intrinsic bandwidth

$$\frac{\Delta\omega}{\omega_1} = \frac{\Delta\lambda}{\lambda_1} \sim \frac{1}{N_u}. \tag{2.54}$$

This is illustrated in Figure 2.6(a). As the observation angle ϕ increases from zero, the wavelength is "red"-shifted because

$$\frac{\lambda_1(\phi)-\lambda_1(0)}{\lambda_1(0)} = \frac{\gamma^2\phi^2}{1+K^2/2} = \frac{\lambda_u}{2\lambda_1(0)}\phi^2 > 0. \tag{2.55}$$

For $\Delta\phi \sim 1/\gamma$, the normalized bandwidth is broad like that of a single bending magnet, $|\Delta\lambda/\lambda| \sim O(1)$, and the spectral distribution integrated over the angle looks like the left-hand side of Figure 2.6(b). However, if we introduce a pinhole that has an angular

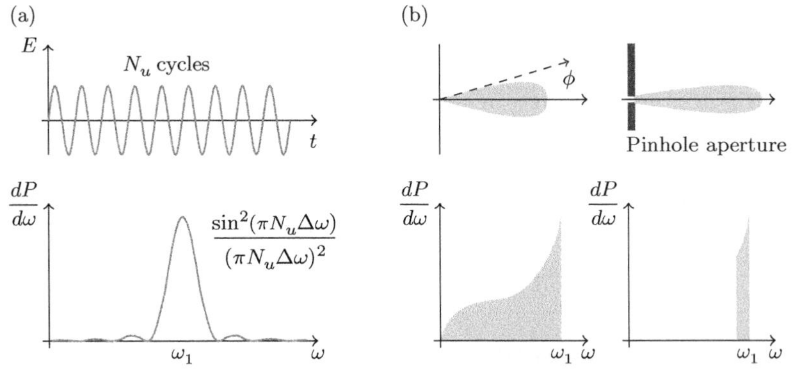

Figure 2.6 (a) Temporal and spectral profile of undulator radiation. (b) Spectral distributions before and after a pinhole. Adapted from Ref. [14].

acceptance $\Delta\phi \ll 1/\gamma$, the spectrum is narrowed to its intrinsic width $\sim 1/N_u$ as shown on the right-hand side of Figure 2.6(b). For typical undulators with $N_u \sim 100$, the source is quasi-monochromatic with a normalized bandwidth $\Delta\omega/\omega_1 \sim 1$ percent.

We can obtain an approximate expression for the natural angular divergence of undulator radiation by determining how small the pinhole must be to observe the intrinsic $1/N_u$ bandwidth. We use Equation (2.55) to write

$$\frac{\Delta\lambda}{\lambda} = \frac{\gamma^2\phi^2}{1+K^2/2} = \frac{\lambda_u}{2\lambda_1(0)}(\Delta\phi)^2 \le \frac{1}{N_u}. \tag{2.56}$$

Defining the RMS angular divergence $\sigma_{r'} \equiv \Delta\phi/2$, (2.56) leads to

$$\Delta\phi \le \frac{1}{\gamma}\sqrt{\frac{1+K^2/2}{N_u}} \sim \sqrt{\frac{2\lambda_1}{L_u}} \quad \Rightarrow \quad \sigma_{r'} = \sqrt{\frac{\lambda_1}{2L_u}}, \tag{2.57}$$

where $L_u = \lambda_u N_u$ is the length of the undulator. Hence, we identify the angular divergence of the undulator source of length L_u with $\sqrt{\lambda_1/2L_u}$. The frequency-integrated angular distribution of synchrotron radiation typically has a width $\sim 1/\gamma$ due to the relativistic effect, but Equation (2.57) indicates that the angular distribution of undulator radiation is further reduced by a factor of $N_u^{-1/2}$ within the frequency bandwidth $1/N_u$. As we have derived it, $\sigma_{r'}$ in (2.57) can only be considered an approximate characterization of the angular divergence, although we will use $\sigma_{r'}$ later when we approximate the undulator radiation by a Gaussian field.

To determine the effective transverse size Δx of a radiation source extended over the length L_u, consider Figure 2.7. Assuming that the angle $\Delta\phi \ll 1$, the diagram in Figure 2.7 shows that the effective source size of undulator radiation is given by

$$\Delta x \sim \Delta\phi\frac{L_u}{2} \sim \sqrt{\frac{\lambda_1 L_u}{2}}. \tag{2.58}$$

Considering (2.57) and (2.58), we see that the product of the source size and divergence is limited by the diffraction law

$$\Delta\phi\Delta x \sim \lambda_1. \tag{2.59}$$

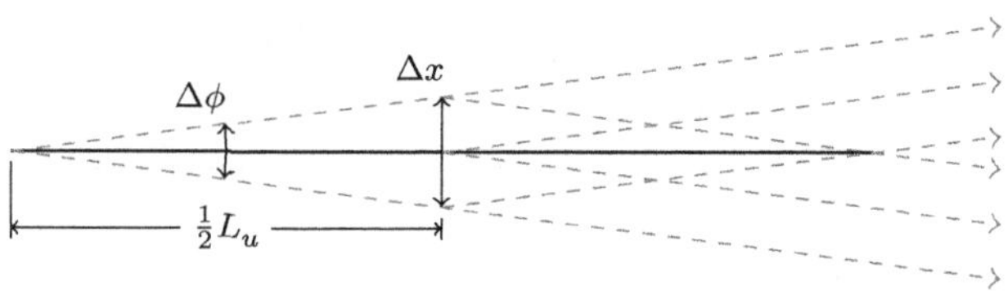

Figure 2.7 Effective source size and angular divergence of undulator radiation.

We will derive and discuss these properties in a more systematic, formal manner in the following section.

2.4.2 Paraxial Analysis of Undulator Radiation

We begin our mathematical investigation of the properties of undulator radiation using the paraxial wave equation; for a collection of N_e electrons, the paraxial equation (2.10) becomes

$$\tilde{\mathcal{E}}_\omega(\boldsymbol{\phi}; z) = \int_{-\infty}^{z} dz' \sum_{j=1}^{N_e} \frac{e[\boldsymbol{\beta}_j(z') - \boldsymbol{\phi}]}{4\pi\epsilon_0 c\lambda^2} e^{ik[ct_j(z') - z' - \boldsymbol{\phi}\cdot\boldsymbol{x}_j(z') + \boldsymbol{\phi}^2(z'-z)/2]}. \tag{2.60}$$

Equation (2.60) applies if the variation of the envelope $\tilde{\mathcal{E}}_\omega$ occurs over a much longer scale length than that set by the resonant wavelength λ_1, which will typically be satisfied provided that $N_u \gg 1$. In this section we will assume that the electron beam is ideal, characterized by a single energy γ_r and directed along the optical axis (finite energy spread and emittance will be discussed in Section 2.4.5). Using the transverse velocity of an electron propagating along the undulator axis (2.45), we find that the $\hat{x}$-component of the paraxial wave equation (2.60) is now

$$\tilde{\mathcal{E}}_\omega(\boldsymbol{\phi}; z) = \int_{0}^{z} dz' \sum_{j=1}^{N_e} \frac{e[K\cos(k_u z') - \phi_x]}{4\pi\epsilon_0 \gamma_r c\lambda^2} e^{ik[ct_j(z') - z' - \boldsymbol{\phi}\cdot\boldsymbol{x}_j(z') + \boldsymbol{\phi}^2(z'-z)/2]}, \tag{2.61}$$

where we will find the resonant energy γ_r shortly.

Before going any further to simplify Equation (2.61), we first note that our primary concern here is the field profile in the central cone where $\gamma\phi \lesssim 1/\sqrt{N_u}$, so that in the following we drop the charge density term $\sim \gamma\phi_x$. Next, we note that the integral in (2.61) involves the phase associated with the particle trajectory, and so that the integral largely cancels to zero unless this phase is approximately constant. To find the appropriate slowly varying (resonant) source current, we use the ideal reference trajectory (2.48) to write

$$\begin{aligned}
\cos(k_u z)e^{ik[ct_j(z) - z - \boldsymbol{\phi}\cdot\boldsymbol{x}_j(z)]} &= \cos(k_u z)\exp\left[ik\left(\tfrac{1+K^2/2}{2\gamma_r^2}z + ct_j(0)\right)\right] \\
&\quad \times \exp\left[ik\left(\tfrac{K^2}{8\gamma_r^2 k_u}\sin(2k_u z) - \tfrac{K\phi_x}{\gamma_r k_u}\sin(k_u z)\right)\right] \\
&\approx e^{ickt_j(0)} \sum_n J_n\left(\tfrac{K^2 k}{8\gamma_r^2 k_u}\right) \\
&\quad \times \cos(k_u z)\exp\left[i\left(\tfrac{1+K^2/2}{2\gamma_r^2}\right)kz + 2ink_u z\right].
\end{aligned} \tag{2.62}$$

The second line follows from neglecting the angular contribution which scales $\sim 2\gamma_r\phi_x \lesssim 2/\sqrt{N_u}$ in the central cone, which is consistent with our previous dropping of the space-charge term from (2.61), and from applying the Jacobi–Anger identity

$$e^{ix\sin\theta} = \sum_{n=-\infty}^{\infty} J_n(x)e^{in\theta}. \tag{2.63}$$

Inspection of Equation (2.62) shows that the phase of the undulator source current is given by

$$\sum_n \exp\left[ik_u z\left(\frac{k}{k_u}\frac{1+K^2/2}{2\gamma_r^2} \mp 1 + 2n\right)\right], \tag{2.64}$$

where the $\pm ik_u z$ comes from writing $\cos(k_u z)$ in exponential notation. Hence, the phase is approximately constant when λ/λ_u equals $(1+K^2/2)/2\gamma_r^2$ divided by an odd integer; in other words, we find that in the central cone the phase (2.62) is stationary and the spectrum is peaked only for the odd harmonics

$$\lambda_h \equiv \frac{1+K^2/2}{2\gamma_r^2}\frac{\lambda_u}{h} \quad \text{for } h = 2n-1,\ n = 1,2,3,\ldots, \tag{2.65}$$

where (2.65) defines the on-axis (odd) harmonic wavelength $\lambda_h \equiv \lambda_{2n-1}(\phi = 0)$ in terms of the resonant energy γ_r. The on-axis spectrum has only odd harmonics because the particle trajectory, although not sinusoidal, is antisymmetric about its half-period $\lambda_u/2$, which means that its Fourier expansion has no even components. This symmetry is broken off-axis, and all harmonics are present. We additionally define the scaled frequency $\nu \equiv \omega/\omega_1 \equiv k/k_1$ and the scaled frequency difference from resonance $\Delta\nu \equiv (k - k_h)/k_1$, in terms of which we have

$$\frac{\omega}{ck_u} = \frac{k}{k_u} = \frac{k_1}{k_u}\nu = \frac{2\gamma_r^2}{1+K^2/2}\nu = \frac{2\gamma_r^2}{1+K^2/2}(h+\Delta\nu). \tag{2.66}$$

We use the slowly-varying phase (2.64) in the current of the paraxial equation, and simplify the right-hand side using the Bessel function identity $J_{-n}(x) = (-1)^n J_n(x)$, the resonance condition (2.66), and by considering frequencies near the resonance condition $\nu = 2n - 1 + \Delta\nu = h + \Delta\nu$ with $\Delta\nu \ll 1$. We find that within the central cone the undulator field satisfies

$$\tilde{\mathcal{E}}_\omega(\boldsymbol{\phi}; L_u) = \sum_{h\ \text{odd}} \frac{eK[\text{JJ}]_h}{8\pi\epsilon_0\gamma_r c\lambda^2} \sum_{j=1}^{N_e} e^{i\omega t_j(0)} \int_0^{L_u} dz\ e^{i\Delta\nu k_u z} e^{ik[\boldsymbol{\phi}^2(z-L_u)/2 - \boldsymbol{\phi}\cdot\boldsymbol{x}_j(z)]}, \tag{2.67}$$

where we have introduced the harmonic Bessel function factor

$$[\text{JJ}]_h = (-1)^{(h-1)/2}\left[J_{(h-1)/2}\left(\frac{hK^2}{4+2K^2}\right) - J_{(h+1)/2}\left(\frac{hK^2}{4+2K^2}\right)\right]. \tag{2.68}$$

The solution (2.67) has reduced the equation for the undulator field in the central cone to a sum over odd harmonics of the fundamental; these harmonics are well-separated in frequency and can be treated essentially independently when $\gamma\,|\phi| \lesssim 1/\sqrt{N_u}$. For an ideal electron beam with $\boldsymbol{x}_j(z) = 0$ the resulting integrand of (2.67) is $\propto e^{iaz}$, and the integral over z is elementary; we find that the radiation field at the end of the undulator is

$$\tilde{\mathcal{E}}_\omega(\boldsymbol{\phi}; L_u) = \sum_{h\text{ odd}} \frac{eKL_u[\text{JJ}]_h}{8\pi\epsilon_0 c\lambda^2\gamma_r} \exp\left[i\pi N_u\left(\Delta\nu + \tfrac{h(\gamma_r\phi)^2}{1+K^2/2}\right)\right]$$
$$\times \frac{\sin\left[\pi N_u\left(\Delta\nu + \tfrac{h(\gamma_r\phi)^2}{1+K^2/2}\right)\right]}{\pi N_u\left(\Delta\nu + \tfrac{h(\gamma_r\phi)^2}{1+K^2/2}\right)} \sum_{i=1}^{N_e} e^{i\omega t_j(0)}. \tag{2.69}$$

The spectral-angular shape of the undulator field in Equation (2.69) is dominated by the sinc function. Peaks of $\mathcal{E}_\omega(\boldsymbol{\phi}; L_u)$ correspond to maxima of the sinc function, which occur at frequencies such that

$$\omega_h(\phi) = h\omega_1\left[1 - \frac{\gamma_r^2\phi^2}{1+K^2/2}\right]. \tag{2.70}$$

Hence, as discussed in the previous section, the resonant frequency is lower (red-shifted) off-axis. In addition, setting the argument of the sinc function equal to π gives a measure of the width. The spectral width at fixed angle and the angular width at fixed frequency scale, respectively, as

$$\frac{\Delta\omega}{\omega} \sim \frac{1}{hN_u} \qquad \sigma_{r'} \sim \frac{1}{\gamma}\sqrt{\frac{1+K^2/2}{hN_u}}, \tag{2.71}$$

precisely as we found in the previous section. The former of these relations can be understood if we realize that the spectral shape results from the Fourier transform of a wave that is N_u cycles long. Since the number of cycles is fixed, so is the bandwidth $\Delta\omega$, which in turn implies that near the harmonic frequency $\omega \sim h\omega_1$ the normalized bandwidth $\Delta\omega/\omega_h \sim 1/hN_u$.

If we consider the longitudinal physics in the time domain, we know that at every undulator period the electrons slip one radiation wavelength behind their spontaneous emission. Hence, the radiation field from any electron is confined to the "slippage distance" $N_u\lambda_1$, and the correlation time of the undulator radiation

$$\sigma_\tau \leq 2\pi N_u/\omega_1 = 2\pi/\Delta\omega, \tag{2.72}$$

which is typically much shorter that the electron bunch duration T; for X-rays at 1 Å and $N_u \sim 10^2$, $\sigma_\tau \lesssim 0.1$ fs, while the electron bunch in storage rings is typically larger than a few ps. Therefore, the generated undulator radiation is temporally chaotic light with many longitudinal modes ($M_L \sim 10^4$–10^5). While the field is chaotic, the relative fluctuation in the number of photons per pulse is small: from Equation (1.110) we have $\sigma_{\mathcal{N}_{\text{ph}}}/\mathcal{N}_{\text{ph}} = 1/\sqrt{M} \leq 1/\sqrt{M_L} \lesssim 10^{-2}$.

Because the randomly phased electrons radiate incoherently, the undulator radiation power and flux scale linearly with the number of particles. Mathematically, the absolute square of the field

$$\left|\tilde{\mathcal{E}}_\omega(\boldsymbol{\phi}; L_u)\right|^2 \sim \sum_{j,k}^{N_e} e^{-i\omega[t_j(0)-t_k(0)]} = N_e + \sum_{j\neq k}^{N_e} e^{-i\omega[t_j(0)-t_k(0)]} \to N_e. \tag{2.73}$$

Thus, the photon flux from a beam of N_e electrons moving along the optical axis is merely N_e times the single particle photon flux; the FEL differs in that it includes the interaction between the radiation and the particle, so that Equation (2.67) depends on the dynamical time $t_j(z)$ and the solution (2.69) no longer applies.

Returning to the spontaneous emission, the photon flux can be found by inserting the electric field (2.69) into the power formula (1.72). We use the current $I = eN_e/T$, the resonance condition (2.66), and the fine structure constant $\alpha \equiv e^2/(4\pi\epsilon_0\hbar c) \approx 1/137$ to simplify the resulting expression, finding

$$\frac{d\mathcal{F}}{d\omega d\boldsymbol{\phi}} = \alpha N_u^2\gamma_r^2\frac{I}{e}\frac{1}{\omega}\left\{\frac{\sin\left[\pi N_u\left(\Delta\nu + \frac{h(\gamma_r\phi)^2}{1+K^2/2}\right)\right]}{\pi N_u\left(\Delta\nu + \frac{h(\gamma_r\phi)^2}{1+K^2/2}\right)}\right\}^2 \frac{h^2K^2[\mathrm{JJ}]_h^2}{(1+K^2/2)^2}, \tag{2.74}$$

where we recall that h is an odd integer (i.e., $h = 2n - 1$ for $n = 1, 2, \ldots$). The function $h^2K^2[\mathrm{JJ}]_h^2/(1 + K^2/2)^2$ is plotted in Figure 2.8(a) for reference. Along the optical axis, the photon flux can be expressed in practical units as

$$\left.\frac{d\mathcal{F}}{d\omega d\boldsymbol{\phi}}\right|_{\boldsymbol{\phi}=0} = 1.74 \times 10^{14} N_u^2(\gamma mc^2)^2[\mathrm{GeV}]I[\mathrm{A}] \times \frac{h^2K^2[\mathrm{JJ}]_h^2}{(1+K^2/2)^2}\,\frac{\text{photons}}{\text{s}\cdot\text{mrad}\cdot(0.1\%\ \text{BW})}, \tag{2.75}$$

where the chosen 0.1 percent bandwidth (BW) is a standard convention in the synchrotron radiation community. The total flux in the central cone defined by $\phi \lesssim 1/(\gamma\sqrt{N_u})$ is obtained by integrating (2.74) over the solid angle. Approximating the angular distribution in the vicinity of the odd harmonic h by a Gaussian, we have

$$\frac{d\mathcal{F}_h}{d\omega} = \left.\frac{d\mathcal{F}_h}{d\omega d\boldsymbol{\phi}}\right|_0 2\pi\sigma_{r'}^2 = \left.\frac{d\mathcal{F}_h}{d\omega d\boldsymbol{\phi}}\right|_0 2\pi\frac{\lambda_h}{2L_u} = \left.\frac{d\mathcal{F}_h}{d\omega d\boldsymbol{\phi}}\right|_0 2\pi\frac{1+K^2/2}{4h\gamma_r^2 N_u}, \tag{2.76}$$

where we generalize the Gaussian angular width (2.57) to the harmonic h as indicated. Inserting the flux density along the optical axis [Equation (2.74) with $\boldsymbol{\phi} = 0$] yields the total spontaneous flux

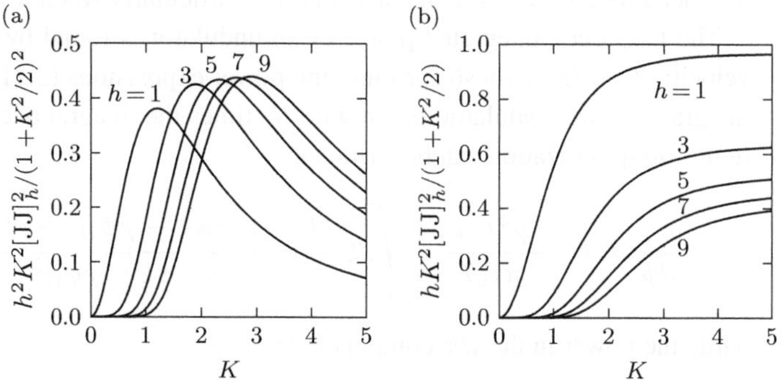

Figure 2.8 The dimensionless undulator flux functions for the first nine harmonics plotted against the deflection parameter K. (a) graphs the function of h and K that appears in the expression for the spectral flux density (2.74), while (b) shows the function relevant to the total photon flux in the central cone equation (2.77). Adapted from Ref. [1].

$$\frac{d\mathcal{F}_h}{d\omega} = \frac{\pi}{2}\alpha N_u \frac{I}{e}\frac{\Delta\omega}{\omega}\frac{hK^2[\mathrm{JJ}]_h^2}{1+K^2/2}. \tag{2.77}$$

As we mentioned earlier, the characteristic spectral width for the photon flux in the central cone is given by $\Delta\omega/\omega \lesssim 1/N_u$, and we plot the function $hK^2[\mathrm{JJ}]_h^2/(1+K^2/2)$ in Figure 2.8(b). The photon flux can be simply understood in the following way. Since the deflection angle per period is $\sim K/\gamma$, each electron emits roughly αK photons every period, and of order $\alpha K N_u$ photons over the length of the undulator. Thus, an electron beam of current I radiates $\sim \alpha K N_u (I/e)$ photons per second per unit bandwidth. Since the total number of emitted photons within the characteristic bandwidth $\Delta\omega/\omega \sim 1/N_u$ is independent of N_u, the spectral density (and brightness) is proportional to the number of undulator periods. In practical units, the undulator photon flux in the central cone is

$$\frac{d\mathcal{F}_h}{d\omega} = \frac{1}{2} 1.43\times 10^{14} N_u I[\mathrm{A}] \frac{h[\mathrm{JJ}]_h^2}{1+K^2/2}\frac{\text{photons}}{\text{s}\cdot(0.1\%\ \text{BW})}. \tag{2.78}$$

The flux formula above is one-half of those quoted in the previous publications [15, 16]. This is because we have computed the flux at $\omega = h\omega_1$; more flux is produced at slightly lower frequencies, becoming twice that given by (2.77–2.78) at approximately $\omega = h\omega_1[1-(hN_u)^{-1}]$. For these frequencies, however, the angular distribution is not maximized in the forward direction and therefore the Gaussian approximation used in the present treatment does not apply.

2.4.3 Frequency Integrated Power

Up to now our discussion of undulator radiation has largely focused on its frequency characteristics. We showed how interference effects from the periodic magnetic structure result in an X-ray flux that is concentrated into harmonics of the fundamental resonant wavelength. This is in stark contrast to the broad spectrum produced by bending magnets, so that both the X-ray brightness and the spectral flux associated with undulator radiation is much higher than what would be produced by just summing $2N_u$ independent bending magnets of the same strength. In contrast to this, in this section we show that the frequency-integrated power from an undulator is more similar to that produced by $2N_u$ bending magnet sources, particularly when $K \gtrsim 1$.

The frequency-integrated power of an undulator is found by inserting the undulator velocity $\beta_x = (K/\gamma)\cos(ck_u\zeta)$ into the power expressions (2.21) and (2.22). Replacing integral over the undulator length with N_u times the integral over one period, the power in the x/σ-polarization can be written as

$$\frac{dP_\sigma}{d\boldsymbol{\phi}} = N_u \frac{e^2\gamma^4K^2k_u}{2\pi\epsilon_0 T}\frac{1}{\pi}\int_0^{2\pi} d\xi\, \frac{[1-(K\cos\xi-\gamma\phi_x)^2+\gamma^2\phi_y^2]^2\sin^2\xi}{[1+(K\cos\xi-\gamma\phi_x)^2+\gamma^2\phi_y^2]^5}, \tag{2.79}$$

while the power in the y/π component is

$$\frac{dP_\pi}{d\boldsymbol{\phi}} = N_u \frac{e^2\gamma^4K^2k_u}{2\pi\epsilon_0 T}\frac{1}{\pi}\int_0^{2\pi} d\xi\, \frac{(K\cos\xi-\gamma\phi_x)^2\gamma^2\phi_y^2\sin^2\xi}{[1+(K\cos\xi-\gamma\phi_x)^2+\gamma^2\phi_y^2]^5}. \tag{2.80}$$

The power density on-axis (where $\phi_x = \phi_y = 0$) can be obtained analytically by setting $e^{i\xi}$ equal to the complex variable z, $e^{-i\xi} = 1/z$, $d\xi = -idz/z$, and interpreting the integration in (2.79) as a contour integral about the unit circle defined by $|z| = 1$. Since the contour is closed, the integral is given by a sum over the residues. The contour encloses two, five-fold degenerate poles at $\pm i(\sqrt{1+K^2}-1)/K$, so that the total power on-axis is [17]

$$\left.\frac{dP_\sigma}{d\boldsymbol{\phi}}\right|_{\boldsymbol{\phi}=0} = \left[\frac{7e^2Kk_u\gamma^4}{64\pi\epsilon_0}\frac{I}{e}\right]2N_u\frac{K\left(K^6+\frac{24}{7}K^4+4K^2+\frac{16}{7}\right)}{(1+K^2)^{7/2}}. \tag{2.81}$$

Using the definition of K, we can show that the expression in square brackets from (2.81) is the same as that for a bending magnet whose magnetic field equals the peak undulator field B_0, Equation (2.35). The factor $2N_u$ accounts for the $2N_u$ poles in an undulator that is N_u period long, while the final function of K accounts for interference effects. This function increases linearly $\sim 16K/7$ when $K^2 \ll 1$, and is practically unity for $K \gtrsim 1$. In practical units of W/mrad^2, the peak, on-axis power density is

$$\begin{aligned}\left.\frac{dP}{d\boldsymbol{\phi}}\right|_{\boldsymbol{\phi}=0}[\text{W/mrad}^2] &\approx 10.84N_uB_0[\text{T}](\gamma mc^2)^4[\text{GeV}]I[\text{A}]\\ &\times\frac{K\left(K^6+\frac{24}{7}K^4+4K^2+\frac{16}{7}\right)}{(1+K^2)^{7/2}}.\end{aligned} \tag{2.82}$$

The total power density at arbitrary angles is of course the sum of (2.79) and (2.80). Inspection of these expressions shows that the vertical angular width $\Delta\phi_y \sim 1/\gamma$ as expected. On the other hand, the angular width along the horizontal direction is roughly given by the maximum angular deflection of the electron, $\Delta\phi_x \sim K/\gamma$. Finally, integrating the angular power densities over solid angle, we find that the power contained in the σ/x-polarization is seven times that in the π/y-polarization as was the case for bending magnet radiation, and the sum of the two is

$$P = \frac{e^2\gamma^2}{12\pi\epsilon_0}\frac{I}{e}\left(\frac{eB_0}{mc}\right)^2L_u, \tag{2.83}$$

The total power produced by an undulator is precisely that of a bending magnet if we identify the arc length $\rho\Delta\phi_x$ from (2.38) with the undulator length L_u, and replace the bending magnetic field with its RMS value in an undulator, $B^2 \to B_0^2/2$. In practical units we have

$$P[\text{kW}] \approx 0.63\,(\gamma mc^2)^2[\text{GeV}]B_0^2[\text{T}]I[\text{A}]L_u[\text{m}]. \tag{2.84}$$

2.4.4 Polarization Control

Planar undulators like those that we have discussed produce linearly polarized radiation. The field is linearly polarized along x when viewed by an observer in the horizontal plane, while in general the polarization direction is a complicated function of the observation angle [18]. The field is linearly polarized because of the symmetry of the electron trajectory within each period. To see this, consider observing the undulator radiation from a position that is some distance perpendicular to the oscillatory motion of the electron, i.e., at some angle along the y direction of Figure 2.5. From this vantage point, the electron rotates in a counter-clockwise fashion for the first half of an undulator period producing elliptically polarized radiation. In the next half-period the electron rotation is in the opposite direction, which gives rise to elliptically polarized radiation with opposite helicity. Because the motion in the two half-periods is symmetric, the ellipticity cancels and the resulting field is linearly polarized.

The argument above suggests a way to produce elliptically polarized radiation from a specially designed planar undulator. The idea is to arrange an electron trajectory that is periodic but in which the symmetry within each period is broken. A periodic magnetic field such as the one shown in Figure 2.9 will lead to such a trajectory, which in turn will generate elliptically polarized radiation when viewed at an angle transverse to the wiggle plane (i.e., along y). In addition to producing elliptically polarized radiation off-axis, such an undulator will also produce non-vanishing, even harmonics in the forward direction, since it is the symmetric motion that also precludes even harmonics from the Fourier expansion of the motion. It turns out, however, that neither the strength of the on-axis even harmonics nor the deviation from linear polarization for the off-axis radiation is significant unless the asymmetry is extreme. In a wiggler where $K \gg 1$ and the orbit deflection is large, such an effect could be made appreciable.

Nevertheless, polarization control is typically more easily and efficiently accomplished by arranging magnet designs that vary in x, y, and z. For example, the electron wiggle motion in a helical undulator [9, 10] follows a helix, and the resulting X-rays are circularly polarized. In addition, since the transverse velocity $\beta_x = (K/\gamma)\cos(k_u z)$ and $\beta_y = (K/\gamma)\sin(k_u z)$, the longitudinal velocity has no oscillating part,

$$\left.\frac{v_z}{c}\right|_{\text{helical}} = \sqrt{1 - \frac{1}{\gamma^2} - \boldsymbol{\beta}_\perp^2(z)} \approx 1 - \frac{1+K^2}{2\gamma^2}. \tag{2.85}$$

Since $v_z = \bar{v}_z$, the on-axis emission in a helical undulator is only at the fundamental, and harmonics are only present off-axis. However, helical undulators are not often used, partly because most applications that make use of X-ray polarization require the ability to vary the polarization from linear to elliptical to circular and back. In what follows we briefly discuss a few undulator designs that can deliver on-axis X-rays of variable polarization.

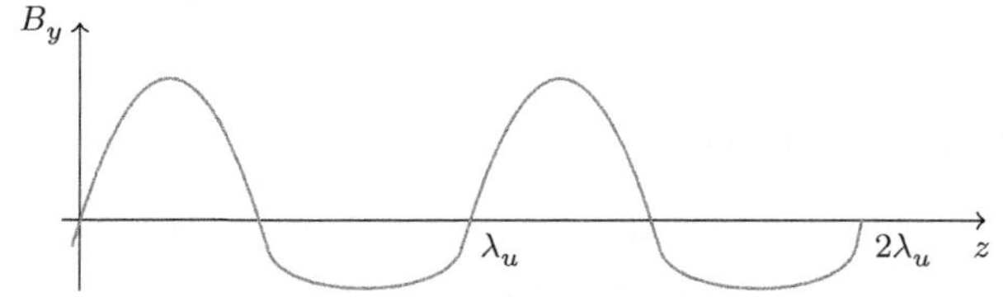

Figure 2.9 Schematic of magnetic field configuration for a non-sinusoidal undulator that can produce even harmonics radiation in the z direction and elliptical polarization in directions $\phi_y \neq 0$.

Variable Polarization Undulator

One way to vary the polarization is by controlling the current in a specially designed electromagnetic undulator. However, the limited field strength and power requirements typically make electromagnetic undulators less attractive then their permanent magnet counterparts, particularly for hard X-rays. For this reason there have been several proposed designs of permanent magnet undulators capable of dynamically varying the X-ray polarization. We mention two types of such undulators here.

In storage rings, where the electron beam is much larger in the horizontal direction than in the vertical direction, radiation devices must stay clear of the horizontal plane. For these applications, the Advanced Planar Polarized Light Emitter (APPLE) device [19] is the most commonly used undulator for variable polarization. The APPLE undulator is constructed using four standard Halbach-type magnet rows, as shown in Figure 2.10. The rows on opposite corners make up a pair that is kept fixed with respect to one another, and the polarization is controlled by varying the magnetic phase between the two pairs. As an example, in the configuration shown in Figure 2.10 the pairs differ in phase by one-half period, which ideally results in equal x and y components of the magnetic field and circularly polarized output. If the pair comprising the lower back and upper front rows is moved along z by $-\lambda_u/2$, the magnetic fields on the top and bottom match, the B-field is directed along y just like a "traditional" planar undulator, and the output X-rays are linearly polarized along x. On the other hand, if the lower back/upper front pair is moved by $+\lambda_u/2$ along z, the B_y components cancel on-axis and the electron motion and radiated field is linearly polarized along y. Note that this type of undulator consists of planar magnet arrays that are entirely above and below the electron beam, which is therefore compatible with elliptical vacuum chambers that are used in third-generation storage rings.

Another type of variable polarization undulator is called the Delta undulator [20], so named because the permanent magnet blocks look like the Greek letter Δ. The Delta undulator is similar to an APPLE device in that it has four magnetic rows, but the Delta has its magnetic rows placed symmetrically above, below, and to the sides of the gap. Hence, by restricting the Delta to circular vacuum vessels, it can place its magnetic poles

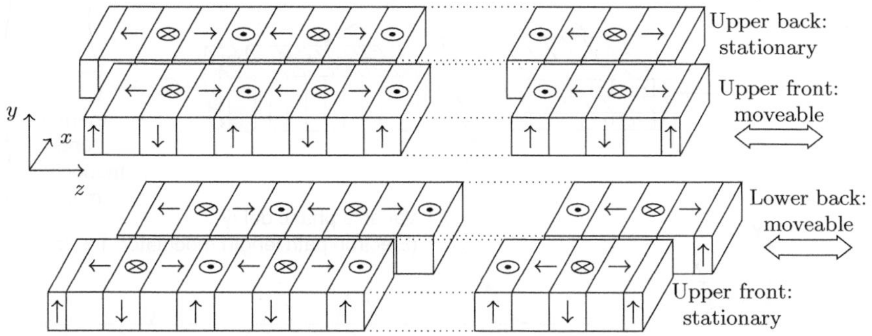

Figure 2.10 Schematic drawing of APPLE-2 undulator, showing the coordinate system used in the text. The arrow for each magnet block represents the magnetization direction. Adapted from Ref. [19].

closer (on average) to the beam axis. This results in a helical field strength that is approximately a factor of 1.7 larger than that of an APPLE-2 undulator. Much like the APPLE device, the Delta undulator operates in fixed gap mode and tunes the photon energy and the polarization state by moving the four magnetic blocks longitudinally with respect to each other. While the circular vacuum chamber is typically not suitable for storage rings, it can be used for FELs that use electron beams with essentially round transverse cross-sections. The increased magnetic field on axis allows for a higher undulator parameter K and hence a larger photon energy tuning range.

Crossed Undulator

An alternative method to produce undulator radiation of arbitrary polarization uses a pair of planar undulators oriented at right-angles to each other [21, 22], as shown in Figure 2.11. The radiation amplitude from these so-called crossed undulators is a linear superposition of two parts: one linearly polarized along the x direction and another linearly polarized along the y direction. In between the two undulators is a variable field magnet that controls the delay of the electrons and therefore the relative phase between the radiation from each undulator; for one delay the output is linearly polarized as indicated by Figure 2.11, and by changing the delay one can quickly vary the polarization to right circular, the mutually orthogonal linear, and/or left circular. For this device to work, it is necessary to use a monochromator with a sufficiently small bandpass, so that the wave trains from the two undulators are stretched and overlapped. Also, the angular divergence of the electron beam should be sufficiently small; otherwise, the fluctuation in the relative phase limits the achievable degree of polarization. A similar arrangement of two helical coils producing variable polarization was described by Alferov *et al.* [23].

2.4.5 Undulator Brightness and the Effects of the Electron Beam Distribution

Previously, we computed the undulator radiation emitted from a collection of ideal, resonant electrons moving along the optical axis. The particles in realizable electron

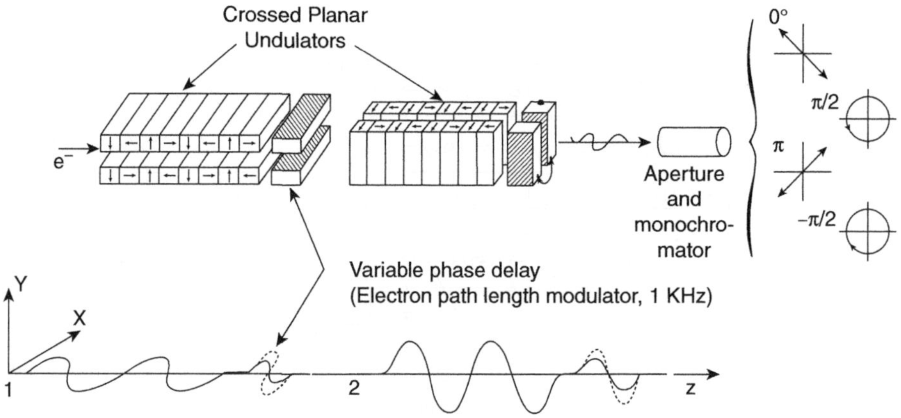

Figure 2.11 Schematic of a pair of crossed undulators, which could be used to produce variably polarized radiation. Reproduced from Ref. [22].

beams have coordinates that are distributed in phase space, with a variation in particle energy and arrival time, and a spread in transverse coordinates and angles characterized by its emittance. Here, we are interested in determining the source strength of undulator radiation including the effect of the e-beam distribution; this may be quite different from that due to N_e ideal electrons.

Our calculation will make use of the brightness function of undulator radiation, which we introduced for general paraxial fields in Chapter 1.2.3. First, we compute the radiation brightness produced by an ideal electron, namely, the brightness associated with the solution to (2.67). We then derive the brightness convolution theorem, which relates the single particle brightness to that produced by a distribution of electrons. Finally, we apply the brightness convolution formula to obtain the undulator radiation from both an electron beam with nonzero energy spread, and from a beam with finite transverse emittance.

We begin by calculating the radiation brightness from an ideal electron. The angular representation of the electric field at the end of the undulator is given by solving (2.67) with $h = 1$. The resulting field can then be propagated a distance ℓ along z by multiplying by the phase $e^{-ik\phi^2\ell/2}$. We will find it convenient to model the radiation as originating from a fictitious planar source located at the midpoint of the undulator. Thus, we propagate the solution at the end of the undulator a distance $\ell = -L_u/2$ by multiplying by the phase $e^{ik\phi^2 L_u/4}$, yielding

$$\mathcal{E}_\nu(\boldsymbol{\phi}; z = L_u/2) = \frac{eK[\mathrm{JJ}]e^{i\omega t_j}}{8\pi\epsilon_0\gamma_r c\lambda^2}e^{-ik\phi^2 L_u/4}\int_0^{L_u} dz\, e^{ikz\phi^2/2}e^{i(\nu-1)k_u z}. \tag{2.86}$$

The undulator brightness from a single electron can be found by inserting (2.86) into the definition (1.75). Choosing to defer the integrations over the undulator length and recalling that $\mathcal{B}$ is real, we find that the brightness of undulator radiation from an ideal electron is given by

$$\begin{aligned}\mathcal{B}_\nu^0(\boldsymbol{x},\boldsymbol{\phi}) = \frac{d\mathcal{F}^0/d\omega}{(\lambda/2)^2}\int_0^{L_u} dp \int_{-p}^{L_u-p} dq\, \frac{1}{\pi L_u q} \\ \times \sin\left\{\frac{2k}{q}\left[\boldsymbol{x} + \left(p + \tfrac{q-L_u}{2}\right)\boldsymbol{\phi}\right]^2 - \left[\frac{k\phi^2}{2} + (\nu-1)k_u\right]q\right\},\end{aligned} \tag{2.87}$$

where the single electron flux $\mathcal{F}^0$ is given by (2.77) at harmonic $h = 1$ and with I/e equal to the reciprocal time $1/T$. On resonance ($\nu = 1$) the integral over q can be evaluated analytically in terms of sine and cosine integrals, from which it is easy to show that at the origin $\mathcal{B}(\mathbf{0},\mathbf{0}) = \mathcal{F}^0/(\lambda/2)^2$. This implies that the wave is transversely coherent as was discussed in Chapter 1.2.4, which is not surprising since one might expect that the field produced by a single point-source to be completely correlated in the transverse plane. We plot the undulator brightness from an ideal reference electron when the transverse coordinates and angles are parallel in Figure 2.12. It is characterized by a large central

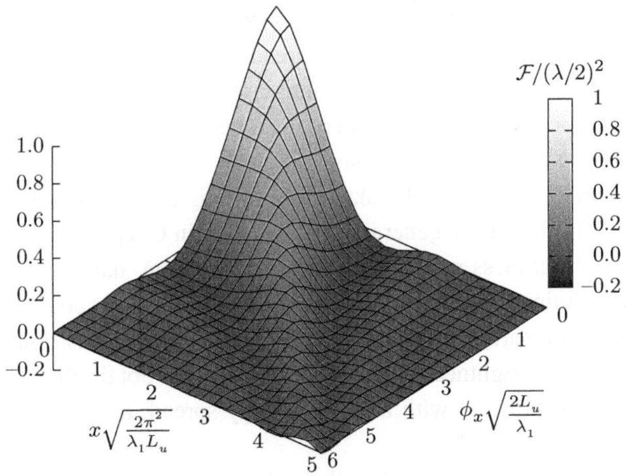

Figure 2.12 Brightness function $\mathcal{B}_1^0(\boldsymbol{x}, \boldsymbol{\phi})$ for undulator radiation from an ideal electron. We plot $\mathcal{B}$ in the x-ϕ_x plane (so that $\boldsymbol{x}$ and $\boldsymbol{\phi}$ are in the "same direction"), and have scaled x by σ_r and ϕ by $\sigma_{r'}$ for clarity. Adapted from Ref. [24].

peak in phase space, together with oscillations along the coordinate axes and a broad, slowly-decaying tail.

The brightness convolution theorem shows how the brightness from a beam of electrons can be found from the single particle $\mathcal{B}_\nu^0$ and the electron distribution function f. We will derive the theorem in a rather general setting that considers any synchrotron radiation device employing a magnetic field with no transverse gradient. In this context we will show how the radiation field generated by an electron with arbitrary initial coordinates is related to that produced by the ideal, reference electron. We then use this result to prove the brightness convolution theorem, and finally apply the theorem to a planar undulator.

We imagine we have an insertion device that begins at $z = 0$ and produces synchrotron radiation over the length L_u. Propagating the paraxial solution (2.12) at $z = L_u$ back to the radiator midpoint by multiplying the paraxial solution at $z = L_u$ by $e^{ikL_u/4}$, we find that the field due to a single electron is

$$\begin{aligned}\mathcal{E}_{\omega,j}(\boldsymbol{\phi}) &= e^{ik\phi^2 L_u/4} e^{-ik\phi^2 L_u/2} \int_0^{L_u} dz\, \frac{e[\boldsymbol{\beta}_j(z) - \boldsymbol{\phi}]}{4\pi\epsilon_0 c\lambda^2} e^{ik[ct_j(z) - z - \boldsymbol{\phi}\cdot\boldsymbol{x}_j(z) + \phi^2 z/2]} \\ &= e^{-ik\phi^2 L_u/4} \int_0^{L_u} dz\, \frac{e[\boldsymbol{\beta}_j(z) - \boldsymbol{\phi}]}{4\pi\epsilon_0 c\lambda^2} e^{ik[S_j(\boldsymbol{\phi},z) + \phi^2 L_u/4]} \end{aligned} \tag{2.88}$$

at the virtual source $z = L_u/2$. Here, we write $S_j(\boldsymbol{\phi}, z)$ in the particle phase as

$$\begin{aligned} S_j(\boldsymbol{\phi}, z) = &\int_{L_u/2}^{z} dz' \left(c\frac{dt_j}{dz'} - 1 - \boldsymbol{\phi}\cdot\boldsymbol{\beta}_j(z') + \frac{1}{2}\phi^2 \right) \\ &+ ct_j(L_u/2) - L_u/2 - \boldsymbol{\phi}\cdot\boldsymbol{x}_j(L_u/2), \end{aligned} \tag{2.89}$$

from which we can eliminate the time coordinate by using

$$c\frac{dt_j}{dz'} - 1 = \frac{1}{\sqrt{1-\gamma_j^{-2}-\boldsymbol{\beta}_j^2}} - 1 \approx \frac{1}{2\gamma_j^2} + \frac{1}{2}\boldsymbol{\beta}_j^2. \tag{2.90}$$

Now, we assume that the accelerating force due to the radiation device is purely magnetic and that we can approximate the magnetic field as being independent of the transverse direction (neglecting any gradient and/or focusing forces). Then, γ_j is constant and the particle's transverse angles are given by

$$\boldsymbol{\beta}_j(z) = \left[\frac{K_x(z)}{\gamma_j} + x_j'(L_u/2)\right]\hat{x} + \left[\frac{K_y(z)}{\gamma_j} + y_j'(L_u/2)\right]\hat{y}. \tag{2.91}$$

Hence, if we denote kS_0 to be the phase associated with the reference particle whose coordinates at the radiator center are $(\boldsymbol{x}, \boldsymbol{x}') = (\boldsymbol{0}, \boldsymbol{0})$ and $(t, \gamma) = (t_0, \gamma_0)$, in general we have

$$kS_j(\boldsymbol{\phi}, z) \approx k(1-2\eta_j)S_0(\boldsymbol{\phi} - \boldsymbol{x}_j', z) + kc(t_j - t_0) - k\boldsymbol{\phi}\cdot\boldsymbol{x}_j, \tag{2.92}$$

where the coordinates $(\boldsymbol{x}_j, \boldsymbol{x}_j', t_j, \eta_j)$ again refer to their values at the middle of the radiator. Inserting (2.92) into (2.88) and setting $t_0 = 0$, we find that the spontaneous radiation $\mathcal{E}_{\nu j}$ produced by a generic electron is related to that of the ideal, reference electron $\mathcal{E}_\nu^0$ via

$$\begin{aligned}\boldsymbol{\mathcal{E}}_{\nu,j}(\boldsymbol{\phi}) &= e^{ickt_j}e^{-ik\boldsymbol{\phi}\cdot\boldsymbol{x}_j}\int_0^{L_u} dz\,\frac{e[K(z) + (\boldsymbol{x}_j' - \boldsymbol{\phi})]}{4\pi\epsilon_0 c\lambda^2}e^{ik(1-\eta_j)S_0[(\boldsymbol{\phi}-\boldsymbol{x}_j'),z]}\\ &= e^{ickt_j}e^{-ik\boldsymbol{\phi}\cdot\boldsymbol{x}_j}\boldsymbol{\mathcal{E}}^0_{\nu-2\eta_j)}(\boldsymbol{\phi} - \boldsymbol{x}_j').\end{aligned} \tag{2.93}$$

The first two factors represent the phase difference due to the particle arrival time and finite electron beam size, respectively. Additionally, the electric field generated by the particle j is now centered about the frequency $\nu - 2\eta_j$ which is attributable to the dependence of the resonant condition on energy, while the angle $(\boldsymbol{\phi} - \boldsymbol{x}_j')$ can be interpreted as a shift of the optical axis for particles making finite transverse angles with respect to z. In physical space the virtual source is given by the Fourier transformed expression

$$E_\nu(\boldsymbol{x}) = e^{ickt_j}e^{ik(\boldsymbol{x}-\boldsymbol{x}_j)\cdot\boldsymbol{x}_j'}E_\nu(\boldsymbol{x} - \boldsymbol{x}_j). \tag{2.94}$$

The main assumption required to arrive at the translation formulas (2.93) and (2.94) is that the magnetic field is independent of $\boldsymbol{x}$, so that the particle oscillates along a nominally straight line through the device. However, the B-field of a planar undulator has a magnetic minimum along the axis, so that there will be some natural focusing over the length scale $\sim \lambda_u\gamma/K$. Hence, to apply Equations (2.93) or (2.94) to a planar undulator requires that it is much shorter than the focal length, $N_u \ll \gamma/K \sim 10^3$.

The total field is obtained by summing the contributions (2.93) over the N_e electrons, while inserting the resulting expression into the definition (1.75) results in the total undulator brightness from an electron beam. The expression for $\mathcal{B}$ can be divided into two sums: one containing N_e diagonal terms and the other the remaining $N_e(N_e - 1)$ as follows:

$$\mathcal{B}_\nu(\boldsymbol{x}, \boldsymbol{\phi}) = \sum_{j=1}^{N_e} \mathcal{B}^0_{\nu-2\eta_j}(\boldsymbol{x} - \boldsymbol{x}_j, \boldsymbol{\phi} - \boldsymbol{x}'_j) + \left\langle \sum_{j \neq k} e^{i\omega(t_j - t_k)}[\cdots] \right\rangle. \tag{2.95}$$

We compute the ensemble average using the single particle probability function $f(\boldsymbol{x}_j, \boldsymbol{x}'_j, \eta_j, t_j)$ assuming that the electrons are uncorrelated, so that the probability of finding an electron at a given point in phase space is independent of the positions of all the other particles. To show that the second term can generally be neglected, we assume that the particle beam is a Gaussian in t with a RMS beams duration σ_e, so that

$$f(\boldsymbol{x}_j, \boldsymbol{x}'_j, \eta_j, t_j) = f(\boldsymbol{x}_j, \boldsymbol{x}'_j, \eta_j) e^{-t_j^2/2\sigma_e^2}. \tag{2.96}$$

As we showed in Section 1.2.6, in this case the second sum in (2.95) is a factor $\sim N_e e^{-\omega^2\sigma_e^2} \ll 1$ smaller than the diagonal term in (2.95). Furthermore, although we rigorously derived this assuming a Gaussian temporal profile, the general conclusion holds provided the electron beam does not have significant Fourier content (bunching) at the wavelength of interest. This is a more formal justification of our incoherent argument in the previous section, which led to the conclusion that the total flux is N_e times the single particle flux. Dropping the second sum in (2.95), we arrive at the brightness convolution formula [24]

$$\mathcal{B}_\nu(\boldsymbol{x}, \boldsymbol{\phi}) = N_e \int d\boldsymbol{x}_j d\boldsymbol{x}'_j d\eta_j \, \mathcal{B}^0_{\nu-2\eta_j}(\boldsymbol{x} - \boldsymbol{x}_j, \boldsymbol{\phi} - \boldsymbol{x}'_j) f(\boldsymbol{x}_j, \boldsymbol{x}'_j, \eta_j), \tag{2.97}$$

where the index j is meant to refer to any single electron, since we have assumed that the particle probability distribution functions are identical and independent. As mentioned previously, the brightness from a Gaussian distributed electron beam will be everywhere positive if the emittance $\varepsilon_x \geq \lambda/4\pi$ [25]; for more general distributions f this general rule seems to at least approximately apply. Related work can also be found in, e.g., the Refs. [26, 27].

The convolution theorem (2.97) is the mathematical statement that in order to maximize the brightness of the spontaneous radiation one would like to match the electron beam phase space distribution function to that of the radiation. We will illustrate these ideas over the next few pages with two simple but useful examples.

First, we consider an electron beam with finite energy spread. Our primary reason for doing this is that we will derive similar results in a more general context when we discuss FEL radiation. Nevertheless, including the finite energy spread is important when the variation of the resonant frequency $\Delta\nu \sim \sigma_\eta$ becomes of order the normalized bandwidth of undulator radiation $\sim 1/2\pi N_u$. For simplicity we will assume that the angular spread of the electron beam is much smaller than that of the undulator radiation. In this case, i.e., when $\sigma_{x'} \ll \sigma_{r'}$, we can approximate the electron probability distribution function by $f(\boldsymbol{x}_j, \boldsymbol{x}'_j, \eta_j) = f(\boldsymbol{x}_j, \eta_j)\delta(\boldsymbol{x}'_j)$ in the convolution theorem (2.97). We evaluate the angular power density by integrating the resulting expression over the transverse dimension and multiplying by the photon energy $\hbar\omega$:

$$\frac{dP}{d\omega d\boldsymbol{\phi}} = N_e\hbar\omega\int d\boldsymbol{x}\int d\boldsymbol{x}_j d\eta_j\, f(\boldsymbol{x}_j,\eta_j)\mathcal{B}^0_{\nu-2\eta_j}(\boldsymbol{x}-\boldsymbol{x}_j,\boldsymbol{\phi})$$
$$= N_e\hbar\omega\int d\boldsymbol{y}\int d\boldsymbol{x}_j d\eta_j\, f(\boldsymbol{x}_j,\eta_j)\mathcal{B}^0_{\nu-2\eta_j}(\boldsymbol{y},\boldsymbol{\phi}), \tag{2.98}$$

where $\boldsymbol{y} \equiv \boldsymbol{x} - \boldsymbol{x}_j$. If the electron distribution is uncorrelated in position and energy the integration over $\boldsymbol{x}_j$ is trivially unity, while the integral in $\boldsymbol{y}$ can be evaluated once we insert the undulator brightness (2.87) as follows:

$$\int d\boldsymbol{y}\;\mathcal{B}^0_\nu(\boldsymbol{y},\boldsymbol{\phi}) = \frac{d\mathcal{F}^0/d\omega}{(\lambda/2)^2}\int_0^{L_u} dp\int_{-p}^{L_u-p} dq\;\frac{ie^{i[k\phi^2/2+(\nu-1)k_u]q}}{\pi L_u q}\left(-\frac{2\pi iq}{4k}\right)$$
$$= \frac{d\mathcal{F}^0}{d\omega}\frac{L_u}{\pi\lambda}\left\{\frac{\sin\!\left[\left(\frac{1}{2}k\phi^2+\Delta\nu k_u\right)L_u/2\right]}{\left(\frac{1}{2}k\phi^2+\Delta\nu k_u\right)L_u/2}\right\}^2. \tag{2.99}$$

Combining the two Equations (2.98)–(2.99), we find after some manipulations that the power spectral density in the forward direction is

$$\left.\frac{dP}{d\omega d\boldsymbol{\phi}}\right|_{\boldsymbol{\phi}=\boldsymbol{0}} = mc^2N_u^2\gamma^2\frac{I}{I_A}\frac{K^2[JJ]^2}{(1+K^2/2)^2}$$
$$\times\int d\eta_j\, f(\eta_j)\left\{\frac{\sin[(\eta_j-\Delta\nu/2)k_uL_u]}{(\eta_j-\Delta\nu/2)k_uL_u}\right\}^2, \tag{2.100}$$

where the Alfvén current $I_A \equiv emc^2/\alpha\hbar \equiv 4\pi\epsilon_0 mc^3/e \approx 17$ kA. The spectral density is broadened and reduced due to the variation of the resonance condition with electron energy, with the effect becoming significant when the energy spread $\sigma_\eta \equiv \sigma_\gamma/\gamma \gtrsim 1/2\pi N_u$. Equation (2.100) will appear again when we discuss self-amplified spontaneous radiation of FELs.

Our second application of the brightness convolution theorem is both instructive for the study of undulator radiation itself and also can be used to begin to show the dramatic improvement in radiation brightness that can be had using FELs. In this case, we focus on the transverse distribution of the electron beam, ignoring the spectral/energy dependence. To obtain simple analytic results, we approximate the undulator brightness (2.87) on resonance by a Gaussian function in both position and angle

$$\mathcal{B}^0_1(\boldsymbol{x},\boldsymbol{\phi}) \approx \frac{d\mathcal{F}^0/d\omega}{(\lambda/2)^2}\exp\left(-\frac{\boldsymbol{x}^2}{2\sigma_r^2}-\frac{\boldsymbol{\phi}^2}{2\sigma_{r'}^2}\right), \tag{2.101}$$

where the angular divergence is given by (2.57), while the physical size is chosen to satisfy $\sigma_r\sigma_{r'} = \lambda/4\pi$:

$$\sigma_{r'} = \sqrt{\frac{\lambda_1}{2L_u}} \qquad \sigma_r = \frac{\sqrt{2\lambda_1 L_u}}{4\pi}. \tag{2.102}$$

Since the undulator intensity profile is not Gaussian, however, these definitions for σ_r and $\sigma_{r'}$ are not uniquely defined. Indeed, other choices that differ by numerical factors also have utility as discussed in, e.g., [3, 28, 29].

Returning to our calculation, we will also assume that the electron beam distribution is a Gaussian, with

$$f(\boldsymbol{x}_j, \boldsymbol{x}_j') = \frac{1}{(2\pi)^2 \sigma_x \sigma_y \sigma_{x'} \sigma_{y'}} \exp\left(-\frac{x_j^2}{2\sigma_x^2} - \frac{y_j^2}{2\sigma_y^2} - \frac{x_j'^2}{2\sigma_{x_j'}^2} - \frac{y'^2}{2\sigma_{y'}^2} \right), \tag{2.103}$$

in which case the convolution (2.97) is trivial. The undulator brightness for a mono-energetic beam is

$$\mathcal{B}_1(\boldsymbol{x}, \boldsymbol{\phi}) = \frac{N_e d\mathcal{F}^0/d\omega}{(2\pi)^2 \Sigma_x \Sigma_y \Sigma_{x'} \Sigma_{y'}} \exp\left(-\frac{x^2}{2\Sigma_x^2} - \frac{y^2}{2\Sigma_y^2} - \frac{\phi_x^2}{2\Sigma_{x'}^2} - \frac{\phi_y^2}{2\Sigma_{y'}^2} \right), \tag{2.104}$$

where the convolved RMS widths are defined in each direction via $\Sigma_{x,y}^2 \equiv \sigma_{x,y}^2 + \sigma_r^2$ and $\Sigma_{x',y'}^2 \equiv \sigma_{x',y'}^2 + \sigma_{r'}^2$. The invariant source strength is given by the brightness at the phase space origin, $\mathcal{B}(\mathbf{0}, \mathbf{0})$. To simplify notation we drop the subscript 1 and denote the total ideal undulator photon flux $N_e \mathcal{F}^0 \equiv \mathcal{F}$, which we computed for a beam of current I in (2.77). In this case, two simple limits of the invariant source brightness can be readily obtained from (2.104) as follows:

Emittance dominated regime: $\sigma_{x,y} \gg \sigma_r$ and $\sigma_{x',y'} \gg \sigma_{r'}$

$$\mathcal{B}(\mathbf{0}, \mathbf{0}) = \frac{N_e d\mathcal{F}^0/d\omega}{(2\pi)^2 \sigma_x \sigma_{x'} \sigma_y \sigma_{y'}} = \frac{d\mathcal{F}/d\omega}{(2\pi)^2 \varepsilon_x \varepsilon_y}, \tag{2.105}$$

Radiation dominated regime: $\sigma_r \gg \sigma_{x,y}$ and $\sigma_{r'} \gg \sigma_{x',y'}$

$$\mathcal{B}(\mathbf{0}, \mathbf{0}) = \frac{N_e d\mathcal{F}^0/d\omega}{(2\pi)^2 \sigma_r^2 \sigma_{r'}^2} = \frac{d\mathcal{F}/d\omega}{(\lambda/2)^2}. \tag{2.106}$$

We see that $\mathcal{B}(\mathbf{0}, \mathbf{0})$ is maximized for Equation (2.106) when the beam emittance is much less than the minimal radiation emittance $\lambda/4\pi$. On the other hand, the brightness is a factor $\lambda/4\pi\varepsilon_x = \varepsilon_r/\varepsilon_x$ smaller for the "large" emittance beam of (2.105), which is related to the fact that the photon flux in the emittance dominated regime comprises of order $\varepsilon_x \varepsilon_y/\varepsilon_r^2 \gg 1$ transverse modes. Note that although the brightness (photons per unit phase space volume) is quite different, the total photon flux, obtained by integrating over all of phase space, is equal to $d\mathcal{F}/d\omega$ in both instances.

2.4.6 From Undulator Radiation to Free-Electron Lasers

This chapter has presented some basic physics and important results for synchrotron radiation sources, with a primary focus on undulator radiation. Undulators in modern synchrotron light sources provide X-ray radiation whose brightness is orders of magnitude larger than other traditional sources. An increase in the brightness by many more orders of magnitude yet can be obtained with free-electron lasers. Here, we discuss how this can be. To reiterate, the brightness (sometimes referred to as the brilliance) is one of the more important figures of merit for radiation sources. It is defined by

$$\mathcal{B} = \frac{\text{spectral flux}}{\text{transverse phase space area}}, \tag{2.107}$$

which is, as we have defined things, equal to the invariant value of the brightness function at the phase space origin, $\mathcal{B}(\mathbf{0},\mathbf{0})$. In the previous section we showed that the transverse phase space area relevant for undulator radiation in the Gaussian approximation is given by

$$\left(\sqrt{2\pi}\,\Sigma_x\right)\left(\sqrt{2\pi}\,\Sigma_{x'}\right)\left(\sqrt{2\pi}\,\Sigma_y\right)\left(\sqrt{2\pi}\,\Sigma_{y'}\right), \tag{2.108}$$

where Σ is again the convolved width in each phase space dimension, e.g., $\Sigma_x = \sqrt{\sigma_x^2 + \sigma_r^2}$. Plugging the photon flux (2.77) into the expression (2.104), at the origin the brightness of undulator radiation is

$$\mathcal{B} = \frac{\pi\alpha N_u (I/e)(\Delta\omega/\omega)}{(2\pi\,\Sigma_x\Sigma_{x'})(2\pi\,\Sigma_y\Sigma_{y'})}\frac{K^2[JJ]^2}{1+K^2/2}. \tag{2.109}$$

The standard units are Σ_x in mm, $\Sigma_{x'}$ in mrad, and I/e in #/sec, while the conventional bandwidth $(\Delta\omega/\omega) = 0.1$ percent, so that brightness is typically expressed in terms of "# of photons/[(mm)^2 (mrad)^2 (sec) (0.1 % BW)]."

The characteristics of a typical "third generation" light source are as follows:

$$\Sigma_x\Sigma_{x'} \approx \varepsilon_x \sim 10^{-2}\ \text{mm-mrad}, \qquad I/e \sim 100\ \text{mA}/e \approx 10^{18}/\text{sec},$$
$$\Sigma_y\Sigma_{y'} \approx \varepsilon_y \sim 10^{-4}\ \text{mm-mrad}, \qquad \alpha N_u \sim 100/137 \approx 1,$$

and $(\Delta\omega/\omega) = 10^{-3}$ by convention. It therefore follows that the typical brightness of undulator radiation is

$$\mathcal{B} = \frac{10^{18}\times 10^{-3}}{(2\pi)^2\times 10^{-6}} \sim 10^{20}\,\frac{\text{photons/sec in }0.1\%\text{ BW}}{(\text{mm})^2\ (\text{mrad})^2}. \tag{2.110}$$

When we consider FEL radiation, the brightness increases by a large factor due to the following considerations:

1. Transverse enhancement: FEL radiation is approximately coherent transversely, so that the phase space area $\Sigma_x\Sigma_{x'}\Sigma_y\Sigma_{y'} \to (\lambda/4\pi)^2$. For 1.5 Å radiation, this gives an enhancement factor over a typical third-generation storage ring of

$$\frac{\varepsilon_x\varepsilon_y}{(\lambda/4\pi)^2} \sim \frac{(10^{-9}\ \text{m-rad})(10^{-11}\ \text{m-rad})(4\pi)^2}{(1.5\times 10^{-10}\ \text{m-rad})^2} \approx 10^2. \tag{2.111}$$

2. Temporal enhancement: to achieve the high peak currents required by the FEL process, the bunch length is squeezed by bunch compressors to $\lesssim 100$ fs, which is $\gtrsim 2$ orders of magnitude shorter than that of third-generation sources.
3. Phase-coherence enhancement: in a FEL, electrons in one coherence length radiate together. The intensity enhancement is

$$N_{l_\text{coh}} = \#\text{ of electrons in one coherence length } l_\text{coh} \approx 10^6. \tag{2.112}$$

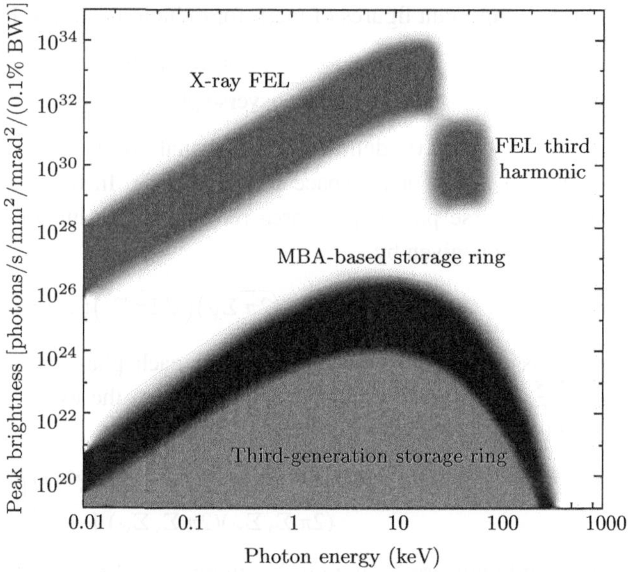

Figure 2.13 Peak brightness envelopes of third- and fourth-generation light sources. MBA-based storage rings have the potential to increase the brightness by about two orders of magnitude over traditional third-generation sources. FELs can increase the brightness another eight orders of magnitude.

Together, the peak brightness of X-ray FELs can be more than ten orders of magnitude larger than that of "third generation" light sources! A radiation source with this scale of improvement is often referred to as a "fourth generation" source. We present a representative comparison of the peak spectral brightness curves for some modern X-ray light sources in Figure 2.13. To similarly compare the average brightness requires knowing the repetition rate of the X-ray pulses, and we make such a comparison in Figure 2.14 including an FEL driven by a superconducting linac operating at 1 MHz.

2.5 Future Directions of Synchrotron Radiation Sources

In the next chapter we will begin our discussion of free-electron lasers which, as we just mentioned, can enhance the X-ray's brightness by ten orders of magnitude over traditional synchrotron radiation sources. Will this then make third-generation storage rings obsolete? We do not think so. In fact, we believe that their ability to produce high levels of average flux, impressive levels of stability, and to accommodate many experiments simultaneously will make third-generation storage rings an important scientific tool for many years to come. Additionally, we think that further development of those X-ray sources for higher brightness and increased agility will continue for the foreseeable future.

2.5.1 Multi-Bend Achromat Lattices for Smaller Storage Ring Emittances

One way to improve X-ray brightness employs a so-called "ultimate" storage ring that is designed to decrease the emittance by one to two orders of magnitude from that in

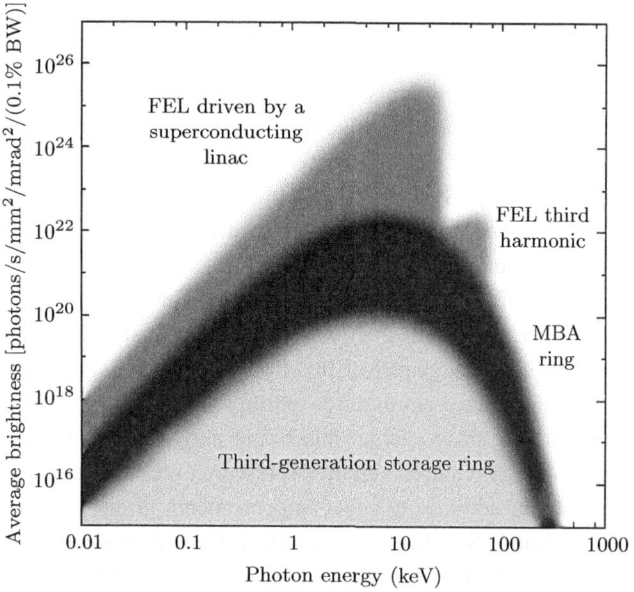

Figure 2.14 Average brightness envelopes of third- and fourth-generation light sources. The average brightness depends on the repetition rate, which for an FEL driven by a normal conducting rf-linac is ~100 Hz. Here, we assume that the upper bound of the FEL average brightness is set by a superconducting linac operating near 1 MHz.

third-generation rings. The emittance in an electron storage ring is determined by the balance of two physical processes characteristic of spontaneous emission – diffusion and damping. Damping is due to the fact that higher energy electrons on average radiate more than lower energy electrons, so that the trajectories of all electrons tend to converge to that of the reference electron. On the other hand, diffusion arises from electron recoil after the emission of a discrete photon, since the emission is a stochastic event described by quantum mechanics. The equilibrium emittance is achieved when the damping and diffusion processes balance, with ε_x scaling as the product of the diffusion constant and the damping time. Hence, the electron beam emittance can be reduced if the diffusion is suppressed.

Unfortunately, diffusion in the horizontal plane (understood by convention to be the plane in which the design orbit lies) cannot be completely eliminated because dipole bending magnets are required to form a closed trajectory. The horizontal emittance will therefore be nonzero. The vertical emittance, on the other hand, can be much, much smaller. In the ideal case, the uncertainty of photon emission in the vertical direction is limited to the opening angle $\sim 1/\gamma$, so that the diffusion along y is significantly reduced. In practice, however, the vertical emittance is dictated by some coupling of horizontal motion into vertical, which leads to a much larger ε_y than its theoretical minimum. This small x-y coupling is typically intentionally applied and controlled, but some amount of coupling is always present due to misaligned magnets and other errors.

Although diffusion in the horizontal plane will always lead to a non-vanishing ε_x, its effect can be limited to the "achromatic bending" sections that have a certain number M of bending magnets interspersed with suitably designed focusing quadrupoles. The smaller the bending angle of each dipole, the smaller will be the diffusion from the achromat, and the smaller will be the equilibrium emittance. Since the total bending angle from all achromats should add up to 360 degrees, the total number of bending magnets should be large to reduce the diffusion and therefore to achieve a small emittance. The $M = 2$ achromat, referred to as the double bend achromat, was first proposed for the National Synchrotron Light Source (NSLS) facility at Brookhaven [30]. Since then, all storage rings for synchrotron radiation facilities have adopted either double bend or triple bend ($M = 3$) achromats.

The use of multi-bend achromats (MBAs) with $M \gg 3$ was proposed as a straightforward generalization of the double bend achromat in 1995 [31]. An MBA-based storage ring can offer a large payoff in terms of emittance reduction, since the emittance scales as $1/M^3$ for a fixed circumference. Implementing an MBA-based ring, on the other hand, is not simple, since a larger number of magnets must fit in a given length. Overcoming this challenge required significant research into the design and manufacture of compact magnets, in addition to other improvements in accelerator technology. The MAX-IV ring in Lund, Sweden, is currently in the commissioning phase and will soon be the first $M = 7$ MBA storage ring [32]. It is designed to achieve a horizontal emittance of 0.33 nm, which is an order of magnitude smaller than any other third-generation source. Presently, several other synchrotron radiation facilities have plans to either upgrade their existing rings with an MBA lattice or possibly build an entirely new MBA ring. The predicted horizontal emittance may be as small as a few tens of picometers, resulting in undulator radiation brightness that is approximately two orders of magnitude greater than is presently available at third-generation rings.

One additional advantage of MBA lattices is that they can be designed to provide essentially round electron beams while maintaining high X-ray brightness. This is done by coupling the horizontal and vertical degrees of freedom such that $\varepsilon_x \approx \varepsilon_y$. While this can in principle be done at any storage ring, the total damping and diffusion implies that $\varepsilon_x + \varepsilon_y$ is a constant,[3] so that making a round beam in a third-generation storage ring dramatically degrades X-ray brightness. The round X-ray beams that would be produced by round electron beams at an MBA can often be better matched to X-ray optical instruments. In addition, round e-beams minimize the electron density and the related Coulomb scattering rate, which improves the lifetime of the beam. A recent series of articles that review MBA-based storage rings can be found in Ref. [33].

As we mentioned in the opening of this section, the low emittance rings based on multi-bend achromats have sometimes been referred to as ultimate storage rings. This terminology gives the impression that advances in storage ring design will stop after realization of the $M = 7$ MBA-based rings, but further developments may yet occur.

2.5.2 Energy Recovery Linacs

In a storage ring, the beam is at equilibrium. Hence, the effort to improve X-ray brightness by decreasing the emittance largely becomes a quest to reduce the diffusion

[3] It is only true that $\varepsilon_x + \varepsilon_y$ is constant if the damping times in x and y are the same; this statement is approximately correct provided the damping rates are similar.

associated with bending. An alternative strategy employs a linear accelerator to generate high-energy electrons, in which the electron beam properties are not in equilibrium and bending can be largely avoided. One promising way to do this for X-ray generation uses an energy recovery linac (ERL). An ERL begins with a high-brightness photocathode electron gun that produces low-emittance electron bunches at up to GHz repetition rates. These bunches are then accelerated by a superconducting linac up to energies of a few GeV, at which point they can be used to generate intense X-rays in insertion devices. After X-ray production, the high-power, high-energy e-beam is guided along a closed loop back to the linac. There, the beam is decelerated by the linac back to the injection energy, and safely disposed of in a beam dump. When the beam is decelerated in the appropriate rf phase, its kinetic energy is converted back to electromagnetic energy. Hence, the rf energy spent to accelerate the electrons is mostly recovered, which gives the energy recovery linac its name. It is possible to design an ERL such that the electrons are accelerated and then decelerated over multiple circuits through linac, which reduces the length and cost of the superconducting accelerator.

An ERL has the potential to provide several different operating modes, including a high-intensity mode that uses a large charge per bunch, a high-coherence mode characterized by an ultra-small emittance, and an ultra-short pulse mode. Such agility is difficult to achieve with a storage ring, where the beam properties are in a stable equilibrium; producing ultra-short X-ray pulses is particularly challenging for storage rings, so that ERLs have a marked advantage in this case. On the other hand, obtaining the stability levels of a storage ring on an essentially single-pass machine like an ERL remains a significant challenge. A review of recent R&D efforts toward an ERL can be found in [34].

2.5.3 Superconducting Undulators

Our look at what the future may hold for advanced synchrotron radiation sources would not complete without mentioning the recent progress in superconducting undulators. Over the last few decades, permanent magnet technology has dominated undulator design, since it has provided the best way to construct high-field, short (few cm) period devices with a gap large enough for the electron beam. The performance of permanent magnet undulators has made some progress since the original Halbach design; for example, in-vacuum undulators eliminate the beam pipe and permit a smaller magnetic gap, while cryogenic devices can achieve larger magnetic fields. As we have previously mentioned, electromagnetic undulators that employ normally conducting coils cannot compete with permanent magnet-based devices in terms of compactness or high-field capability. An electromagnetic undulator making use of superconducting wires, however, can in principle achieve large enough current densities such that the generated magnetic fields will outperform permanent magnet devices. Building practical superconducting undulators has required a number of technical and engineering issues to be overcome; nevertheless, successful devices using NbTi superconducting wires have been operating at ANKA since 2005 [35] and at the APS since 2013 [36]. We expect more to be installed and brought online in the near future.

The primary advantage of superconducting undulators is their higher field strength, which means that one can employ a shorter period λ_u for a given K, and fit more undulator periods into a given length. Alternatively, one can use the increase in K at fixed λ_u to significantly enhance output at higher harmonics. Both of these approaches particularly benefit hard X-ray output, so that one can consider more moderate energy electron beams and more compact, less costly synchrotron radiation facilities.

The improvements in magnetic field for a given pole-to-pole gap g and undulator wavelength λ_u can be evaluated using two fitting formulas developed in Ref. [37]. For a niobium–titanium planar SCU, the peak field

$$B_0[\mathrm{T}] \approx (0.48534 + 0.41611\lambda_u[\mathrm{cm}] - 0.039932\lambda_u^2[\mathrm{cm}]) \times \exp\left[-\frac{\pi(2g-\lambda_u)}{2\lambda_u}\right] \quad [\mathrm{NbTi}], \tag{2.113}$$

while an SCU using niobium–tin wires follows

$$B_0[\mathrm{T}] \approx (0.68115 + 0.64105\lambda_u[\mathrm{cm}] - 0.060986\lambda_u^2[\mathrm{cm}]) \times \exp\left[-\frac{\pi(2g-\lambda_u)}{2\lambda_u}\right] \quad [\mathrm{Nb_3Sn}]. \tag{2.114}$$

Equations (2.113) and (2.114) are the SCU counterparts to the permanent magnet Halbach formula given in Equation (2.40), and they are valid for a period length λ_u between 0.8 and 3.6 cm and at a coil current that is 80 percent of the critical current. We plot a comparison of the predicted K vs. λ_u behavior for two different gap sizes in Figure 2.15. The plot shows that a NbTi-based undulator nearly triples the K parameter for a given g and λ_u, or can be built with a smaller period and/or gap while achieving the same K. Note that g is the magnetic gap, and that inner diameter of the beam vacuum chamber is typically 2 to 4 mm smaller than g.

Finally, superconducting undulator technology will also continue to evolve; for example, Nb_3Sn wires hold the promise of even higher current densities and K values as

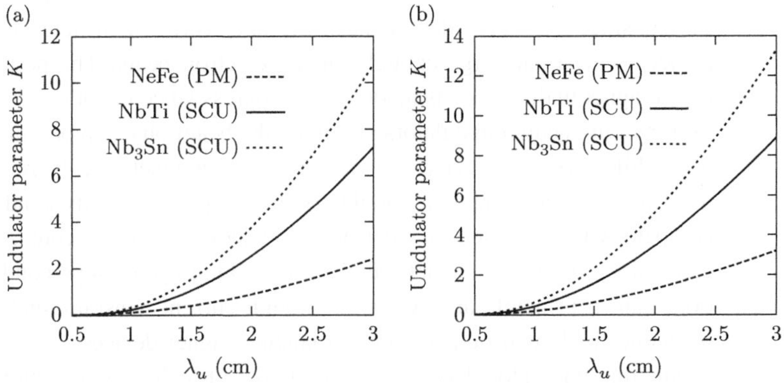

Figure 2.15 Comparison of superconducting (SCU) and permanent magnet (PM) undulator parameter K as a function of wavelength λ_u. (a) plots K assuming the full magnetic gap $g = 9$ mm, while (b) sets $g = 7$ mm. The K of the NbTi-based SCU is about three times stronger than the PM device at the same gap and λ_u, while Nb_3Sn is ~50 percent stronger than NbTi.

shown in Figure 2.15, but additional challenges remain. Eventually, magnets employing a high Tc superconductor may revolutionize undulator construction, perhaps in an electron storage ring based on an MBA design with $M \gg 7$.

2.5.4 Laser Undulator

The synchrotron sources described in this book typically call for multi-GeV electron beams, so that conventionally they have been built using rather long particle accelerators in large and costly facilities. Perhaps the best way to reduce the size and cost of these X-ray sources is to decrease the accelerator length, which may be achieved by either increasing the accelerating gradient or by finding ways to make use of lower energy electrons. While attaining ever higher accelerating gradients is a very active area of research, it does not directly affect X-ray production and will not be further discussed here. Rather, in this section we describe one promising scheme that can produce intense X-rays with relatively low energy electrons, which works by replacing the usual magnetic undulator with a laser; because the laser wavelength can be $\sim 10^{-3}$–10^{-4} times smaller than magnetostatic undulators, one can consider electron beam energies that are up to 100 times lower. Various proposals for compact X-ray sources using laser undulators have been made over the years (see, e.g., [38, 39, 40, 41, 42]), and some experimental results include [43, 44, 45].

X-ray sources that employ laser undulators have come to be known as "inverse Compton sources," although we do not think that this is a suitable name. First, Compton scattering applies when the recoil of the electron is important, while the effect described here is simple Thomson scattering in the rest frame of the electron. Second, it is not obvious to us what the "inverse" applies to. Hence, we will eschew the name "inverse Compton source," even though it appears to be here to stay.

Now, we return to treat the physics of a laser undulator by considering a plane wave laser that is directed along $\hat{k} = \boldsymbol{k}/k_L$. In this case the magnetic field can be derived from the electric field using $\boldsymbol{B} = (\hat{k} \times \boldsymbol{E})/c$, and the Lorentz force law becomes

$$\frac{d}{dt}(\gamma mc\boldsymbol{\beta}) = -e\left[\boldsymbol{E} + \boldsymbol{\beta} \times (\hat{k} \times \boldsymbol{E})\right] = -e\left[\boldsymbol{E} + (\boldsymbol{\beta} \cdot \boldsymbol{E})\hat{k} - (\boldsymbol{\beta} \cdot \hat{k})\boldsymbol{E}\right]. \tag{2.115}$$

We orient our axes such that the laser undulator is polarized along $\hat{x}$ and directed as shown in Figure 2.16. Hence, $\boldsymbol{E} = E_0 \sin(\omega_L t - \boldsymbol{k} \cdot \boldsymbol{x})\hat{x}$ with $\boldsymbol{k} = k_L(0, \sin\Psi, -\cos\Psi)$ and $\omega_L = ck_L$ the laser frequency. Then, the $\hat{x}$ component of the force law (2.115)

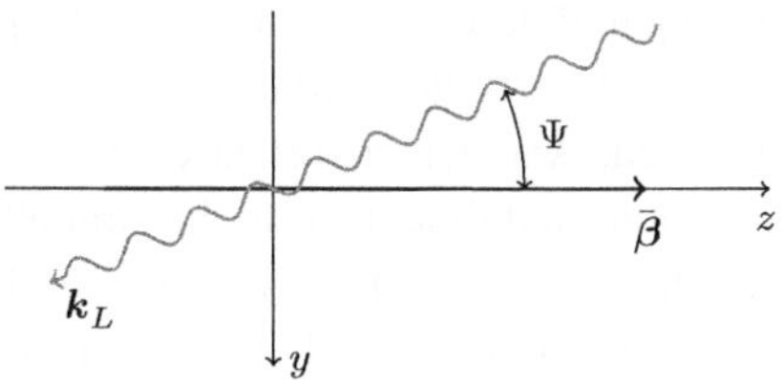

Figure 2.16 Geometry for the laser undulator that is directed to collide with an electron beam at the angle Ψ.

becomes

$$\frac{d}{dt}(\gamma mc\beta_x) = -eE_0 \sin(ck_L t - \boldsymbol{k}\cdot\boldsymbol{x})\left[1 - (\hat{k}\cdot\boldsymbol{\beta})\right]$$
$$= \frac{eE_0}{ck_L}\frac{d}{dt}\cos(ck_L t - \boldsymbol{k}\cdot\boldsymbol{x}). \tag{2.116}$$

This equation expresses the conservation of canonical momentum along the horizontal direction, and can easily be integrated to obtain

$$\beta_x = \frac{eE_0}{\gamma mc^2 k_L}\cos(ck_L t - \boldsymbol{k}\cdot\boldsymbol{x}). \tag{2.117}$$

The equations for the transverse "wiggle velocity" in the laser field equations (2.115) and (2.117) are very similar to their magnetic counterparts (2.44) and (2.45). From this we see that the deflection parameter for the laser undulator is

$$K_L = \frac{eE_0}{mc^2 k_L} = \sqrt{\frac{2r_e I_{\text{laser}}}{\pi mc^3}}\lambda_L, \tag{2.118}$$

where I_{laser} is the intensity of the laser and the classical radius of the electron $r_e \equiv e^2/(4\pi\varepsilon_0 mc^2) \approx 2.818 \times 10^{-15}$ m. In other literature K_L is often referred to as the peak dimensionless vector potential a_L. In terms of convenient quantities we have $K_L \approx 8.9 \times 10^{-10}\lambda_L[\mu\text{m}]\sqrt{I_{\text{laser}}[\text{W/cm}^2]}$, and $K_L < 1$ except for the most intense, high-power lasers.

In addition to the deflection parameter, we can find the effective undulator wavelength by replacing $t \approx z/\bar{\beta}_z$ in Equation (2.117). Then, we find that the transverse velocity oscillates with $\cos[k_L(1/\bar{\beta}_z + \cos\Psi)z - k_L y \sin\Psi]$, so that the laser field acts like an undulator with wavelength

$$\lambda_u \to \frac{\bar{\beta}_z \lambda_L}{1 + \bar{\beta}_z \cos\Psi}. \tag{2.119}$$

All the properties of a laser-based X-ray source can be derived from the corresponding expressions that we derived for undulator radiation by making the wavelength substitution (2.119) and using the deflection parameter K_L. In particular, the resonant wavelength for synchrotron radiation production in a laser field is

$$\lambda = \frac{1 + K_L^2/2}{2\gamma^2}\frac{\lambda_L}{1 + \bar{\beta}_z \cos\Psi}. \tag{2.120}$$

Laser-based synchrotron sources offer the possibility of generating very short wavelength X-ray photons, or of producing intense X-rays at a given wavelength with a significantly lower energy electron beam. Hence, one can significantly reduce the cost and size of the required accelerator. However, the promise of a compact source comes with a cost: because the e-beam emittance $\varepsilon_x = \varepsilon_{x,n}/\gamma$, the X-ray coherence and brightness will be reduced. Another important consideration is that the laser envelope will vary in all dimensions and time. Hence, the strength parameter K_L will not be constant over the entire interaction, and the resonance condition (2.120) will vary accordingly. While this affect is not significant when $K_L \ll 1$, in FEL applications that typically call for $K_L^2 > 0.1$ it can be quite important. Nevertheless, there are some proposals for which laser-based FELs may be possible; see, e.g., Refs. [46, 47].

References

[1] K.-J. Kim, "Characteristics of synchrotron radiation," in *Proc. US Particle Accelerator School*, ser. AIP Conference Proceedings, M. Month and M. Dienes, Eds., no. 184. New York: AIP, p. 565, 1989.

[2] H. Wiedemann, *Synchrotron Radiation*. Berlin, Germany: Springer-Verlag, 1993.

[3] A. Hofmann, *The Physics of Synchrotron Radiation*. Cambridge, UK: Cambridge University Press, 2004.

[4] G. A. Schott, *Electromagnetic Radiation*. Cambridge, UK: Cambridge University Press, 1912.

[5] J. Schwinger, "On the classical radiation of accelerated electrons," *Phys. Rev.*, vol. 75, p. 1912, 1949.

[6] R. P. Feynman, R. B. Leighton, and M. Sands, *The Feynman Lecture on Physics*. Reading, Mass: Addison-Wesley, 1963.

[7] L. D. Landau and E. M. Lifshitz, *The Classical Theory of Fields*, 4th ed., vol. 2, ser. Course of Theoretical Physics. London: Pergamon, 1979 (Translated from the Russian).

[8] J. D. Jackson, *Classical Electrodynamics*, 2nd ed. New York: Wiley, 1975.

[9] D. F. Alferov, Y. A. Bashmakov, and E. G. Bessonov, "Undulator radiation," *Zh. Tekh. Fiz.*, vol. 43, p. 2126, 1973, translated from Russian in Sov. Phys. Tech. Phys., 18:1336, 1974.

[10] B. M. Kincaid, "A short-period helical wiggler as an improved source of synchrotron radiation," *J. Appl. Phys.*, vol. 48, p. 2684, 1977.

[11] G. Geloni, E. L. Saldin, E. A. Schneidmiller, and M. V. Yurkov, "Paraxial green's functions in synchrotron radiation theory," DESY, Tech. Rep. 05-032, 2005.

[12] K. Halbach, "Physical and optical properties of rare earth cobalt magnets," *Nucl. Instrum. Methods Phys. Res.*, vol. 187, p. 109, 1981.

[13] —, "Permanent magnetic undulators," *J. Phys. Colloques C1*, vol. 44, pp. C1–211, 1983.

[14] D. Attwood, *Soft X-rays and Extreme Ultraviolet Radiation*. Cambridge, UK: Cambridge University Press, 1999.

[15] S. Krinsky, "Undulators as sources of synchrotron radiation," *IEEE Trans. Nucl. Sci.*, vol. 30, p. 3078, 1983.

[16] K.-J. Kim, in *X-ray Data Booklet*, D. Vaughn, Ed., no. 490. New York: Lawrence Berkeley Laboratory, 1986.

[17] —, "Angular distribution of undulator power for an arbitrary deflection parameter K," *Nucl. Instrum. Methods Phys. Res., Sect. A*, vol. 246, p. 67, 1986.

[18] H. Kitamura, "Polarization of undulator radiation," *Japanese J. Appl.Phys.*, vol. 19, p. 2185, 1980.

[19] S. Sasaki, "Analyses for a planar variably-polarizing undulator," *Nucl. Instrum. Methods Phys. Res., Sect. A*, vol. 347, p. 83, 1994.

[20] A. Temnykh, "Delta undulator for Cornell energy recovery linac," *Phys. Rev. ST-Accel. Beams.*, vol. 11, p. 120702, 2008.

[21] M. B. Moiseev, M. M. Nikitin, and N. I. Fedosov, "Change in the kind of polarization of undulator radiation," *Sov. Phys. J.*, vol. 21, p. 332, 1978.

[22] K.-J. Kim, "A synchrotron radiation source with arbitrarily adjustable elliptical polarization," *Nucl. Instrum. Methods Phys. Res.*, vol. 219, p. 425, 1984.

[23] D. F. Alferov, Y. A. Bashmakov, and E. G. Bessonov, "Generation of circularly-polarized electromagnetic radiation," *Zh. Tekh. Fiz.*, vol. 46, p. 2392, 1976, translated from Russian in Sov. Phys. Tech. Phys., 21:1408, 1976.

[24] K.-J. Kim, "Brightness, coherence, and propagation characteristics of synchrotron radiation," *Nucl. Instrum. Methods Phys. Res., Sect. A*, vol. 246, p. 71, 1986.

[25] N. D. Cartwright, "A non-negative Wigner-type distribution," *Physica*, vol. 83, p. 210, 1976.

[26] G. Geloni, E. L. Saldin, E. A. Schneidmiller, and M. V. Yurkov, "Transverse coherence properties of X-ray beams in third-generation synchrotron radiation sources," *Nucl. Instrum. Methods Phys. Res., Sect. A*, vol. 588, p. 463, 2008.

[27] I. V. Bazarov, "Synchrotron radiation representation in phase space," *Phys. Rev. ST-Accel. Beams*, vol. 15, p. 050703, 2012.

[28] H. Onuki and P. Elleaume, *Undulators, Wigglers, and Their Applications*. London, UK: CRC Press, 2003.

[29] R. R. Lindberg and K.-J. Kim, "Compact representations of partially coherent undulator radiation suitable for wave propagation," *Phys. Rev. ST-Accel. Beams*, vol. 18, p. 090702, 2015.

[30] R. Chasman, G. K. Green, and E. M. Rowe, "Preliminary design of a dedicated synchrotron radiation facility," in *Proceedings of the 1975 Particle Accelerator Conference*, Washington, DC, 1975, p. 1765.

[31] D. Einfeld, J. Schaper, and M. Plesko, "Design of a diffraction limited light source (DIFL)," in *Proceedings of the 1995 Particle Accelerator Conference*, Dallas, TX, p. 177, 1995.

[32] S. C. Leemann, A. Åndersson, M. Eriksson, L. J. Lindgren, E. Wallén, J. Bengtsson, and A. Streun, "Beam dynamics and expected performance of Sweden's new storage-ring light source: MAX IV," *Phys. Rev. ST-Accel. Beams.*, vol. 12, p. 12070, 2009.

[33] *J. of Synchrotron Rad.*, vol. 21, no. 5, Special Issue on Diffraction-Limited Storage Rings and New Science Opportunities, 2014.

[34] D. H. Bilderback, J. D. Brock, D. S. Dale, K. D. Finkelstein, M. A. Pfeifer, and S. M. Gruner, "Energy recovery linac (ERL) coherent hard X-ray sources," *New J. Phys.*, vol. 12, p. 035011, 2010.

[35] S. Casalbuoni, M. Hagelstein, B. Kostka, R. Rossmanith, M. Weisser, E. Steffens, A. Bernhard, D. Wollmann, and T. Baumbach, "Generation of X-ray radiation in a storage ring by a superconductive cold-bore in-vacuum undulator," *Phys. Rev. ST-Accel. Beams.*, vol. 9, p. 010702, 2006.

[36] Y. Ivanyushenkov *et al.*, "Development and operating experience of a short-period superconducting undulator at the advanced photon source," *Phys. Rev. ST-Accel. Beams.*, vol. 18, p. 040703, 2015.

[37] S. H. Kim, "A scaling law for the magnetic fields of superconducting undulators," *Nucl. Instrum. Methods Phys. Res., Sect. A*, vol. 546, p. 604, 2005.

[38] R. H. Milburn, "Electron scattering by an intense polarized photon field," *Phys. Rev. Lett.*, vol. 10, p. 75, 1963.

[39] P. Sprangle, A. Ting, E. Esarey, and A. Fisher, "Tunable, short pulse hard X-rays from a compact laser synchrotron source," *J. Appl. Phys.*, vol. 72, p. 5032, 1992.

[40] K.-J. Kim, S. Chattopadhyay, and C. V. Shank, "Generation of femtosecond X-rays by 90° Thomson scattering," *Nucl. Instrum. Methods Phys. Res., Sect. A*, vol. 341, p. 351, 1994.

[41] Z. Huang and R. D. Ruth, "Laser-electron storage ring," *Phys. Rev. Lett.*, vol. 80, p. 976, 1998.

[42] W. Graves, J. Bessuille, P. Brown, S. Carbajo, V. Dolgashev, K.-H. Hong, E. Ihloff, B. Khaykovich, H. Lin, K. Murari, E. Nanni, G. Resta, S. Tantawi, L. Zapata, F. Kärtner, and D. Moncton, "Compact X-ray source based on burst-mode inverse Compton scattering at 100 khz," *Phys. Rev. ST-Accel. Beams*, vol. 17, p. 120701, 2014.

[43] R. W. Schoenlein, W. P. Leemans, A. H. Chin, P. Volfbeyn, T. E. Glover, P. Balling, M. Zolotorev, K.-J. Kim, S. Chattopadhyay, and C. V. Shank, "Femtosecond X-ray pulses at 0.4 å generated by 90° Thomson scattering," *Science*, vol. 274, p. 236, 1996.

[44] W. J. Brown, S. G. Anderson, C. P. J, Barty, S. M. Betts, R. Booth, J. K. Crane, R. R. Cross, D. N. Fittinghoff, D. J. Gibson, F. V. Hartemann, E. P. Hartouni, J. Kuba, G. P. Le Sage, D. R. Slaughter, A. M. Tremaine, A. J. Wootton, P. T. Springer, and J. B. Rosenzweig, "Experimental characterization of an ultrafast Thomson scattering X-ray source with three-dimensional time and frequency-domain analysis," *Phys. Rev. ST-Accel. Beams*, vol. 7, p. 060702, 2004.

[45] H. Shimizu, M. Akemoto, Y. Arai, S. Araki, A. Aryshev, M. Fukuda, S. Fukuda, J. Haba, K. Hara, H. Hayano, Y. Higashi, Y. Honda, T. Honma, E. Kako, Y. Kojima, Y. Kondo, K. Lekomtsev, T. Matsumoto, S. Michizono, T. Miyoshi, H. Nakai, H. Nakajima, K. Nakanishi, S. Noguchi, T. Okugi, M. Sato, M. Shevelev, T. Shishido, T. Takenaka, K. T., J. Urakawa, K. Watanabe, S. Yamaguchi, A. Yamamoto, Y. Yamamoto, K. Sakaue, S. Hosoda, H. Iijima, M. Kuriki, R. Tanaka, A. Kuramoto, M. Omet, and A. Takeda, "X-ray generation by inverse Compton scattering at the superconducting RF test facility," *Nucl. Instrum. Methods Phys. Res., Sect. A*, vol. 772, p. 26, 2015.

[46] J. C. Gallardo, R. C. Fernow, R. Palmer, and C. Pellegrini, "Theory of a free-electron laser with a gaussian optical undulator," *IEEE J. Quantum Electron.*, vol. 24, p. 1557, 1988.

[47] J. E. Lawler, J. Bisognano, R. A. Bosch, T. C. Chiang, M. A. Green, K. Jacobs, T. Miller, R. Wehlitz, D. Yavuz, and R. C. York, "Nearly copropagating sheared laser pulse FEL undulator for soft X-rays," *J. Appl. Phys. D*, vol. 46, p. 325501, 2013.

3 Basic FEL Physics

This chapter provides a somewhat qualitative introduction to free-electron laser physics. After a brief introduction to how FELs fit in with other sources of coherent radiation and how FEL amplification can work, we derive in Section 3.2 the 1D equations of motion for the electrons in an FEL. We find that the electrons move in a pendulum-type potential formed by the transverse undulator and radiation fields. We then discuss the FEL particle dynamics in the limit that the gain is small in Section 3.3, for which we can approximate the electric field as constant so that we do not need Maxwell's equations; this regime of operation is most suited to oscillator FELs that we cover in more detail in Chapter 7. We compute the FEL gain in the small-signal limit, and qualitatively describe the FEL when the radiation power is large. In Section 3.4 we include the self-consistent evolution of the radiation, which extends our equations to include high-gain FELs that experience exponential gain. We show how the basic physics of small signal gain and exponential growth can be derived in terms of three collective variables, and how the FEL Pierce parameter ρ plays a central role in determining the output characteristics. Finally, we make a brief introduction to self-amplified spontaneous radiation, which we will describe more thoroughly in Chapter 4.

3.1 Introduction

3.1.1 Coherent Radiation Sources

Powerful, coherent sources are familiar for wavelengths longer than 1 mm – the microwave devices – and also for wavelengths between a few microns and approximately 0.1 μm – lasers based on atomic and molecular transitions. Microwave devices (including magnetrons, klystrons, and, increasingly, solid-state devices) have found numerous applications including radar, accelerating structures, and food preparation. The applications of lasers operating in the IR, visible, and near UV are incredibly numerous and diverse: uses vary from precise measurements of time and distance to cutting and etching on both large and nano-scales to addressing single atoms for quantum computing.

Although the applications stated above cover a wide range of activities, there remains a vast spectral region, including both the THz ($\lambda \sim 0.1$ mm) and X-ray ($\lambda \lesssim 40$ Å) wavelengths, for which coherent radiation is not readily available. Free-electron lasers

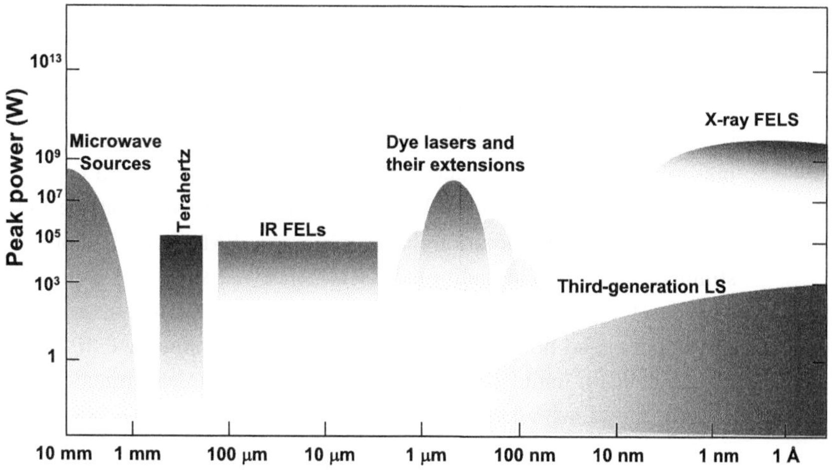

Figure 3.1 Power of tunable sources of radiation.

(FELs) are coherent radiation sources based on radiation from "free" electrons rather than electrons bound in atomic and molecular systems. Invented by John Madey [1] in 1971 and subsequently demonstrated at Stanford University [2], FELs can in principle operate at any arbitrary wavelength, limited only by the energy and quality of the electron beam that is produced by accelerators. FELs can therefore be used to fill gaps in the spectral regions where coherent sources have not otherwise been available. Figure 3.1 schematically illustrates a number of radiation sources, including both the characteristic wavelength and power produced. In the infrared (1 μm $<$ radiation wavelength $<$ 1 mm), FEL oscillators employing a variety of accelerators including Van de Graafs, microtrons, and rf linacs have been constructed and are currently in operation. FEL oscillators in storage rings (with beam energies ranging from 500 MeV to 1 GeV) produce coherent visible and UV radiation. The shortest wavelength achieved in a storage ring FEL is about 200 nm, which is presently limited by the mirror reflectivity at these short wavelengths.

Generating coherent radiation at shorter wavelengths has been primarily pursued using high-gain FELs that do not require mirrors. Devices generating intense UV and vacuum ultraviolet (VUV) radiation were first designed and built in the late 1990s and early 2000s, which paved the way for the planning and construction of numerous X-ray FEL facilities. Since 2005, FLASH in Hamburg has operated a soft X-ray FEL for scientific users, while the era of hard X-ray FELs began in 2009 with the Linac Coherent Light Source (LCLS) at SLAC [3]. X-ray FEL facilities in Japan (SACLA) [4] and Italy (FERMI) have since been commissioned and operated for users, while other FELs are in various stages of construction/development (e.g., the European XFEL in Germany [5], the XFEL in Pohang, Korea [6], and the SwissFEL at PSI, Switzerland [7]). X-ray FELs are sometimes referred to as fourth-generation light sources to distinguish them from the third-generation sources that use high-brightness storage rings to generate quasi-coherent spontaneous radiation from magnetic insertion devices such as bending magnets, wigglers, and undulators.

3.1.2 What Is an FEL?

Thus far, we have studied the generation of spontaneous electromagnetic (EM) radiation by "free" electrons in an undulator, which is analogous to the spontaneous emission by bound electrons in atoms. For undulators, the spontaneous radiation is produced by the prescribed motion in an external magnetic field. Under certain favorable conditions, the radiated field interacts with the electrons in such a way that the induced acceleration produces additional radiation that amplifies the EM wave. This process is analogous to stimulated emission in traditional lasers. Since FELs are not tied to any electronic levels, however, they can in principle be built to produce radiation at any wavelength.

To understand how radiation amplification is possible, we first consider the electron trajectory in an undulator. The transverse motion is determined very accurately by the magnetic field, with the transverse velocity as we found in (2.45):

$$v_x = \frac{Kc}{\gamma}\cos(k_u z). \tag{3.1}$$

In addition to the motion from the undulator, the electron is acted upon by the spontaneous radiation field that we model as an electromagnetic wave propagating colinearly with the particle beam,

$$\boldsymbol{E}(z,t) = \hat{x}E_0 \sin(kz - \omega t + \phi), \qquad \omega = ck = \frac{2\pi c}{\lambda}. \tag{3.2}$$

The rate of energy transfer from the radiation field to the electron is given by the incremental work

$$\boldsymbol{F}\cdot\boldsymbol{v} = -e\boldsymbol{E}\cdot\boldsymbol{v} = -\frac{eE_0Kc}{\gamma}\cos(k_u z)\sin(kz - \omega t + \phi) \neq 0, \tag{3.3}$$

so that energy can be exchanged between the field and the particles. The electron is accelerated if $\boldsymbol{F}\cdot\boldsymbol{v} > 0$ (referred to as an inverse FEL), while it is decelerated if $\boldsymbol{F}\cdot\boldsymbol{v} < 0$. In the latter case, the particle kinetic energy is converted into electromagnetic energy, and the field strength of the wave is increased.

In general, the interaction cannot be sustained since the electromagnetic wave propagates ahead of the electrons. However, there is another way for the energy transfer to continue. Recall that the fundamental wavelength of the undulator radiation $\lambda = (1 + K^2/2)\lambda_u/2\gamma^2$ is equal to the distance that light propagates ahead of an electron during the time the electron traverses one undulator period λ_u. This is equivalent to the statement that the electron slips one λ behind the EM wave while traveling one undulator period. Hence, due to the periodicity of the system, the interaction can continue. We illustrate this in Figure 3.2. On the left is an electron whose transverse velocity due to the undulator field is initially directed along $-\hat{x}$, which we assume is parallel to the radiation field as indicated by the arrow. Thus, this electron is initially in a decelerating phase of the wave, and it radiates its kinetic energy into field energy. After one undulator period, the EM wave has advanced one radiation period and the situation is identical as shown on the right. In between the energy transfer is always in the same direction; for example, at $\lambda_u/2$ both the electron velocity and the field are pointed along $+\hat{x}$, so that $\boldsymbol{F}\cdot\boldsymbol{v} < 0$. Furthermore, if the interaction phase is such that the electrons are on average

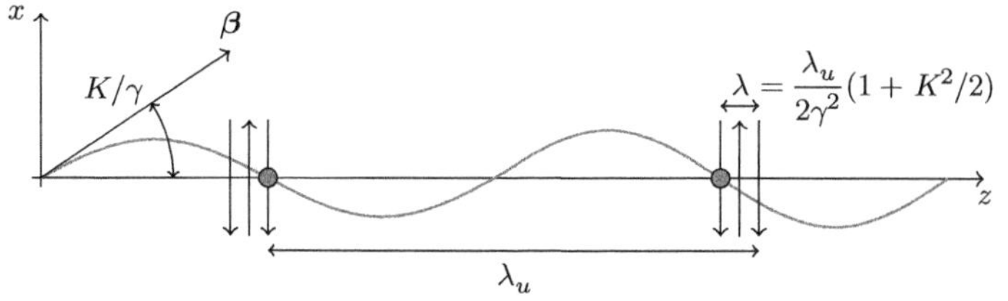

Figure 3.2 After an electron (gray dot) travels one undulator period λ_u of the sinusoidal trajectory (in gray), a plane wave (represented by alternating vertical arrows) overtakes the electron by one resonant wavelength λ. Thus, the undulator radiation carrying this resonant wavelength can exchange energy with the electron over many undulator periods.

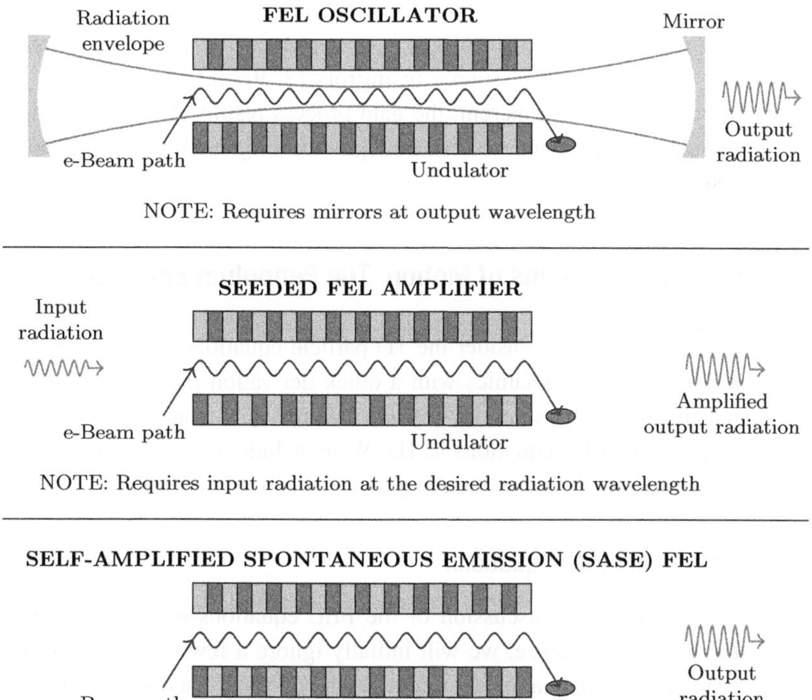

Figure 3.3 Schematic FELs operating in three different modes: oscillator, amplifier, and self-amplified spontaneous emission.

decelerated by the field, then the electromagnetic wave gains energy and amplification occurs.

As with any laser, free-electron lasers can have several operating modes. Figure 3.3 illustrates the operating principle of an FEL oscillator, amplifier, and one based upon self-amplified spontaneous emission. Oscillators use mirrors to trap the radiation, so that the field is built up over many amplification passes through the undulator. FEL oscillators have thus far been built in the IR to UV spectral range where mirrors of sufficient quality are readily available. With advances in multilayer mirrors for soft X-rays and hard X-ray mirrors based on Bragg crystal optics, however, FEL oscillators covering the X-ray portion of the spectrum can also be considered.

FELs can also act as linear amplifiers that will magnify radiation whose central frequency is close to the undulator resonance condition. Since this requires a source at the relevant wavelength, this mode of operation is most easily realized in spectral regions where other devices already exist. Extending FEL amplifiers to other wavelengths typically requires advanced techniques such as harmonic generation, so that applying them at X-ray energies is quite challenging. For this reason, the first X-ray FELs that have been planned and built are based on self-amplified spontaneous emission (SASE), where for the FEL system the "spontaneous emission" refers to the undulator radiation that we described previously. A SASE FEL amplifies the spontaneous undulator radiation generated by shot noise, so that it can produce intense, quasi-coherent radiation without either an external source or mirrors [8, 9, 10, 11]. While the resulting radiation is not completely coherent, the gain process results in a source whose brightness can be approximately ten orders of magnitude brighter than traditional synchrotron light sources.

3.2 Electron Equations of Motion: The Pendulum Equations

In this section we consider the 1D particle equations for the FEL. First, we introduce the longitudinal dynamics with a quick derivation that omits certain technical details but illustrates the essential physics. We then fill in the specifics and write down the longitudinal FEL equations in 1D. We conclude this section by explaining the particle coordinates and their motion in phase space in more detail.

3.2.1 Derivation of the Equations

We begin with a discussion of the FEL equations designed to show the essence of the derivation. Hence, we will initially ignore a few complexities, and clean up these issues shortly thereafter. According to the opening discussion of this chapter, the energy exchange due to the electron-radiation interaction in an undulator is given by

$$\begin{aligned}\frac{d\gamma}{dt} &= -\frac{eE_0Kc}{mc^2\gamma}\cos(k_uz)\sin(kz-\omega t+\phi)\\ &= -\frac{eE_0Kc}{2mc^2\gamma}\Big\{\underbrace{\sin\left[(k+k_u)z-\omega t+\phi\right]}_{\sim\,\text{particle phase }\theta}+\underbrace{\sin\left[(k-k_u)z-\omega t+\phi\right]}_{\text{neglect, will justify later}}\Big\},\end{aligned} \tag{3.4}$$

where we have identified the particle phase $\theta \sim (k+k_u)z - \omega t + \phi$. We do not write equality here because the "correct" particle phase should also account for the longitudinal oscillations in the undulator; this is one of the complexities that we temporarily ignore but will rectify with Equation (3.14).

The phase θ is often referred to as the ponderomotive phase, since it is the phase relevant to the effective longitudinal potential originating from the combined motion in the undulator and electromagnetic fields. Inspection of Equation (3.4) shows that the particle energy oscillates with little net change if θ evolves secularly, while the energy can decrease (or increase) significantly if the particle phase is nearly constant. The time rate of change in the phase θ is given by

$$\frac{d\theta}{dt} = (k + k_u)v_z - ck. \tag{3.5}$$

In Equation (3.5) we will neglect the oscillating part of the longitudinal velocity by making the replacement

$$v_z \to \bar{v}_z = c\left(1 - \frac{1 + K^2/2}{2\gamma^2}\right) \tag{3.6}$$

with $\bar{v}_z$ is the average longitudinal velocity. Plugging this into the phase equation (3.5) and using the fact that $k_u/k \ll 1$, we find

$$\frac{d\theta}{dt} = ck\left(\frac{k_u}{k} - \frac{1 + K^2/2}{2\gamma^2}\right). \tag{3.7}$$

In order to have a stationary phase ($d\theta/dt = 0$) and significant energy exchange, we need

$$\frac{k_u}{k} = \frac{\lambda}{\lambda_u} = \frac{1 + K^2/2}{2\gamma^2}, \tag{3.8}$$

which is exactly the undulator resonance condition. Thus, we find that the phase is nearly constant and the system approximately periodic (in the sense that the radiation slips one λ_1 ahead of the particles every λ_u) when the resonance condition (3.8) is satisfied. In this case, there can be a sustained interaction between the wave and the electron beam which in turn can lead to significant amplification of the electromagnetic field. Since γ is now a function of time, we use γ_r to denote the energy that is resonant with an EM field whose wavelength is λ_1, i.e.,

$$\frac{\lambda_1}{\lambda_u} \equiv \frac{1 + K^2/2}{2\gamma_r^2}. \tag{3.9}$$

We further introduce the (assumed small) normalized electron energy deviation from resonance

$$\eta \equiv \frac{\gamma - \gamma_r}{\gamma_r} \ll 1. \tag{3.10}$$

Writing the Lorentz factor as $\gamma = \gamma_r(1 + \eta)$, the phase equation for $\eta \ll 1$ becomes

$$\frac{d\theta_j}{dt} = 2k_u c\eta_j, \tag{3.11}$$

where for the moment we have included the subscripts j to emphasize that these are individual particle coordinates. The energy equation (3.4) is

$$\frac{d\eta_j}{dt} = \frac{1}{\gamma_r}\frac{d\gamma_j}{dt} = -\frac{eE_0Kc}{2\gamma_j\gamma_r mc^2}\sin\theta_j. \tag{3.12}$$

For a constant E_0, which is a good approximation for low-gain FELs that are characterized by small EM amplification, these two equations completely determine the evolution of the electron energy η and the ponderomotive phase θ. Their structure is identical to the equations describing the physical pendulum, and they are therefore referred to as the pendulum equations after W. Colson's graduate work at Stanford University [12].

While the preceding derivation illustrates the basic physics of the 1D particle dynamics, a few revisions are in order to obtain the equations of motion in their final form. First, it is both conventional and convenient to use z instead of t as the independent variable via $cdt \approx dz$. Next, we justify neglecting the second term from Equation (3.4), which is equivalent to ignoring oscillating terms that have a large phase advance over one undulator period. To do this properly, we recall from Section 2.4.1 that the longitudinal velocity v_z of an electron in a planar undulator also has an oscillatory part, i.e.,

$$v_z = \bar{v}_z - \frac{cK^2}{4\gamma^2}\cos(2k_u z) \approx \bar{v}_z - \frac{ck_u K^2}{k_1(2+K^2)}\cos(2k_u z). \tag{3.13}$$

The oscillations of v_z imply that the particle time coordinate also has fast oscillations, which contradicts our assumption that the phase θ_j is slowly-varying. To remedy this problem, we define θ_j in terms of the slowly-varying average particle time $\bar{t}_j$, which is obtained by subtracting off the oscillating part of t_j:

$$\begin{aligned}\theta_j(z) &\equiv (k_1 + k_u)z - ck_1\bar{t}_j(z)\\ &\equiv (k_1 + k_u)z - ck_1\left[t_j(z) - \frac{K^2}{\omega_1(4+2K^2)}\sin(2k_u z)\right].\end{aligned} \tag{3.14}$$

Differentiating the phase with respect to z now yields

$$\frac{d\theta_j}{dz} = (k_1 + k_u) - c\frac{d\bar{t}_j}{dz} = (k_1 + k_u) - \frac{c}{\bar{v}_z}, \tag{3.15}$$

which, upon inserting the average velocity (3.6) and expanding for $\gamma \gg 1$ and $\eta \ll 1$ results in Equation (3.11) with $cdt \to dz$. We now use the slowly varying phase to manipulate the energy equation as

$$\begin{aligned}&\cos(k_u z)\sin(k_1 z - \omega_1 t + \phi)\\ &= \frac{e^{ik_u z} + e^{-ik_u z}}{4i}\left\{e^{i\theta}e^{-ik_u z}\exp\left[-\tfrac{iK^2}{4+2K^2}\sin(2k_u z)\right] - c.c.\right\}\\ &= \frac{e^{i\theta}}{4i}\sum_n J_n\left(\frac{K^2}{4+2K^2}\right)\left[e^{-2ink_u z} + e^{-2i(n+1)k_u z}\right] - c.c..\end{aligned} \tag{3.16}$$

The terms with $n \neq 0$ from the first term in the sum and $n \neq -1$ from the second sum have a phase whose derivative

$$\frac{d}{dz}(\theta + 2nk_u z) \approx 2nk_u \quad \text{with} \quad n \neq 0, \tag{3.17}$$

meaning that their phase increases by $4n\pi$ per undulator period. Such a fast oscillation tends to average to zero, and cannot support a sustained interaction. This is the justification for retaining only the resonant term in (3.4), and is often referred to as "wiggle averaging" over the fast oscillations in the undulator.[1] Keeping the two slowly varying terms from (3.16) introduces the following change to the energy equation

$$K \longrightarrow K[\mathrm{JJ}], \quad \text{with } [\mathrm{JJ}] \equiv J_0\left(\frac{K^2}{4+2K^2}\right) - J_1\left(\frac{K^2}{4+2K^2}\right). \tag{3.18}$$

This factor of [JJ] looks similar to that derived for the paraxial wave equation (2.61); in fact, both arise from the electron's longitudinal oscillations at $2k_u$ in a planar undulator, which modifies the averaged coupling between the particles and the field. In a helical undulator, on the other hand, $v_z = \bar{v}_z$ (see Equation (2.85) and preceding discussion), and there is no [JJ]. In addition, in a helical undulator the peak magnetic field equals its RMS value, so that the FEL equations applicable in a helical device can be found from their planar undulator counterparts using

$$\text{helical undulator :} \qquad [\mathrm{JJ}] \to 1 \qquad K \to K\sqrt{2}. \tag{3.19}$$

Including the derived modifications and corrections, the 1D particle equations for the FEL are now given by

$$\frac{d\theta}{dz} = 2k_u\eta, \qquad \frac{d\eta}{dz} = -\frac{\epsilon}{2k_u L_u^2}\sin\theta, \tag{3.20}$$

where we have introduced the dimensionless field strength

$$\epsilon = \frac{eE_0K[\mathrm{JJ}]}{\gamma_r^2 mc^2}k_u L_u^2 \tag{3.21}$$

for later convenience.

We use Figure 3.4 to illustrate the physical meaning of the variables in the pendulum equations. In short, we have

z: The independent variable giving the location inside the undulator.
$t(z)$: The time an electron arrives at z.
$\theta(z)$: The ponderomotive phase defined by

$$\begin{aligned}\theta = (k_1 + k_u)z - \omega_1\bar{t} + \text{const.} &= \omega_1\left[\frac{k_1 + k_u}{\omega_1}z - \bar{t}(z)\right] + \text{const.}\\ &= \omega_1\left[z/\bar{v}_z - \bar{t}(z)\right] + \text{const.}\end{aligned} \tag{3.22}$$

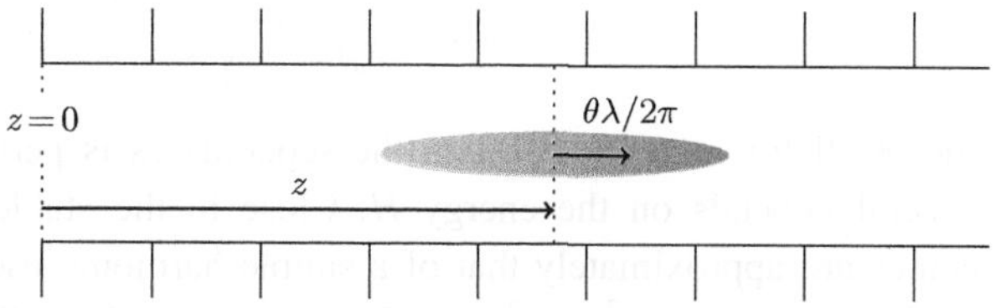

Figure 3.4 Schematic showing the variables defined in the FEL pendulum equations.

[1] The astute reader will notice that the previously neglected term actually contributes the slowly varying piece $\sim J_1$ due to the longitudinal oscillations.

The reference electron enters the undulator at $z = 0$ when $t = 0$, so that $z/\bar{v}_z$ is the time that this reference electron arrives at the location z. If we define the reference electron to be at the center of the bunch, than θ is the electron longitudinal position relative to the bunch center in units of $\lambda_1/2\pi$. For an electron bunch with the pulse length 300 fs $\sim$ 0.1 mm, the phase spread at the radiation wavelength $\lambda_1 = 1$ nm is $2\pi \times 10^5$, a very large number.

The pendulum equations (3.20) describe the motion of electrons in the so-called "ponderomotive potential" due to the combined undulator and radiation fields [12]. The ponderomotive potential gives rise to a bunching force that moves at the average beam velocity $\bar{v}_z$ and is slowly varying and approximately periodic in θ.

3.2.2 Motion in Phase Space

For a constant ϵ and therefore constant electric field, the pendulum motion is restricted by the constant of motion H:

$$H = k_u\eta^2 - \frac{\epsilon}{2k_uL_u^2}(\cos\theta - 1). \tag{3.23}$$

It turns out that H is also the pendulum Hamiltonian, from which the equations of motion can be obtained as shown in Appendix A. Each dashed line in Figure 3.5 is a possible trajectory in the (θ, η) phase space that is characterized by a different but constant value of H. The trajectories joining the two unstable fixed points at $(\theta = \pm\pi, \eta = 0)$ are known as the separatrices, for which $H = \epsilon/k_uL_u^2$ and thus

$$\eta = \pm\frac{\sqrt{\epsilon}}{k_uL_u}\sqrt{(1+\cos\theta)/2} = \pm\frac{\sqrt{\epsilon}}{k_uL_u}\cos(\theta/2). \tag{3.24}$$

The separatrix is shown as a thick dashed line in Figure 3.5. Outside the separatrices the trajectories are unbounded, and the motion can be identified with a pendulum rotating about its axis in one direction. Inside the separatrices the particle exhibits periodic oscillations about the stable fixed point $(0, 0)$, which corresponds to the vibration motion of the pendulum. The stable region between the separatrices is referred to as the ponderomotive bucket, since for a constant electric field the particles are trapped in this region. The maximum value $\eta_{\max}$ on the separatrix, which defines the bucket height, is given by

$$\eta_{\max} = \frac{\sqrt{\epsilon}}{k_uL_u} = \sqrt{\frac{eE_0K[JJ]}{k_u\gamma_r^2mc^2}}. \tag{3.25}$$

The oscillatory motion between the separatrices is periodic with a frequency that in general depends on the energy H. Close to the stable fixed point, however, the dynamics are approximately that of a simple harmonic oscillator; for $|\theta| \ll 1$ Equations (3.20) reduce to $d^2\theta/dz^2 = -\Omega_s^2\theta$. The natural oscillation frequency is given by the synchrotron wavenumber Ω_s, with z_s its period:

$$\Omega_s = \frac{\sqrt{\epsilon}}{L_u} = \sqrt{\frac{eE_0K[JJ]k_u}{\gamma_r^2mc^2}} \qquad z_s \equiv \frac{2\pi}{\Omega_s}. \tag{3.26}$$

3.3 Low-Gain Regime

In the low-gain regime we assume that the change in the electric field due to its interaction with the electrons is small. Hence, in this case we can derive the FEL gain using only the electron equations of motion and energy conservation. To determine the detailed field evolution of course requires Maxwell's equations, to which we will return when we discuss the high-gain regime in Sections 3.4 and 4.3.

Before we get into the mathematics of the low-gain regime, we pause to give a brief overview of the FEL pendulum dynamics, assuming that the electron beam is initially monoenergetic with its energy satisfying the resonance condition. In this case, the beam first develops an energy modulation at the resonant wavelength as it interacts with the undulator and radiation fields. As the evolution progresses, the energy modulation leads to a change in the relative longitudinal position that depends on the electron's initial conditions, as can be inferred from Figure 3.5. This in turn results in a density modulation (microbunching) of the beam at the resonant wavelength. If the particle energy is initially on resonance as assumed, than the net energy exchange between the particles and the EM field is zero because equal numbers of electrons gain and lose energy. In order for a low-gain FEL to amplify the radiation, one must choose the initial beam energy to be somewhat higher than that given by the resonance condition, so that more electrons lose energy to the field than are accelerated by it. Now, in the next section we turn to deriving these results in a more quantitative fashion.

3.3.1 Derivation of Gain

We will solve the pendulum equations (3.20) in the low-gain regime where the amplification in a single pass through the undulator is small, so that $\epsilon \approx$ constant.

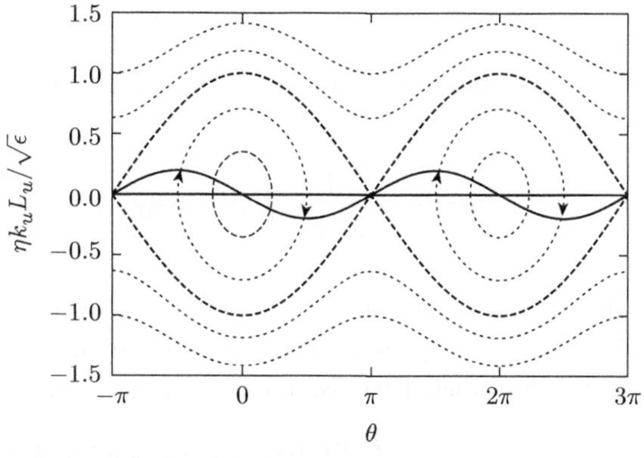

Figure 3.5 Electron motion in the longitudinal phase space (θ, η) due to the presence of a resonant EM wave in the undulator. The dashed lines are the phase-space trajectories, with the thicker ones denoting the separatrix. The straight, thick black line at $\eta = 0$ plots the initial positions of the electron beam, assuming that the electrons populate all longitudinal phases. Initial evolution in the undulator leads to an energy modulation that is approximately sinusoidal, which we also show. As the evolution progresses, the electron distribution develops sharper kinks near $\theta = 2n\pi$ for integer n, which implies that the density of the electron beam also becomes modulated/microbunched.

In addition, we simplify the analysis by solving for the electron motion perturbatively under the assumption that the electrons do not significantly rotate in the pendulum potential. This assumption is equivalent to assuming that the synchrotron period times the undulator length is small, or that the dimensionless field strength $\epsilon \equiv (\Omega_s L_u)^2 \ll 1$. We then develop the perturbation equations by expanding the phase space variables as

$$\theta = \theta_0(z) + \epsilon\theta_1(z) + \epsilon^2\theta_2(z) + \ldots$$
$$\eta = \eta_0(z) + \epsilon\eta_1(z) + \epsilon^2\eta_2(z) + \ldots.$$

Inserting the expansion above into the equations of motion (3.20) yields

$$\frac{d\theta_m}{dz} = 2k_u\eta_m, \tag{3.27}$$

$$\frac{d\eta_0}{dz} + \epsilon\frac{d\eta_1}{dz} + \epsilon^2\frac{d\eta_2}{dz} = -\frac{\epsilon}{2k_u L_u^2}(\sin\theta_0 + \epsilon\theta_1\cos\theta_0), \tag{3.28}$$

where (3.27) is the θ equation at arbitrary order m, while (3.28) is the energy equation through $O(\epsilon^2)$. At zeroth order, we have

$$\frac{d\eta_0}{dz} = 0 \qquad\qquad \frac{d\theta_0}{dz} = 2k_u\eta_0$$
$$\Rightarrow \eta_0 = \text{const} \qquad\qquad \Rightarrow \theta_0(z) = 2k_u\eta_0 z + \phi_0, \tag{3.29}$$

with (ϕ_0, η_0) defined to be the initial particle coordinates in phase space. At lowest order the particles move in straight lines with constant velocity. The interaction with the field comes in with the first-order equations

$$\frac{d\theta_1}{dz} = 2k_u\eta_1, \tag{3.30}$$

$$\frac{d\eta_1}{dz} = -\frac{1}{2k_u L_u^2}\sin\theta_0\,. \tag{3.31}$$

Putting the zeroth-order solutions (3.29) into Equation (3.31) and solving for η_1, we have

$$\eta_1(z) = -\frac{1}{2k_u L_u^2}\int_0^z dz'\,\sin\theta_0(z') = \frac{\cos\theta_0(z) - \cos\phi_0}{4k_u^2 L_u^2\eta_0}. \tag{3.32}$$

Note that the average of the first-order energy deviation over the initial electron phases vanishes, i.e., $\langle\eta_1(z)\rangle_{\phi_0} = 0$. Thus, there is no energy exchange between the particles and field at this order. There is, however, energy modulation, since

$$\epsilon\eta_1(z) = -\frac{eE_0K[\mathrm{JJ}]k_u L_u^2}{\gamma_r^2 mc^2}\frac{\cos(2k_u\eta_0 z + \phi_0) - \cos\phi_0}{4k_u^2 L_u^2\eta_0}$$
$$\approx -\frac{eE_0K[\mathrm{JJ}]}{2\gamma_r^2 mc^2}\frac{\sin(2k_u\eta_0 z)\sin(\phi_0)}{2k_u\eta_0} \approx -\frac{eE_0K[\mathrm{JJ}]}{2\gamma_r^2 mc^2} z\sin\phi_0 \tag{3.33}$$

for $2k_u\eta_0 z = 4\pi N_u(\gamma_0 - \gamma_r)/\gamma_r \ll 1$. While here we view the modulation in η as a precursor to FEL gain, a variety of longitudinal phase space manipulation techniques take advantage of this phenomenon by employing an optical laser as the driving EM field. We mention a few of these related to harmonic generation in Chapter 6.

The FEL gain appears at second order in ϵ when the energy modulation evolves into a density modulation that can radiate coherently. First, we solve for θ_1 from Equation (3.30)

$$\frac{d\theta_1}{dz} = \frac{\cos\theta_0(z) - \cos\phi_0}{2k_u L_u^2 \eta_0}$$
$$\Rightarrow\ \theta_1(z) = \frac{\sin\theta_0(z) - \sin\phi_0}{(2k_u L_u \eta_0)^2} - \frac{z\cos\phi_0}{2k_u L_u^2 \eta_o}. \tag{3.34}$$

Thus, the second-order energy equation is

$$\frac{d\eta_2}{dz} = -\frac{\theta_1(z)}{2k_u L_u^2}\cos\theta_0(z)$$
$$\Rightarrow\ \eta_2(L_u) = -\int_0^{L_u} dz\left[\frac{\cos\theta_0(\sin\theta_0 - \sin\phi_0)}{(2k_u L_u^2)(2k_u L_u \eta_0)^2} - \frac{z\cos\theta_0\cos\phi_0}{(2k_u L_u^3)(2k_u L_u \eta_0)}\right]. \tag{3.35}$$

Taking the average over initial phases ϕ_0 yields

$$\langle\cos\theta_0\sin\theta_0\rangle_{\phi_0} = \frac{1}{2}\langle\sin 2\theta_0\rangle_{\phi_0} = 0,$$
$$\langle\cos\theta_0\sin\phi_0\rangle_{\phi_0} = \left\langle\big[\cos(2k_u\eta_0 z)\cos\phi_0 - \sin(2k_u\eta_0 z)\sin\phi_0\big]\sin\phi_0\right\rangle_{\phi_0}$$
$$= \frac{1}{2}\sin(2k_u\eta_0 z),$$
$$\langle\cos\theta_0\cos\phi_0\rangle_{\phi_0} = \frac{1}{2}\cos(2k_u\eta_0 z).$$

Hence,

$$\langle\eta_2(L_u)\rangle_{\phi_0} = -\frac{1}{2k_u L_u^2}\int_0^{L_u} dz\left[\frac{\sin(2k_u\eta_0 z)}{(2k_u L_u \eta_0)^2} - \frac{z\cos(2k_u\eta_0 z)}{L_u(2k_u L_u \eta_0)}\right]. \tag{3.36}$$

Writing the scaled initial electron deviation from resonance as $x = k_u\eta_0 L_u$, it is straightforward to obtain

$$\langle\eta_2(L_u)\rangle_{\phi_0} = \frac{2x\sin x\cos x - 2\sin^2 x}{8k_u L_u x^3} = \frac{1}{16k_u L_u}\frac{d}{dx}\left(\frac{\sin x}{x}\right)^2. \tag{3.37}$$

Therefore, the average net change in the electron beam energy at second order is given by

$$\langle\Delta\eta\rangle = \langle\epsilon^2\eta_2(L_u)\rangle = -\frac{e^2E_0^2K^2[\mathrm{JJ}]^2}{4\gamma_r^4(mc^2)^2}\frac{k_u L_u^3}{4}g(x), \tag{3.38}$$

where we have introduced the normalized gain function

$$g(x) = -\frac{d}{dx}\left(\frac{\sin x}{x}\right)^2. \tag{3.39}$$

We have written the energy change (3.38) in a suggestive manner in terms of $g(x)$ to make a deep connection between the FEL gain (stimulated emission) derived here and the undulator radiation (spontaneous emission) discussed in the previous chapter. To make this connection explicit, we recall from Equation (2.100) that in the forward direction the spectrum of undulator radiation from a monoenergetic e-beam $f(\eta) = \delta(\eta - \eta_0)$ is

$$\left.\frac{dP}{d\omega d\phi}\right|_{\phi=0} \propto \mathcal{S}(\omega, \eta_0) \propto \left\{\frac{\sin[k_u L_u(\eta_0 - \Delta\nu/2)]}{k_u L_u(\eta_0 - \Delta\nu/2)}\right\}^2. \tag{3.40}$$

Hence, we find that the energy change (gain/stimulated emission) is related to the spontaneous emission via $\langle\Delta\eta\rangle \propto g(\eta_0) \propto \frac{d}{d\eta_0}\mathcal{S}(\omega, \eta_0)$; explicitly, in terms of the change in Lorentz factor γ we have

$$\langle\Delta\gamma\rangle = \frac{\pi\lambda^2 cT}{N_e}\frac{\epsilon_0 E_0^2}{(mc^2)^2}\frac{\partial}{\partial\gamma}\left.\frac{dP}{d\omega d\phi}\right|_{\phi=0}. \tag{3.41}$$

For our particular case of a straight undulator with constant K the spectrum $dP/d\omega$ is a function of $\eta - \Delta\nu/2 = (\gamma - \gamma_r)/\gamma_r - (\omega - \omega_1)/2\omega_1$, so that in addition to (3.41) we also have

$$\langle\Delta\eta\rangle \propto -\frac{d}{d\omega}\mathcal{S}(\omega, \eta_0). \tag{3.42}$$

The fact that the spectral gain curve can be deduced from the spontaneous radiation spectrum equation (3.41) was originally derived as Madey's second theorem in Ref. [1]. Madey's first theorem relates the ensemble averaged energy loss $\langle\epsilon^2\eta_2\rangle \equiv \langle\Delta\eta\rangle$ with the induced energy spread $\langle(\epsilon\eta_1)^2\rangle \equiv \langle\eta^2\rangle$. Using the expression for the first-order energy change (3.32) we can easily show that

$$\langle\eta^2\rangle = \frac{e^2E_0^2K^2[\mathrm{JJ}]^2}{4\gamma_r^4(mc^2)^2}\frac{L_u^2}{2}\left(\frac{\sin x}{x}\right)^2 \quad\Rightarrow\quad \langle\Delta\eta\rangle = \frac{1}{2}\frac{\partial}{\partial\eta_0}\langle\eta^2\rangle. \tag{3.43}$$

The collection of results (3.41), (3.42), (3.43) is often loosely referred to as Madey's theorem, although (3.41) and (3.43) are the actual statements of the second and first theorem, respectively. Here, we derived Equations (3.41) and (3.43) for the specific case of an ideal undulator. Nevertheless, Madey's theorems are quite general results that can be proven in the low-gain limit using very few assumptions [13, 14]. In addition, Madey's theorem is of practical use, since it is typically significantly easier to compute the spontaneous emission than the gain.

Returning to the actual FEL gain, we plot the gain function $g(x)$ in Figure 3.6. The gain reaches its maximum of $g(x) \approx 0.54$ when the normalized detuning $x \equiv 2\pi N_u\eta_0 \approx 1.3$, so that for an undulator N_u periods long, the optimal initial energy offset is

$$\left.\frac{\gamma_0 - \gamma_r}{\gamma_r}\right|_{\substack{\text{max}\\\text{gain}}} = \frac{1.3}{2\pi N_u} \approx \frac{1}{5N_u}. \tag{3.44}$$

The RMS width of the gain curve around its maximum is of order unity with respect to x, which translates to a variation in beam energy $\Delta\eta \sim 1/6N_u$. Using the resonance condition, the frequency width of the gain curve is

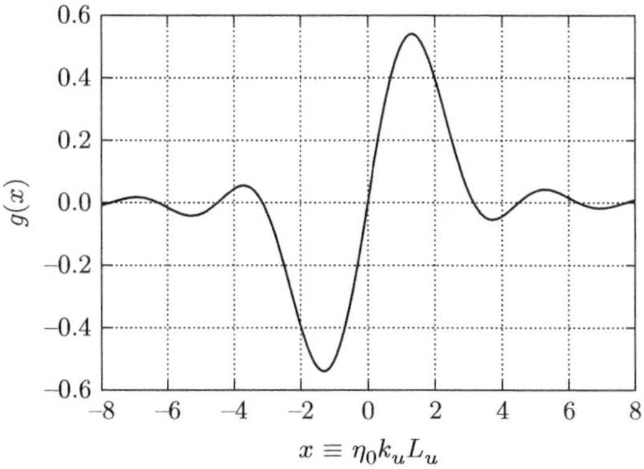

Figure 3.6 The low-gain function $g(x)$ of Equation (3.39). A positive value of $g(x)$ corresponds to a net energy loss for the electrons and thus gain for the radiation signal.

$$\frac{\sigma_\omega}{\omega_r} \sim 2\Delta\eta \sim \frac{1}{3N_u}. \tag{3.45}$$

Thus, we see that the characteristic bandwidth of a low-gain FEL scales inversely with the number of undulator periods in a manner similar to that of the spontaneous radiation.

The net decrease in particle energy results in an increase in the electromagnetic field energy. To compute this energy exchange, recall that the energy density of an EM plane wave is

$$u = \frac{\epsilon_0}{2}\left(\boldsymbol{E}^2 + c^2\boldsymbol{B}^2\right) = \frac{\epsilon_0}{2}E_0^2, \tag{3.46}$$

while the FEL interaction changes the electron energy density by an amount

$$\Delta u = mc^2\langle\Delta\gamma\rangle\frac{N_e}{cT\mathscr{A}_{\text{tr}}} = mc^2\langle\Delta\gamma\rangle\frac{I}{ec}\frac{1}{\mathscr{A}_{\text{tr}}}, \tag{3.47}$$

where $\mathscr{A}_{\text{tr}}$ is the relevant transverse cross sectional area, I is the beam current, and the average energy change $\langle\Delta\gamma\rangle$ is obtained from (3.38). The change in electromagnetic energy density equals $-\Delta u$, so that the relative FEL gain, $G \equiv -\Delta u/u$, is given by

$$G = -\frac{2\pi\lambda_1^2}{mc^2\mathscr{A}_{\text{tr}}}\frac{\partial}{\partial\gamma}\frac{dP}{d\omega d\phi}\bigg|_{\phi=0} = -\frac{I}{I_A}\frac{4\pi^2K^2[\text{JJ}]^2}{(1+K^2/2)^2}\frac{\gamma N_u^3\lambda_1^2}{\mathscr{A}_{\text{tr}}}g(x). \tag{3.48}$$

Here $I_A \equiv ec/r_e \equiv 4\pi\epsilon_0 mc^3/e \approx 17045$ A is the Alfvén current. The appropriate transverse area $\mathscr{A}_{\text{tr}}$ for the 3D interaction will be determined by some convolution of the radiation and electron beam sizes. If we assume that the electron and radiation beams have Gaussian profiles with widths σ_x and σ_r at their respective waists, we have

$$\mathscr{A}_{\text{tr}} \to 2\pi(\sigma_x^2 + \sigma_r^2), \qquad \sigma_r^2 = \frac{\lambda_1}{4\pi}Z_R, \tag{3.49}$$

where Z_R is the radiation Rayleigh length. The replacement of $\mathscr{A}_{\text{tr}}$ shown above is physically reasonable, and in Section 5.4 we will derive a more rigorous 3D gain formula that contains this simple extension in the appropriate limit. The Rayleigh length is usually chosen to be one-half the undulator length, $Z_R = L_u/2 = \lambda_u N_u/2$, while the optimal overlap of the electron and radiation beams occurs when their cross-sectional areas are equal,

$$\mathscr{A}_{\text{tr}} \approx 2 \times 2\pi\sigma_r^2 = \lambda_1 Z_R = \lambda_1 \lambda_u N_u/2. \tag{3.50}$$

In this case, the gain is

$$G = 8\pi^2 \frac{I}{I_A} \frac{K^2[\text{JJ}]^2}{(1+K^2/2)^2} \frac{N_u^2 \lambda_1}{\lambda_u} g(x). \tag{3.51}$$

We will return to this idea of electron and radiation beam mode matching in a more rigorous fashion when we treat three-dimensional effects in Section 5.4.

Having obtained the linear gain to second order in ϵ, we now add some more details regarding the statement that ϵ is a small parameter. In order to do this, it is convenient to replace the variables z and η by dimensionless quantities whose magnitudes are of order unity. We scale the longitudinal coordinate by the undulator length, introducing the dimensionless propagation distance τ via

$$\tau \equiv z/L_u . \tag{3.52}$$

As for the energy η, we have seen that the bandwidth of the gain curve is of order $1/N_u$, so that any electron falls out of resonance with the FEL when its energy changes by an amount $\sim 1/N_u$. We will verify this result in our subsequent discussion of the low-gain FEL efficiency, from which we will find that the maximum value of $|\eta|$ at saturation is about $1/2N_u$. Therefore, we define the scaled energy $\tilde{\eta}$ by

$$\tilde{\eta} = 2N_u\eta . \tag{3.53}$$

In terms of these dimensionless variables, the pendulum equations (3.20) become

$$\frac{d\theta}{d\tau} = 2\pi\tilde{\eta}, \qquad \frac{d\tilde{\eta}}{d\tau} = -\frac{\epsilon}{2\pi}\sin\theta . \tag{3.54}$$

These equations make it readily apparent that $\epsilon \equiv (\Omega_s L_u)^2$ is the relevant expansion parameter. We can get another interpretation of ϵ by writing it as

$$\epsilon \equiv \frac{(eE_0K[\text{JJ}]/\gamma_r)L_u}{2mc^2\gamma_r/N_u} \ll 1 . \tag{3.55}$$

The quantity (eE_0K/γ_r) is the electromagnetic (ponderomotive) force on the electron in the direction of its motion, so that the numerator in (3.55) is the maximum work the field imparts to an electron while is traverses the length of the undulator (apart from the factor [JJ]). On the other hand, the denominator is twice the maximum energy that an electron will lose at saturation. Thus, the condition (3.55) states that the perturbation expansion is valid when the optical field is much smaller than its value at saturation.

3.3.2 Particle Trapping and Low-Gain Saturation

The gain decreases from the small signal value derived in the previous section when the optical power is large such that (3.55) is no longer satisfied. The decrease in gain occurs when the field amplitude is large enough to trap electrons in the ponderomotive bucket and then rotate them to an absorptive phase where they extract energy from the field.

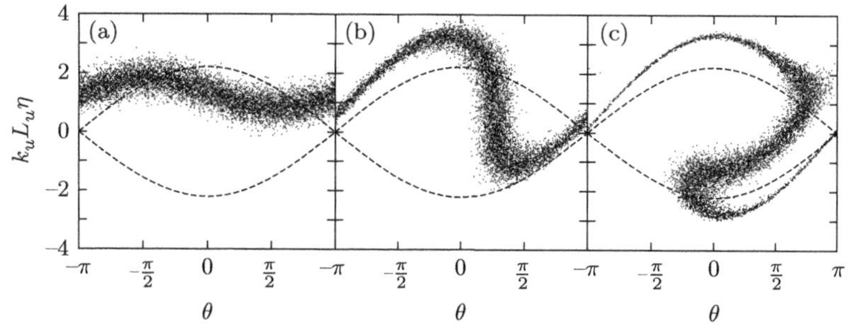

Figure 3.7 Numerical illustrations of longitudinal phase space at the beginning (a), middle (b), and end (c) of the undulator for a saturated FEL oscillator. The dashed lines show the separatrix of the bucket, while the beam is initially centered at the optimal energy for gain defined by $k_u L_u \eta \approx 1.3$.

A full discussion of this "saturation" of the gain requires solving the pendulum equations for arbitrary values of the electric field. Rather than doing this, we can get a feel for the particle motion near saturation by plotting the electron phase space near the beginning, middle, and end of the undulator of a low-gain FEL operating at saturation in Figure 3.7. Upon entering the undulator, the radiation field with power $P_{\text{sat}} - \Delta P$ captures a large fraction of the electrons in the ponderomotive bucket as shown in Figure 3.7(a). The beam quickly develops a modulation in energy. As the evolution proceeds, the energy modulation is converted to density bunching at λ_1 shown in Figure 3.7(b). Up to this point in the evolution, which takes about one half of the undulator length, the net exchange of energy between the electron beam and the optical field is relatively small. In the second half of the undulator, the rotation in phase space results in more electrons in the lower half of the bucket, meaning that significantly more electrons have lost energy to the field than have been accelerated by it (see Figure 3.7(c)). The lost kinetic energy is converted to field energy, which we can estimate by considering an electron located at one-half the height of the separatrix with initial phase space coordinates $(\theta, \eta) = (0, 0.5\eta_{\text{max}})$. It will lose a maximum amount of energy if it rotates to $(\theta, \eta) = (0, -0.5\eta_{\text{max}})$ during the second half of the undulator. Hence, maximum energy exchange occurs if the particle makes one-half an oscillation in the field, or if $\Omega_s L_u \approx \pi$. In view of the formula for the synchrotron period (3.26), maximum energy exchange will occur if

$$\Omega_s L_u = \sqrt{\epsilon} \approx \pi, \tag{3.56}$$

and the electron's contribution to the optical energy is then

$$mc^2 \gamma \eta_{\text{max}} = mc^2 \gamma \frac{\sqrt{\epsilon}}{k_u L_u} = mc^2 \gamma \frac{1}{2N_u}, \tag{3.57}$$

where we use Equation (3.56) to eliminate ϵ. This electron therefore loses a fraction $1/2N_u$ of its energy to the optical field. If we assume that the energy (3.57) equals the average energy loss of all electrons, we obtain the following estimate for the efficiency of an FEL oscillator

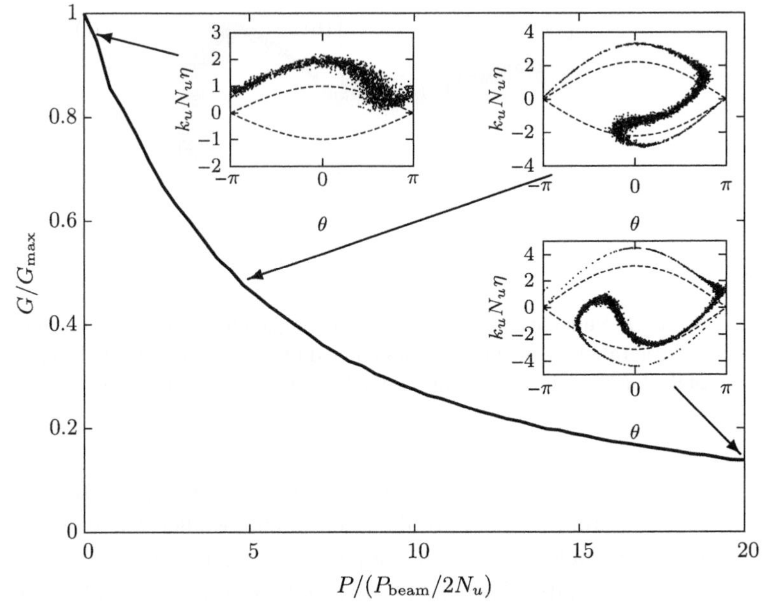

Figure 3.8 Simulation results showing the gain reduction $G/G_{\max}$ as a function of the input power P scaled by $P_{beam}/2N_u$. The insets are the particle phase space (dots) and separatrix (lines) defined by the E-field at the end of the undulator for $P/(P_{beam}/2N_u) = 0.2$, 5, and 20.

$$\Delta P = \text{efficiency} \times P_{\text{beam}} \approx \frac{1}{2N_u} P_{\text{beam}} \,. \tag{3.58}$$

In the above, $P_{\text{beam}} = (I/e)\gamma mc^2$ is the electron beam's kinetic power. At saturation a low-gain device converts $\sim 1/2N_u$ of its kinetic energy into field energy, and the ideal output power of each radiation pulse is $\Delta P \approx P_{\text{beam}}/2N_u$.

We have argued that the FEL gain decreases as the power approaches and exceeds $P_{\text{beam}}/2N_u$ because at this power level the synchrotron period $z_s \sim L_u$ and the particles rotate to an absorptive phase of the ponderomotive potential. While this gain reduction can be determined semi-analytically [15], it is perhaps easier to illustrate the physics by numerically solving the FEL pendulum equations and plotting. We show the results of the gain normalized to its small signal maximum in Figure 3.8. The inset panels show samples of the longitudinal phase space at the end of the undulator for various values of the input power, along with the ponderomotive bucket defined by the optical field at $z = L_u$. When the power is much less than $P_{\text{beam}}/2N_u$ the particles are still predominantly in the decelerating phase, while the evidence of rotation in the ponderomotive bucket is clear when $P = 5(P_{\text{beam}}/2N_u)$. At the largest field power the electrons have significantly rotated in the potential, and the gain is reduced by a large factor.

3.4 High-Gain Regime

The FEL can also act as a high-gain amplifier, in which case the energy exchange during a single pass through the undulator is large and the field amplitude cannot be regarded as a constant. This high-gain regime is particularly important when mirrors are not available to build oscillators, and has been used as the first way to produce intense X-rays from FELs. Here, it is necessary to also consider the field evolution, so that we must study the pendulum equations coupled to the paraxial wave equation for the radiation.

3.4.1 Maxwell Equation

The electromagnetic field equation for an FEL can be derived from the paraxial wave equation (2.10) that we previously employed to study synchrotron radiation. This is to be expected, since the FEL is a natural extension of spontaneous undulator radiation once we include the self-consistent electron motion in the radiation field. We begin our FEL derivation by expressing the paraxial equation (2.10) for N_e source electrons as

$$\left[\frac{\partial}{\partial z} + \frac{ik}{2}\boldsymbol{\phi}^2\right]\tilde{\mathcal{E}}_\omega(\boldsymbol{\phi}; z) = \sum_{j=1}^{N_e} \frac{e[\boldsymbol{\beta}_j(z) - \boldsymbol{\phi}]}{4\pi\epsilon_0 c\lambda^2} e^{ik[ct_j(z)-z]} \times \int d\boldsymbol{x}\; e^{-ik\boldsymbol{\phi}\cdot\boldsymbol{x}}\delta(\boldsymbol{x} - \boldsymbol{x}_j). \tag{3.59}$$

Here, we have rewritten the angular dependence of the current so that we can replace the point-like electron source with a constant charge density in the transverse plane by making the replacement $\delta(\boldsymbol{x} - \boldsymbol{x}_j) \to \mathscr{A}_{\text{tr}}^{-1}$, where $\mathscr{A}_{\text{tr}}$ is the transverse area. Then, in the 1D limit we have

$$\int d\boldsymbol{x}\; e^{-ik\boldsymbol{\phi}\cdot\boldsymbol{x}}\delta(\boldsymbol{x} - \boldsymbol{x}_j) \to \frac{1}{\mathscr{A}_{\text{tr}}}\int d\boldsymbol{x}\; e^{-ik\boldsymbol{\phi}\cdot\boldsymbol{x}} = \frac{\lambda^2}{\mathscr{A}_{\text{tr}}}\delta(\boldsymbol{\phi}), \tag{3.60}$$

and the source is directed entirely in the forward direction. We complete the 1D limit by defining the 1D electric field $\tilde{E}_\omega(z)$ via

$$\tilde{\mathcal{E}}_\omega(\boldsymbol{\phi}; z) = \tilde{E}_\omega(z)\delta(\boldsymbol{\phi}). \tag{3.61}$$

The $\delta(\boldsymbol{\phi})$ enforces the field to be in the forward direction only, which also implies that the spatial representation of the electric field is independent of $\boldsymbol{x}$. We insert the field (3.61) and the transverse electron velocity $\beta_{x,j} = (K/\gamma_j)\cos(k_u z)$ into (3.59), and then integrate over angles to find the 1D field equation

$$\frac{\partial}{\partial z}\tilde{E}_\nu(z) = \frac{1}{2\pi}\sum_{j=1}^{N_e}\frac{eK\cos(k_u z)}{4\epsilon_0 c\mathscr{A}_{\text{tr}}\gamma_j}e^{ik[ct_j(z)-z]}. \tag{3.62}$$

We have assumed that the field $\tilde{E}_\nu(z)$ describes a slowly varying envelope, so that consistency requires that we also identify the slowly varying current in (3.62). As discussed previously, we can do this by introducing the average particle time $\bar{t}_j = t_j - (K^2/8k_u\gamma^2)\sin(2k_u z)$ that subtracts off the oscillatory figure-eight component from t. In terms of the slowly varying ponderomotive phase, we have

$$\begin{aligned} k[ct_j(z) - z] &= \nu\left[\omega_1\bar{t}_j(z) - (k_1 + k_u)z\right] + \nu k_u z + \frac{\nu K^2}{4 + 2K^2}\sin(2k_u z) \\ &= -\nu\theta_j(z) + \Delta\nu k_u z + hk_u z + \nu\xi\sin(2k_u z), \end{aligned} \tag{3.63}$$

where we recall that h is an odd integer identifying the harmonic number, the normalized frequency difference $\Delta\nu \equiv \nu - h \equiv k/k_1 - h$, and we have introduced the shorthand notation $\xi \equiv K^2/(4+2K^2)$. Then, the wave equation (3.62) becomes

$$\frac{\partial}{\partial z}\tilde{E}_\nu(z) = -\frac{eK}{4\epsilon_0 c\mathscr{A}_{\text{tr}}}\frac{1}{2\pi}\sum_{j=1}^{N_e}\frac{1}{\gamma_j}e^{-i\nu\theta_j(z)+i\Delta\nu k_u z} \times \left[e^{i(h-1)k_u z} + e^{i(h+1)k_u z}\right]e^{i\nu\xi\sin(2k_u z)}. \tag{3.64}$$

The envelope $\tilde{E}_\nu(z)$, energy γ_j, and phase $\theta_j(z)$ all vary slowly over one undulator period, as does $e^{i\Delta\nu k_u z}$ if we restrict our attention to small frequency detunings, $|\Delta\nu| \ll 1$. We can extract the slowly varying terms from the second line of Equation (3.64) by averaging over an undulator period λ_u as follows

$$\frac{1}{\lambda_u}\int_0^{\lambda_u} dz\,\left[e^{i(h-1)k_u z} + e^{i(h+1)k_u z}\right]e^{i\nu\xi\sin(2k_u z)} = J_{-(h-1)/2}(\nu\xi) + J_{-(h+1)/2}(\nu\xi) \equiv [\text{JJ}]_h, \tag{3.65}$$

where we have used the Jacobi–Anger identity equation (2.63) to evaluate the integral, from which we find the familiar harmonic Bessel function factor first introduced in (2.68).

We are now in a position to write the frequency domain wave equation for the 1D FEL. However, there are a few notational issues that we would like to simplify. First, we will find it convenient to have the temporal and frequency representations of the field be related by a Fourier transform with respect to the scaled frequency ν. To do this, we write

$$E_x(\boldsymbol{x},t;z) = \int d\omega d\boldsymbol{\phi}\; e^{-i(\omega - k\boldsymbol{\phi}\cdot\boldsymbol{x})}e^{ikz}\tilde{\mathcal{E}}_\omega(\boldsymbol{\phi};z) = \int d\omega\; e^{-i(\omega-kz)}\tilde{E}_\omega(z) = e^{-ih(\omega_1-k_1z)}\int d\nu\; e^{i\Delta\nu\theta}ck_1e^{-i\Delta\nu k_u z}\tilde{E}_\omega(z). \tag{3.66}$$

The integrand contains the slowly varying field, and we can both simplify this and eliminate the phase $e^{i\Delta\nu k_u z}$ from the source current in Equation (3.64) by defining the phase-shifted electric field amplitude

$$E_\nu(z) = ck_1e^{-i\Delta\nu k_u z}\tilde{E}_\omega(z). \tag{3.67}$$

Note that this phase shift must be retained even though $\Delta\nu \ll 1$, since we may also have $k_u z \gg 1$. Finally, the field equation for $E_\nu(z)$ is

$$\left[\frac{\partial}{\partial z} + i\Delta\nu k_u\right]E_\nu(z) = -\frac{ek_1K[\text{JJ}]_h}{4\epsilon_0\gamma_r\mathscr{A}_{\text{tr}}}\frac{1}{2\pi}\sum_{j=1}^{N_e}e^{-i\nu\theta_j(z)} = -\kappa_h n_e\frac{1}{N_\lambda}\sum_{j=1}^{N_e}e^{-i\nu\theta_j(z)}. \tag{3.68}$$

Here, N_λ is the number of electrons in one wavelength, and the harmonic coupling and electron volume density are, respectively,

$$\kappa_h \equiv \frac{eK[\mathrm{JJ}]_h}{4\epsilon_0 \gamma_r} \qquad n_e \equiv \frac{I/ec}{\mathscr{A}_{\mathrm{tr}}} \equiv \frac{N_\lambda}{\lambda_1 \mathscr{A}_{\mathrm{tr}}}. \tag{3.69}$$

Note that while approximating γ_j by γ_r in κ_h is a very good approximation, such a replacement in the particle phase would eliminate the FEL interaction entirely.

In Chapter 4 we will employ Equation (3.68) to fully characterize the 1D FEL dynamics, because we believe that the frequency representation is best suited for theoretical analysis. Nevertheless, there are situations in which the time domain approach is useful. In particular, the time domain equations are well suited for efficient numerical simulation codes (see Appendix B), and in the rest of this chapter we will use the time domain formulation to obtain some basic understanding of the high-gain behavior and its scalings.

The time domain wave equation basically follows from the inverse Fourier transform of (3.68). To make this connection explicit, we use the definitions (3.66) and (3.67) to find that 1D slowly varying envelopes are related by the Fourier transforms

$$E(\theta; z) = \int d\nu\, e^{i\Delta\nu\theta} E_\nu(z), \qquad E_\nu(z) = \frac{1}{2\pi}\int d\theta\, e^{-i\Delta\nu\theta} E(\theta; z). \tag{3.70}$$

Therefore, multiplying (3.68) by $e^{i\Delta\nu\theta}$ and integrating over ν yields

$$\left[\frac{\partial}{\partial z} + k_u \frac{\partial}{\partial \theta}\right] E(\theta; z) = -\kappa_1 n_e \frac{2\pi}{N_\lambda} \sum_{j=1}^{N_e} e^{-i\theta_j(z)} \delta[\theta - \theta_j(z)] \tag{3.71}$$

at the fundamental frequency ω_1.

It may appear that our work is done, but the transverse current in (3.71) is composed of a sum of delta functions that is unfortunately both difficult to treat and apparently in violation of our assumption that E varies slowly. To establish a well-defined, slowly varying current, we average Equation (3.71) over some number of periods in θ. This "slice-averaging" has the same physical significance as our previous assumption that $|\Delta\nu| \ll 1$, and is valid provided the averaging time is much shorter than the characteristic time over which the field amplitude changes. For a high-gain FEL we require the averaging time Δt to be much less than the coherence time, $\Delta t = \Delta\theta/\omega_h \ll t_{\mathrm{coh}}$, which at the fundamental frequency reduces to $\Delta t \ll \lambda_1/(4\pi c\rho)$ or $\Delta\theta \ll 1/2\rho$. The time window over which the beam average is taken is sometimes referred to as an FEL slice.

We average (3.71) over an FEL slice by applying

$$\frac{1}{\Delta\theta} \int_{\theta-\Delta\theta/2}^{\theta+\Delta\theta/2} d\theta' \,\Bigg|_{\text{at fixed } z} \tag{3.72}$$

to both sides. Averaging the left-hand side of Equation (3.71) leaves it unchanged since it is slowly varying, while applying (3.72) to the right-hand side picks out those electrons whose ponderomotive phase θ_j is within the interval $\theta - \Delta\theta/2$ and $\theta + \Delta\theta/2$. In other words, the source for $E(\theta)$ includes the $N_\Delta = N_\lambda(\Delta\theta/2\pi)$ electrons that satisfy

$\left|\theta_j - \theta\right| \leq \Delta\theta/2$ when they arrive at location z. Then, we find that the wave equation in the time domain is

$$\left[\frac{\partial}{\partial z} + k_u \frac{\partial}{\partial \theta}\right] E(\theta; z) = -\kappa_1 n_e \frac{1}{N_\Delta} \sum_{j \in \Delta} e^{-i\theta_j(z)} \tag{3.73}$$

$$= -\kappa_1 n_e \left\langle e^{-i\theta_j(z)} \right\rangle_\Delta. \tag{3.74}$$

The notation in (3.73) denotes that we are to sum over the N_Δ particles within the FEL slice at position z and phase θ. Hence, the average $\langle e^{-i\theta_j}\rangle_\Delta$, which is often referred to as the local bunching factor (or just bunching factor), is a function of both z and θ. For any given z, the bunching factor quantifies the spectral content of the current near the fundamental frequency by a complex number whose magnitude is between 0 and 1.[2] Finally, we note while that the Maxwell equation in the time and frequency domain look quite similar, they differ as follows: the driving current in Fourier version equation (3.68) is a sum over all electrons with the phase $e^{-i\nu\theta_j}$, while the time domain equation (3.73) sums only over those electrons within the FEL time slice using the phase $e^{-i\theta}$.

3.4.2 FEL Equations and Energy Conservation

Let us collect the 1D FEL equations in the time domain. Writing the field equation for the fundamental equation (3.74) and the pendulum equations (3.20) in terms of E, we have

$$\left[\frac{\partial}{\partial z} + k_u \frac{\partial}{\partial \theta}\right] E(\theta; z) = -\kappa_1 n_e \left\langle e^{-i\theta_j} \right\rangle_\Delta, \tag{3.75}$$

$$\frac{d\theta_j}{dz} = 2k_u \eta_j, \tag{3.76}$$

$$\frac{d\eta_j}{dz} = \chi_1 \left(E e^{i\theta_j} + E^* e^{-i\theta_j}\right), \tag{3.77}$$

with

$$\kappa_1 \equiv \frac{eK[\mathrm{JJ}]}{4\epsilon_0 \gamma_r} \qquad \chi_1 \equiv \frac{eK[\mathrm{JJ}]}{2\gamma_r^2 mc^2}. \tag{3.78}$$

Equations (3.75), (3.76), and (3.77) are the central governing equations for a high-gain FEL in 1D. These equations conserve total (particle + field) energy. To show this, we first integrate the electromagnetic energy density u_{EM} over length and multiply the result by the transverse area $\mathscr{A}_{\mathrm{tr}}$ to obtain the field energy

$$U_{\mathrm{EM}} = \frac{\mathscr{A}_{\mathrm{tr}}\lambda_1}{2\pi} \int d\theta\, u_{\mathrm{EM}} = \frac{\mathscr{A}_{\mathrm{tr}}\lambda_1}{2\pi} \int d\theta\, \frac{\epsilon_0}{2}(\boldsymbol{E}^2 + c^2\boldsymbol{B}^2)$$
$$= \frac{\mathscr{A}_{\mathrm{tr}}\lambda_1}{2\pi} \int d\theta\, 2\epsilon_0 \left|E\right|^2. \tag{3.79}$$

[2] Harmonic generalizations of the bunching factor can also be defined as $b_h \equiv \langle e^{-ih\theta_j}\rangle_\Delta$.

Hence, an equation for the electromagnetic field energy can be obtained by multiplying (3.75) by $(\mathscr{A}_{\text{tr}}\lambda_1/\pi)\epsilon_0 E^*$, adding the complex conjugate, and integrating over θ; we find that

$$\begin{aligned}\frac{d}{dz}U_{\text{EM}} &= -\frac{eK[\text{JJ}]}{2\gamma_r}\frac{N_\lambda}{2\pi N_\Delta}\int d\theta \sum_{j\in\Delta} E^* e^{-i\theta_j} + c.c.\\ &= -\frac{eK[\text{JJ}]}{2\gamma_r}\sum_j \frac{e^{-i\theta_j}}{\Delta\theta}\int_{\theta_j-\Delta\theta/2}^{\theta_j+\Delta\theta/2} d\theta\, E^*(\theta) + c.c.\\ &= -\frac{eK[\text{JJ}]}{2\gamma_r}\sum_j E^*(\theta_j)e^{-i\theta_j} + c.c.,\end{aligned} \tag{3.80}$$

where in the last line we assumed that $E(\theta)$ is constant over the length $\Delta\theta$; this assumption is required because of our slice averaging, but is not necessary if one uses the frequency representation (3.68) or the unaveraged equation (3.71). The change in the total kinetic energy is obtained by multiplying (3.77) by $\gamma_r mc^2$ and summing over all electrons,

$$\frac{d}{dz}U_{\text{KE}} = \frac{d}{dz}\sum_j \gamma_r(1+\eta_j)mc^2 = \frac{eK[\text{JJ}]}{2\gamma_r}\sum_j E(\theta_j)e^{i\theta_j} + c.c.. \tag{3.81}$$

Adding Equations (3.80) and (3.81) shows that energy is conserved:

$$\frac{d}{dz}[U_{\text{EM}} + U_{\text{EM}}] = \frac{d}{dz}\left[\sum_j \gamma_r\eta_j mc^2 + \frac{\mathscr{A}_{\text{tr}}\lambda_1}{2\pi}\int d\theta\; 2\epsilon_0\,|E(\theta;z)|^2\right] = 0. \tag{3.82}$$

3.4.3 Dimensionless FEL Scaling Parameter ρ

By expressing the governing equations of physical systems in terms of dimensionless quantities, one can identify important time and length scales and characterize the relevant magnitudes of the physical variables. In this section we cast the FEL equations into dimensionless form and find the fundamental scaling parameter ρ. We will subsequently see that ρ, which is also called the Pierce parameter, characterizes most properties of a high-gain FEL, while the dimensionless beam and radiation variables will give us some sense of the dynamics without any additional computation.

We introduce the as-yet-unspecified parameter ρ by defining the scaled longitudinal coordinate $\hat{z} \equiv 2k_u\rho z$ that leads to the phase equation

$$\frac{d\theta_j}{d\hat{z}} = \hat{\eta}_j \;\text{ for }\; \hat{\eta}_j \equiv \frac{\eta_j}{\rho} \;\text{ (the new "momentum" variable).} \tag{3.83}$$

To simplify the energy equation for $\hat{\eta}_j$, we define the dimensionless complex field amplitude

$$a = \frac{\chi_1}{2k_u\rho^2}E, \tag{3.84}$$

in terms of which the energy equation reduces to

$$\frac{d\hat{\eta}_j}{d\hat{z}} = a(\theta_j,\hat{z})e^{i\theta_j} + a(\theta_j,\hat{z})^* e^{-i\theta_j}. \tag{3.85}$$

Writing the field equation (3.75) in terms of $\hat{z}$ and a, we have

$$\left[\frac{\partial}{\partial\hat{z}} + \frac{1}{2\rho}\frac{\partial}{\partial\theta}\right] a(\theta,\hat{z}) = -\frac{\chi_1}{2k_u\rho^2}\frac{n_e\kappa_1}{2k_u\rho}\langle e^{-i\theta_j}\rangle_\Delta. \tag{3.86}$$

To simplify the field equation, we choose to set the coefficient on the right-hand side of (3.86) to unity. Thus, the dimensionless Pierce parameter ρ must be [11]

$$\rho = \left[\frac{n_e\kappa_1\chi_1}{(2k_u)^2}\right]^{1/3} = \left(\frac{e^2K^2[\mathrm{JJ}]^2 n_e}{32\epsilon_0\gamma_r^3 mc^2 k_u^2}\right)^{1/3} = \left[\frac{1}{8\pi}\frac{I}{I_A}\left(\frac{K[\mathrm{JJ}]}{1+K^2/2}\right)^2\frac{\gamma\lambda_1^2}{2\pi\sigma_x^2}\right]^{1/3}, \tag{3.87}$$

where $I_A = ec/r_e = 4\pi\epsilon_0 mc^3/e \approx 17045$ A is the Alfvén current and we have set the cross-sectional area of the electron beam $\mathscr{A}_{\mathrm{tr}} \to 2\pi\sigma_x^2$ assuming a Gaussian transverse profile.

The scaled FEL equations have all coefficients unity, so that the dimensionless form allows one to make a number of order-of-magnitude estimates regarding the dynamics. First, one may *a priori* expect that the scaled variation $d/d\hat{z} \lesssim 1$. Thus, in the exponential growth regime we may anticipate the 1D gain length $L_{G0} \sim (2k_u\rho)^{-1}$. Additionally, since resonant energy exchange proceeds if the ponderomotive phase is nearly constant, this implies that saturation of the FEL interaction occurs when the scaled energy deviation $\hat{\eta}_j \sim 1$ (or $\eta_j \sim \rho$). At this point we expect that the bunching will approach its maximum value $\left|\langle e^{-i\theta_j}\rangle_\Delta\right| \to 1$, which in turn implies that the maximum scaled amplitude of the radiation $|a| \sim 1$. Furthermore, if we had included the transverse derivatives in the wave equation we would expect

$$\frac{1}{4k_u k_1\rho}\nabla_\perp^2 \to 1. \tag{3.88}$$

Identifying the transverse Laplacian with the radiation size via $\nabla_\perp^2 \sim 1/\sigma_r^2$, we find that the RMS mode size of the laser is roughly given by

$$\sigma_r \sim \sqrt{\frac{\lambda_1}{4\pi}\frac{\lambda_u}{4\pi\rho}}. \tag{3.89}$$

While these arguments are heuristic, they give useful predictions of FEL performance. Besides the observation that the gain length is approximately $\lambda_u/4\pi\rho$, we use the definition (3.84) to translate the scaled radiation amplitude $|a| \to 1$ at saturation to $|E| \to 2k_u\rho^2/\chi_1$, so that the maximum field energy density

$$2\epsilon_0|E|^2 \sim 2\epsilon_0\rho\frac{4k_u^2\rho^3}{\chi_1^2} = 2\epsilon_0\rho\frac{\kappa_1}{\chi_1} = \rho n_e\gamma_r mc^2. \tag{3.90}$$

Because $n_e mc^2\gamma_r$ is the electron energy density, we see that ρ also gives the FEL efficiency at saturation:

$$\rho = \frac{\text{field energy generated}}{\text{e-beam kinetic energy}}. \tag{3.91}$$

To determine the distance at which the FEL gain saturates and $P \sim \rho P_{\text{beam}}$, we consider the motion of the electron in the pendulum potential. Recall from Section 3.2.2 that the period of motion is characterized by the synchrotron wavenumber

$$\Omega_s \equiv \sqrt{\frac{eE_0 k_u K[\text{JJ}]}{\gamma^2 mc^2}} = 2\rho k_u \left|2a_0\right|^{1/2}, \tag{3.92}$$

and that the radiation field gains or loses energy depending on the oscillation phase of the particles. Since the energy exchange to the radiation ends when most of the particles make one-half oscillation in the ponderomotive bucket, we have $\langle\Omega_s\rangle z_{\text{sat}} \approx \pi$, where $\langle\Omega_s\rangle$ is the average value of the synchrotron wavenumber over the FEL length z_{sat}. Taking $\langle\Omega_s\rangle$ to be one-quarter of its maximum value at saturation where $|a_0| \sim 1$, we have $\rho k_u z_{\text{sat}}/\sqrt{2} \sim \pi$, or $z_{\text{sat}} \sim \lambda_u/\rho$. It is interesting to note that the power saturates when the synchrotron wavenumber is roughly equal to the exponential growth rate,

$$P \sim \rho P_{\text{beam}} \quad \Leftrightarrow \quad \Omega_s \sim 2\rho k_u. \tag{3.93}$$

This is to be expected, since when $\Omega_s \sim 2\rho k_u$ the particles can rotate to the accelerating phase of the potential during one growth length, in which case they then extract energy from the field.

Therefore, the FEL (or Pierce) parameter ρ determines the main characteristics of high-gain FEL systems, including

1. Gain length $\sim \lambda_u/4\pi\rho$,
2. Saturation power $\sim \rho\times$(e-beam power),
3. Saturation length $L_{\text{sat}} \sim \lambda_u/\rho$,
4. Transverse mode size $\sigma_r \sim \sqrt{\lambda_1\lambda_u/16\pi^2\rho}$.

In the following sections we will analyze the FEL equations and demonstrate that the dynamics indeed exhibit these simple scalings.

3.4.4 1D Solution Using Collective Variables

In this section, we illustrate the essentials of FEL gain by neglecting the θ dependence of the electromagnetic field. This ignores the propagation (slippage) of the radiation, and is equivalent to assuming that a has only one frequency component. This model will be useful to illustrate the basic physics of the electron beam and radiation field in a high-gain device, but will be insufficient to fully understand the spectral properties of self-amplified spontaneous emission (SASE); we will present a more rigorous discussion of SASE in Section 4.3. The 1D FEL equations ignoring radiation slippage are

$$\frac{d\theta_j}{d\hat{z}} = \hat{\eta}_j \tag{3.94}$$

$$\frac{d\hat{\eta}_j}{d\hat{z}} = ae^{i\theta_j} + a^* e^{-i\theta_j} \tag{3.95}$$

$$\frac{da}{d\hat{z}} = -\langle e^{-i\theta_j}\rangle_\Delta \,. \tag{3.96}$$

These are $2N_\Delta + 2$ coupled first-order ordinary differential equations, $2N_\Delta$ for the particles, and 2 for the complex amplitude a. In general, these can only be solved via computer simulation. However, the system can be linearized in terms of three collective variables [11]:

$$\begin{aligned} &a && \text{(field amplitude)} \\ &b = \langle e^{-i\theta_j}\rangle_\Delta && \text{(bunching factor)} \\ &P = \langle \hat{\eta}_j e^{-i\theta_j}\rangle_\Delta && \text{(collective momentum).} \end{aligned}$$

The equations of motion for the bunching b and the field amplitude a follow directly from Equations (3.94) and (3.96). Differentiating the collective momentum yields

$$\frac{dP}{d\hat{z}} = \left\langle \frac{d\hat{\eta}_j}{d\hat{z}} e^{-i\theta_j}\right\rangle - i\left\langle \hat{\eta}_j^2 e^{-i\theta_j}\right\rangle = a + a^*\left\langle e^{-2i\theta_j}\right\rangle - i\left\langle \hat{\eta}_j^2 e^{-i\theta_j}\right\rangle. \tag{3.97}$$

Note that (3.97) contains additional field variables, and the resulting system of equations is not closed. Nevertheless, these other terms are nonlinear, which we therefore expect to result in negligible higher-order corrections when a, b, and P are much smaller than unity before saturation. Thus, linearizing (3.97) and including the equations for b and a from (3.94) and (3.96) yields the following closed system in the small-signal regime

$$\frac{da}{d\hat{z}} = -b \qquad \text{Bunching produces coherent radiation.} \tag{3.98a}$$

$$\frac{db}{d\hat{z}} = -iP \qquad \text{Energy modulation becomes density bunching.} \tag{3.98b}$$

$$\frac{dP}{d\hat{z}} = a \qquad \text{Coherent radiation drives energy modulation.} \tag{3.98c}$$

These are three coupled first-order equations, which can be reduced to a single third-order equation for a as

$$\frac{d^3 a}{d\hat{z}^3} = ia. \tag{3.99}$$

We solve the linear equation by assuming that the field dependence is $\sim e^{-i\mu\hat{z}}$, which results in the following dispersion relation for μ:

$$\mu^3 = 1. \tag{3.100}$$

This is the well-known cubic equation [16], whose three roots are given by

$$\mu_1 = 1, \qquad \mu_2 = \frac{-1-\sqrt{3}i}{2}, \qquad \mu_3 = \frac{-1+\sqrt{3}i}{2}. \tag{3.101}$$

The root μ_1 is real and gives rise to an oscillatory solution, while μ_2 and μ_3 are complex conjugates that lead to exponentially decaying and growing modes, respectively. Furthermore, the roots obey

$$\sum_{\ell=1}^{3} \mu_\ell = 0, \qquad \sum_{\ell=1}^{3} \frac{1}{\mu_\ell} = \sum_{\ell=1}^{3} \mu_\ell^* = \sum_{\ell=1}^{3} \mu_\ell^2 = 0, \tag{3.102}$$

and the general solution to Equation (3.99) is composed of a linear combination of the exponential solutions:

$$a(\hat{z}) = \sum_{\ell=1}^{3} C_\ell e^{-i\mu_\ell \hat{z}}. \tag{3.103}$$

The three constants C_ℓ are determined from the initial conditions $a(0)$, $b(0)$, and $P(0)$. By differentiating the expression for a and using (3.98), we find

$$a(0) = C_1 + C_2 + C_3, \tag{3.104}$$

$$\left.\frac{da}{d\hat{z}}\right|_0 = -b(0) = -i\left[\mu_1 C_1 + \mu_2 C_2 + \mu_3 C_3\right], \tag{3.105}$$

$$\left.\frac{d^2a}{d\hat{z}^2}\right|_0 = iP(0) = -\left[\mu_1^2 C_1 + \mu_2^2 C_2 + \mu_3^2 C_3\right]. \tag{3.106}$$

Using (3.102), this yields the electromagnetic field evolution as

$$a(\hat{z}) = \frac{1}{3}\sum_{\ell=1}^{3}\left[a(0) - i\frac{b(0)}{\mu_\ell} - i\mu_\ell P(0)\right] e^{-i\mu_\ell \hat{z}}. \tag{3.107}$$

The general solution for the radiation requires all three roots of μ. For long propagation distances, however, the relative importance of the oscillating root μ_1 and decaying root μ_2 becomes insignificant in comparison with the growing solution associated with μ_3. Thus, the radiation field is completely characterized by μ_3 in the exponential growth regime where $\hat{z} \gg 1$, so that

$$a(\hat{z}) \approx \frac{1}{3}\left[a(0) - i\frac{b(0)}{\mu_3} - i\mu_3 P(0)\right] e^{-i\mu_3 \hat{z}}. \tag{3.108}$$

The first term in the bracket describes the coherent amplification of an external radiation signal, while the second and the third term show how modulations in the electron beam density and energy may lead to FEL output. When the source of these modulations is the electron beam shot noise then the exponential growth is called self-amplified spontaneous emission (SASE).

3.4.5 Qualitative Description of Self-Amplified Spontaneous Emission (SASE)

Self-amplified spontaneous emission results from the FEL amplification of the initially incoherent spontaneous undulator radiation [8, 9, 11]. It is of primary importance for FEL applications in wavelength regions where mirrors (and, hence, oscillator configurations) are unavailable.

For our first look at SASE, we use the formula for the radiation in the high-gain regime (3.108) assuming that there is no external field $a(0) = 0$ and that the beam has vanishing energy spread with $P(0) = 0$. In this case, the radiation intensity in the exponential growth regime is

$$\langle |a(\hat{z})|^2 \rangle \approx \frac{1}{9} \langle |b(0)|^2 \rangle e^{\sqrt{3}\hat{z}}. \tag{3.109}$$

Here, the scaled propagation distance $\sqrt{3}\hat{z} = \sqrt{3}(2k_u z\rho) = z/L_{G0}$, and the ideal 1D power gain length is

$$L_{G0} \equiv \frac{\lambda_u}{4\pi\sqrt{3}\rho}. \tag{3.110}$$

The bunching factor at the undulator entrance $\langle |b(0)|^2 \rangle$ derives from the initial shot noise of the beam, which is subsequently amplified by the FEL process. This level of shot noise is determined by the number of particles in the radiation coherence length, and in Section 4.3.2 we will show that

$$\langle |b(0)|^2 \rangle = \left\langle \frac{1}{N_{l_{\text{coh}}}^2} \left| \sum_{j \in l_{\text{coh}}} e^{-i\theta_j} \right|^2 \right\rangle \approx \frac{1}{N_{l_{\text{coh}}}}, \tag{3.111}$$

where $N_{l_{\text{coh}}}$ is the number of electrons in a coherence length l_{coh}. As we will show in the next chapter, the normalized bandwidth of SASE is $\Delta\omega/\omega \sim \rho$, so that the coherence time $t_{\text{coh}} \sim \lambda_1/c\rho$ and the coherence length length $l_{\text{coh}} \sim \lambda_1/\rho$; more precise expressions are given in Equations (4.54) and (4.64). Alternatively, one can recognize the coherence length as the amount the radiation slips ahead of the electron beam in a few gain lengths. Hence, the start-up noise of a SASE FEL is characterized by

$$N_{l_{\text{coh}}} \sim \frac{I}{ec} \frac{\lambda_1}{\rho}. \tag{3.112}$$

Figure 3.9 is a schematic plot that illustrates the initial start-up, exponential growth, and saturation of a SASE FEL. As is clear from the Figure and from the previous

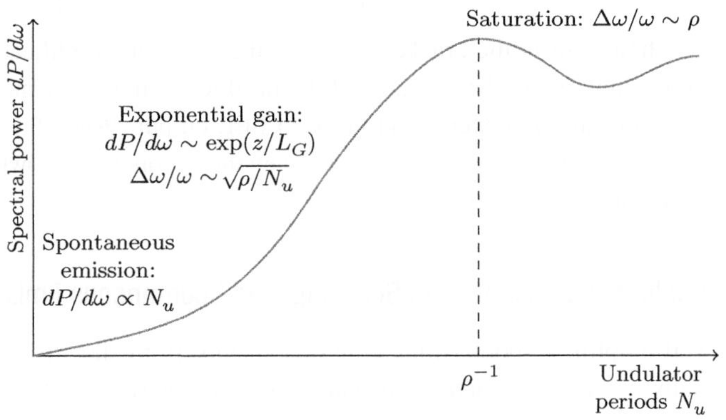

Figure 3.9 Illustration of basic SASE processes. Adapted from Ref. [17].

discussion, ρ plays a fundamental role in the high-gain FEL physics for SASE. While we haven't yet derived all the radiation properties, some of the important ones include

1. Saturation length $L_{\text{sat}} \sim \lambda_u/\rho$;
2. Output power $\sim \rho \times P_{\text{beam}}$;
3. Frequency bandwidth $\Delta\omega/\omega \sim \rho$;
4. 1D power gain length $L_{G0} = \lambda_u/(4\pi\sqrt{3}\rho)$;
5. Transverse coherence: radiation emittance $\varepsilon_r = \lambda/4\pi$;
6. Transverse mode size: $\sigma_r \sim \sqrt{\varepsilon_r L_{G0}}$;
7. For the SASE power $P = P_{\text{in}} \exp(z/L_G)$, the effective noise $P_{\text{in}} \sim \rho\gamma mc^2/N_{l_{\text{coh}}}$.

While these basic scalings and the plot of Figure 3.9 describes the ensemble averaged SASE properties, we should keep in mind that any individual SASE pulse is essentially amplified undulator radiation, and therefore has the same basic power and spectral fluctuations as the chaotic light discussed in Section 1.2.5 (we derive the fluctuation characteristics for SASE explicitly in Sections 4.3.2 and 4.3.3). We can understand the connection of SASE to amplified undulator radiation in another way by considering the undulator energy as computed from the 1D power spectral density,

$$U_{\text{und}} = T\int d\omega d\boldsymbol{\phi}\, \frac{dP}{d\omega d\boldsymbol{\phi}} \xrightarrow{\text{1D}} T\int d\omega\, \frac{\lambda^2}{\mathscr{A}_{\text{tr}}} \frac{dP}{d\omega d\boldsymbol{\phi}}\bigg|_{\boldsymbol{\phi}=\mathbf{0}}, \tag{3.113}$$

where the quantity $\lambda^2/\mathscr{A}_{\text{tr}}$ can be understood as the characteristic angular spread from a source of area $\mathscr{A}_{\text{tr}}$: $\Delta\phi_x\Delta\phi_y \sim \lambda^2/\mathscr{A}_{\text{tr}}$. In the 1D limit this tends to zero and we identify $\delta(\boldsymbol{\phi}) = \mathscr{A}_{\text{tr}}/\lambda^2$, so that

$$\frac{dP}{d\omega}\bigg|_{1D} = \frac{\lambda^2}{\mathscr{A}_{\text{tr}}}\delta(\boldsymbol{\phi})\frac{dP}{d\omega} = \frac{\lambda^2}{\mathscr{A}_{\text{tr}}} \frac{dP}{d\omega\boldsymbol{\phi}}\bigg|_{\boldsymbol{\phi}=\mathbf{0}}. \tag{3.114}$$

The same factor $\lambda^2/\mathscr{A}_{\text{tr}}$ appeared for the 1D limit in (3.60). Inserting the power density in the forward direction equation (2.100) with $f(\eta_j) = \delta(\eta_j)$, we find that

$$\begin{aligned} U_{\text{und}} &= T\left[\frac{\lambda_1^2}{\mathscr{A}_{\text{tr}}}\frac{I}{I_A}\left(\frac{K[\text{JJ}]}{1+K^2/2}\right)^2 \gamma_r^2 mc^2 N_u^2\right]\frac{\omega_1}{\pi N_u}\int dx\left(\frac{\sin x}{x}\right)^2 \\ &= 8\pi\omega_1 T\gamma_r mc^2 N_u\rho^3 \to 8\pi\omega_1 T\gamma_r mc^2\rho^2 \end{aligned} \tag{3.115}$$

at the FEL saturation distance $N_u \approx 1/\rho$. Now, we use (3.115) to rewrite the FEL energy at saturation as

$$U_{\text{FEL}} = N_e\rho\gamma_r mc^2 = \frac{N_e}{\rho\omega_1 T}\frac{U_{\text{und}}}{8\pi} \sim \frac{t_{\text{coh}}N_e}{T}U_{\text{und}} = N_{l_{\text{coh}}}U_{\text{und}} \tag{3.116}$$

$$= \frac{T}{t_{\text{coh}}}N_{l_{\text{coh}}}^2\frac{U_{\text{und}}}{N_e}. \tag{3.117}$$

The first line (3.116) shows that in the forward direction the FEL output at saturation is larger than that of the undulator radiation by the number of particles in a coherence time $N_{l_{\text{coh}}} \gtrsim 10^5$. The second result (3.117) interprets the FEL energy as being proportional to the undulator field energy due to a single electron times the square of the number of electrons in a coherence length times the number of coherent regions T/t_{coh}.

Finally, we would like to emphasize that X-ray FELs based on SASE would not have been realized without incredible improvements in the production, transport, and manipulation of electron beams, since very high brightness electron beams are essential for X-ray FELs. In particular, SASE FELs have been made possible through recent advances in photocathode gun design (see [18] and a review in [19]), and tremendous improvements of rf linac and undulator technology. These advances have made it possible to produce sufficient gain in the undulator for transversely coherent radiation, meaning that the electron beam meets the following criteria:

1. Energy spread $\Delta\gamma/\gamma < \rho$;
2. Emittance $\varepsilon_x \lesssim \lambda/(4\pi)$;
3. Beam size $\sigma_x \gtrsim \sigma_r \sim \sqrt{\frac{\lambda}{4\pi}\frac{\lambda_u}{4\pi\rho}}$ to have 1D scalings approximately apply;
4. High peak current to achieve $\rho \sim 10^{-3}$ and, hence, a reasonable saturation length and power efficiency.

The production and transport of such high-brightness beams is itself a rich subject, but beyond our intended scope; rather, the next two chapters will attempt to explain the physics behind these e-beam requirements, and how they ultimately relate to the performance of advanced X-ray sources.

References

[1] J. M. J. Madey, "Stimulated emission of bremsstrahlung in a periodic magnetic field," *J. Appl. Phys.*, vol. 42, p. 1906, 1971.

[2] D. A. G. Deacon, L. R. Elias, J. M. J. Madey, G. J. Ramian, H. A. Schwettman, and T. I. Smith, "First operation of a free-electron laser," *Phys. Rev. Lett.*, vol. 38, p. 892, 1977.

[3] J. Galayda *et al.*, "Linac Coherent Light Source (LCLS) Conceptual Design Report," SLAC, Report SLAC-R-593, 2002.

[4] T. Shintake *et al.*, 2005, SPring-8 Compact SASE Source Conceptual Design Report, `www-xfel.spring8.or.jp`.

[5] R. Brinkmann *et al.*, "TESLA XFEL: First Stage of the X-ray Laser Laboratory (Technical Design Report, Supplement)," DESY, Report TESLA FEL 2002-09, 2002.

[6] H.-S. Kang, K.-W. Kim, and I.-S. Ko, "Current status of PAL-XFEL project," in *Proceedings of IPAC 2014*, p. 2897, 2014.

[7] R. Ganter *et al.*, "SwissFEL-conceptual design report," Paul Scherrer Institute (PSI), Report, 2010.

[8] A. Kondratenko and E. Saldin, "Generation of coherent radiation by a relativistic electron beam in an ondulator," *Part. Accelerators*, vol. 10, p. 207, 1980.

[9] Y. S. Derbenev, A. M. Kondratenko, and E. L. Saldin, "On the possibility of using a free electron laser for polarization of electrons in storage rings," *Nucl. Instrum. Methods Phys. Res.*, vol. 193, p. 415, 1982.

[10] R. Bonifacio, C. Pellegrini, and L. M. Narducci, "Collective instabilities and high-gain regime free electron laser," in *Free Electron Generation of Extreme Ultraviolet Coherent Radiation*, J. M. J. Madey and C. Pellegrini, Eds., no. 118. SPIE, p. 236, 1984.

[11] —, "Collective instabilities and high-gain regime in a free electron laser," *Opt. Commun.*, vol. 50, p. 373, 1984.

[12] W. B. Colson, "One-body electron dynamics in a free electron laser," *Phys. Lett. A*, vol. 64, p. 190, 1977.

[13] N. M. Kroll, "A note on the Madey gain-spread theorem," in *Free-Electron Generators of Coherent Radiation Vol. 8*. Reading, MA: Addison-Wesley, p. 315, 1982.

[14] S. Krinsky, J. M. Wang, and P. Luchini, "Madey's gain spread theorem for the free-electron laser and the theory of stochastic processes," *J. Appl. Phys.*, vol. 53, p. 5453, 1982.

[15] I. Boscolo, M. Leo, R. A. Leo, G. Soliani, and V. Stagno, "On the gain of the free electron laser (FEL) amplifier for a nonmonoenergetic beam," *IEEE J. Quantum Electron.*, vol. 18, p. 1957, 1982.

[16] N. Kroll and W. McMullin, "Stimulated emission from relativistic electrons passing through a spatially periodic transverse," *Phys. Rev. A*, vol. 17, p. 300, 1978.

[17] K.-J. Kim, "Three-dimensional analysis of coherent amplification and self-amplified spontaneous emission in free electron lasers," *Phys. Rev. Lett.*, vol. 57, p. 1871, 1986.

[18] J. S. Fraser, R. L. Sheffield, and E. R. Gray, "A new high brightness electron injector for free-electron lasers driven by RF linacs," *Nucl. Instrum. Methods Phys. Res., Sect. A*, vol. 250, p. 71, 1986.

[19] I. Ben-Zvi, "Photoinjectors," in *Accelerator Physics, Technology, and Applications*, A. W. Chao, H. O. Moser, and Z. Zhao, Eds. World Scientific, p. 158, 2004.

4 1D FEL Analysis

In this chapter we delve more deeply into the 1D theory of the FEL. The 1D picture is sufficient to understand how an FEL works, since the essential FEL physics is longitudinal in nature. A free-electron laser acts as a linear amplifier in the small signal regime, and we will find that it is most easily analyzed theoretically in the frequency representation. Hence, we begin this section by deriving the Klimontovich equation describing the electron beam in the frequency domain, to which we add the Maxwell equation (3.68). We then apply these equations to the small-gain limit in Section 4.2, finding solutions that generalize those of Section 3.3. We then turn our attention to the high-gain FEL in Section 4.3, showing how the linearized FEL equations can be solved for arbitrary initial conditions using the Laplace transform. In particular, Section 4.3 covers self-amplified spontaneous emission (SASE) in some detail, because SASE provides the simplest way to produce intense X-rays. We derive the basic properties of SASE in the frequency domain, including its initialization from the fluctuations in the electron beam density (shot noise), its exponential gain, and its spectral properties. We then connect our analysis to the time domain picture via Fourier transformation, which helps complete the characterization of SASE's fluctuation properties. The chapter concludes with a discussion of how the FEL gain saturates in Section 4.4. We derive a quasilinear theory that describes the decrease in gain associated with an increase in electron beam energy spread, and show qualitatively how this is related to particle trapping. We also discuss tapering the undulator strength parameter after saturation to further extract radiation energy from the electron beam. Finally, we make a few comments on superradiance, focusing on the superradiant FEL solution associated with particle trapping that can support powers in excess of the usual FEL saturation power.

4.1 Coupled Maxwell–Klimontovich Equations for the 1D FEL

The 1D FEL equations in the frequency domain can be used to obtain a clear understanding of various aspects of the SASE process, including the initial start-up from particle shot noise, the exponential gain, the development of longitudinal coherence, and the effect of the electron beam energy spread. They also qualitatively describes the full physics of X-ray FELs, since the effect of diffraction is smaller for shorter wavelengths. In the previous chapter we derived the spectral wave equation (3.68), so in this section we concentrate on the frequency representation of the Klimontovich equation that describes the dynamics of the electron distribution function.

In our previous analysis we described the electron motion in an FEL using single particle (Newton's) equations, from which we found an approximate collective description. An alternate approach that retains all the generality of the single particle equations while treating the electron beam as a single entity employs a distribution function on phase space. It turns out that this approach is also naturally suited for the frequency representation.

To retain the discrete nature of the electrons, we describe the electron beam using the Klimontovich distribution function in the longitudinal phase space spanned by (θ, η),

$$F(\theta, \eta; z) = \frac{k_1}{I/ec} \sum_{j=1}^{N_e} \delta[\theta - \theta_j(z)]\delta[\eta - \eta_j(z)]. \tag{4.1}$$

Here I/ec is the line density, and F is comprised of sum over all N_e particles in the beam, with each particle contributing a delta function centered about its coordinates in phase space.

The distribution function can be separated into the smooth background $\bar{F}$ and the remainder δF that contains the shot noise and the perturbation due to the FEL interaction. We write this division as

$$F(\theta, \eta; z) = \bar{F}(\eta; z) + \delta F(\theta, \eta; z), \tag{4.2}$$

in which we assume that the smooth background distribution is independent of the phase θ. This corresponds to a uniform bunch (coasting beam) model which is approximately valid for an electron beam that is much longer than the slippage distance over one gain length, $\lambda_1/4\pi\rho$. Its normalization follows from (4.2) to be such that $\int d\eta\, \bar{F} = 1$.

Conservative, Hamiltonian dynamics dictates that the distribution function is conserved along single particle orbits, so that F satisfies the continuity equation

$$\frac{d}{dz}F(\theta, \eta; z) = \frac{\partial F}{\partial z} + \frac{d\theta}{dz}\frac{\partial F}{\partial \theta} + \frac{d\eta}{dz}\frac{\partial F}{\partial \eta} = 0. \tag{4.3}$$

The phase equation $d\theta_j/dz$ is the same as was derived previously; the energy equation (3.77) can be written in terms of $E_\nu(z)$ using Equation (3.70),

$$e^{i\theta_j}E(z, \theta_j) = e^{i\theta_j}\int d\nu\; E_\nu(z)e^{i\Delta\nu\theta_j} = \int d\nu\; E_\nu(z)e^{i\nu\theta_j}. \tag{4.4}$$

Here, recall that $\int d\omega\, E_\omega = \omega_1 \int d\omega\; E_\nu$. If we also include the force due to all odd radiation harmonics, $\chi_1 \to \sum_h \chi_h$, and the Liouville equation (4.3) for $F = \bar{F} + \delta F$ becomes

$$\begin{aligned} &\frac{\partial}{\partial z}\bar{F} + \sum_{h\,\text{odd}} \chi_h \left[\int d\nu\; E_\nu(z)e^{i\nu\theta} + \text{c.c.}\right]\frac{\partial}{\partial \eta}\delta F \\ &+\left[\frac{\partial}{\partial z} + 2k_u\eta\frac{\partial}{\partial \theta}\right]\delta F + \sum_{h\,\text{odd}} \chi_h \left[\int d\nu\; E_\nu(z)e^{i\nu\theta} + \text{c.c.}\right]\frac{\partial}{\partial \eta}\bar{F} = 0, \end{aligned} \tag{4.5}$$

where the integration extends over a small region about $\nu \approx h$.

The second line in Equation (4.5) contains the fluctuations associated with δF and E_ν that during the FEL interaction are dominated by frequencies at the fundamental and, possibly, its odd harmonics. The first line, on the other hand, contains all the terms that vary slowly over the bunch, along with some nonlinear harmonic contributions. Since these two expressions vary over disparate temporal scales, they should separately vanish. The resulting equations can be more cleanly written in frequency space, and therefore we write the frequency representation of the Klimontovich distribution function as

$$F_\nu(\eta; z) = \frac{1}{2\pi}\int d\theta\, e^{-i\nu\theta} F(\theta, \eta; z) = \frac{1}{N_\lambda}\sum_{j=1}^{N_e} e^{-i\nu\theta_j(z)}\delta[\eta - \eta_j(z)], \tag{4.6}$$

where N_λ is the number of electrons in one radiation wavelength λ_1.

Now, we obtain the equation of motion for $\bar{F}$ by averaging Equation (4.5) over the bunch length cT, which effectively eliminates the entire second line along with the fluctuating terms from the first line. In a similar way, the equation for F_ν is found by demanding that the Fourier transform of the second line in (4.5) vanishes; we supply more details of this derivation in Section 4.4. To complete the set of equations we use the definition (4.6) to express the source term of the wave equation (3.68) in terms of F_ν, finding that

$$\frac{\partial}{\partial z}\bar{F}(\eta; z) = -\sum_{h\ \mathrm{odd}} \chi_h \int d\nu \left[E_\nu(z)\frac{\partial F_\nu^*}{\partial \eta} + c.c.\right] \tag{4.7}$$

$$\left[\frac{\partial}{\partial z} + 2i\nu k_u\eta\right] F_\nu(\eta; z) = -\sum_{h\ \mathrm{odd}} \chi_h E_\nu(z)\frac{\partial \bar{F}}{\partial \eta} \tag{4.8}$$

$$\left[\frac{\partial}{\partial z} + i\Delta\nu k_u\right] E_\nu(z) = -\kappa_h n_e \int d\eta\, F_\nu(\eta; z), \tag{4.9}$$

where the harmonic couplings

$$\chi_h \equiv \frac{eK[\mathrm{JJ}]_h}{2\gamma_r^2 mc^2} \qquad \kappa_h \equiv \frac{eK[\mathrm{JJ}]_h}{4\epsilon_0\gamma_r}. \tag{4.10}$$

The three equations (4.7), (4.8), and (4.9) form the basis for our 1D FEL analysis throughout this chapter. Before proceeding, however, we will simplify the FEL equations by assuming that only one radiation harmonic contributes to the force. In the low-gain regime this means that we consider the amplification of a nearly monochromatic field whose frequency is centered at $h\omega_1$ with h odd and positive. This situation is particularly relevant to an FEL oscillator designed to amplify one frequency over many passes through the undulator. On the other hand, our high-gain analysis will typically assume that the FEL operates at the fundamental frequency, since this has the largest amplification; hence, high-gain devices typically operate with $h = 1$.

Next, inspection of Equation (4.7) shows that the change in the smooth distribution depends quadratically on the high-frequency perturbation. Hence, we can neglect the evolution of $\bar{F}$ in the initial stage when both the microbunching F_ν and field E_ν can be considered small. In this limit we set $\bar{F}(\eta; z) \to V(\eta)$, and we find that

$$\left[\frac{\partial}{\partial z} + 2i\nu k_u \eta\right] F_\nu(\eta; z) = -\chi_h \frac{dV}{d\eta} E_\nu(z) \tag{4.11}$$

$$\left[\frac{\partial}{\partial z} + i\Delta\nu k_u\right] E_\nu(z) = -\kappa_h n_e \int d\eta\, F_\nu(\eta; z). \tag{4.12}$$

The Equations (4.11)–(4.12) represent the linearized Maxwell–Klimontovich system for the 1D FEL. The linearized equations do not conserve total energy, which should not surprising since the field energy is a second-order quantity in E_ν. Hence, these equations are limited to the small signal regime before FEL saturation, when $P \ll \rho P_{\text{beam}}$. In the next few sections we will investigate the linear FEL dynamics associated with Equations (4.11) and (4.12), and return to the physics of gain saturation in Section 4.4.

4.2 Pertubative Solution for Small FEL Gain

We have shown that in the linear regime the FEL is governed by the coupled differential equations (4.12) and (4.11). It is easy to see that these are equivalent to the integral equations

$$E_\nu(z) = e^{-i\Delta\nu k_u z}\left[E_\nu(0) - \kappa_h n_e \int_0^z dz'\, e^{i\Delta\nu k_u z'} \int d\eta\, F_\nu(\eta; z')\right] \tag{4.13}$$

$$F_\nu(\eta; z) = e^{-2i\nu k_u \eta z}\left[F_\nu(\eta; 0) - \chi_h \int_0^z dz'\, e^{2i\nu k_u \eta z'} \frac{dV}{d\eta} E_\nu(z')\right]. \tag{4.14}$$

Combining these two, we find that the electric field satisfies

$$\begin{aligned} E_\nu(z) = e^{-i\Delta\nu k_u z}\Bigg[& E_\nu(0) - \kappa_h n_e \int_0^z dz' \int d\eta\, e^{i(\Delta\nu - 2\nu\eta)k_u z'} F_\nu(\eta; 0) \\ & + \chi_h \kappa_h n_e \int_0^z dz'\, e^{i(\Delta\nu - 2\nu\eta)k_u z'} \int_0^{z'} dz''\, e^{2i\nu k_u \eta z''} \frac{dV}{d\eta} E_\nu(z'')\Bigg]. \end{aligned} \tag{4.15}$$

The first term on the right-hand side is the input coherent radiation that propagates unaltered to the point z, while the second term is the spontaneous undulator radiation; we abbreviate the latter as

$$E_\nu^{\text{SR}}(z) \equiv -\kappa_h n_e e^{-i\Delta\nu k_u z} \int_0^z dz' \int d\eta\, e^{i(\Delta\nu - 2\nu\eta)k_u z'} F_\nu(\eta; 0). \tag{4.16}$$

The third term in Equation (4.15) represents the effects of the interaction between the electron beam and the radiation field.

As it stands, (4.15) clearly separates the initial, spontaneous, and amplified parts of the radiation field, but is in general more difficult to solve then (4.12) and (4.11). If the interaction is weak, however, we may replace $E_\nu(z'')$ with the unperturbed field $e^{-i\Delta\nu k_u z''}E_\nu(0)$ in the interaction term. In this case we have a closed-form solution that we can write as

$$E_\nu(z) = E_\nu^{\text{Coh}}(z) + E_\nu^{\text{SR}}(z), \tag{4.17}$$

where in the weak-interaction (low-gain) approximation the coherent part of the field is

$$E_\nu^{\text{Coh}}(z) = e^{-i\Delta\nu k_u z}E_\nu(0) \times \left[1 + \chi_h \kappa_h n_e \int d\eta \int_0^z dz' \int_0^{z'} dz''\, e^{i(\Delta\nu - 2\nu\eta)k_u(z'-z'')}\frac{dV}{d\eta}\right]. \tag{4.18}$$

The integrals over z' and z'' are straightforward, and at the end of the undulator the coherent field can be cast in the form

$$E_\nu^{\text{Coh}}(L_u) = \left\{1 + \frac{j_{C,h}}{8}\int d\eta\; V(\eta)[g(x_{\nu,\eta}) + ip(x_{\nu,\eta})]\right\} e^{-2\pi i\Delta\nu N_u}E_\nu(0). \tag{4.19}$$

The compact solution (4.19) obtains after integrating over η by parts and introducing additional short-hand notation with the constant $j_{C,h}$ and the functions p and g. We have defined $g(x)$ to be the same as the gain function introduced when we first discussed low-gain FEL physics in Section 3.3.1, while we will find that the function $p(x)$ is related to the accompanying phase change of $E_\nu(z)$. Explicitly, these functions are

$$g(x) = -\frac{d}{dx}\left(\frac{\sin x}{x}\right)^2 \qquad p(x) = \frac{d}{dx}\left(\frac{2x - \sin 2x}{2x^2}\right). \tag{4.20}$$

We assume that the frequency is near an odd harmonic, $\nu = h + \Delta\nu$ with h odd and $\Delta\nu \ll 1$, in which case g and p are functions of the argument

$$x_{\nu,\eta} = 2\pi N_u(h\eta - \Delta\nu/2). \tag{4.21}$$

Finally, the dimensionless constant $j_{C,h}$ was introduced by Colson in his low-gain FEL analysis [1], and defined to be

$$j_{C,h} \equiv 4h\chi_h\kappa_h n_e k_u L_u^3 = 4\pi^2 h \frac{e^2 n_e}{4\pi\epsilon_0}\frac{K^2[\text{JJ}]_h^2}{\gamma^3 mc^2}N_u L_u^2 \tag{4.22}$$

$$= 2h(4\pi\rho N_u)^3\frac{[\text{JJ}]_h^2}{[\text{JJ}]^2} \tag{4.23}$$

with $[\text{JJ}] \equiv [\text{JJ}]_1$. The second line shows the relationship between $j_{C,h}$ and the dimensionless Pierce parameter ρ that we introduced in Section 3.4.3. Since $j_{C,h}$ is proportional to the gain when the gain is small, in this limit we also find that $G \propto (\rho N_u)^3$. Again, the low-gain solution (4.19) is valid if $j_{C,h} \ll 1$, which is equivalent to requiring that the undulator length is less then the ideal 1D FEL gain length, $L_u/L_{G0} = 4\pi\sqrt{3}\rho N_u < 1$. This requirement can be relaxed somewhat if the gain is reduced by the energy spread of the electron beam.

To investigate the effects of the electron beam energy spread, we consider the case where the electrons' energy distribution is Gaussian about η_0:

$$V(\eta) = \frac{1}{\sqrt{2\pi}\sigma_\eta} e^{-(\eta-\eta_0)^2/2\sigma_\eta^2}. \tag{4.24}$$

We then write the amplitude formula (4.19) as

$$E_\nu^{\text{Coh}}(L_u) = \left\{1 + \frac{j_{C,h}}{8}\left[\bar{g}(x_0) + i\bar{p}(x_0)\right]\right\} e^{-2\pi i \Delta\nu N_u} E_\nu(0). \tag{4.25}$$

Here, we have defined the integrations

$$\bar{g}(x_0) \equiv \int dx' \, \frac{e^{-x'^2/2(2\pi N_u h\sigma_\eta)^2}}{\sqrt{2\pi}(2\pi N_u h\sigma_\eta)} g(x_0 - x'), \tag{4.26}$$

$$\bar{p}(x_0) \equiv \int dx' \, \frac{e^{-x'^2/2(2\pi N_u h\sigma_\eta)^2}}{\sqrt{2\pi}(2\pi N_u h\sigma_\eta)} p(x_0 - x'), \tag{4.27}$$

with $x_0 \equiv 2\pi N_u(h\eta_0 - \Delta\nu/2)$.

Now, we can easily calculate the gain by considering the radiation energy density $u \propto |E_\nu|^2$. We have

$$G = \frac{u(L_u) - u(0)}{u(0)} \approx \frac{j_{C,h}}{4}\bar{g}(x_0) \tag{4.28}$$

to first order in $j_{C,h}$. This expression generalizes the gain formula (3.48) to include e-beam energy spread; in the limit $\sigma_\eta \ll 1/(2\pi N_u h)$ we have $\bar{g}(x_0) \to g(x_0)$ and (4.28) reproduces (3.48). In other words, the electron beam energy spread can be neglected if its RMS width is much less than that of the spontaneous radiation at the FEL harmonic of interest. On the other hand, if the energy spread $\sigma_\eta \gtrsim 1/(2\pi N_u h)$ then we must account for the fact that electrons with different energies satisfy the FEL resonance condition at different radiation wavelengths. The resulting interference tends to reduce the FEL gain, and this physics is captured mathematically by the convolution (4.26).

There is another convenient way to write $\bar{g}(x_0)$ if the energy distribution is given by the Gaussian (4.24). This expression follows if we defer integrating over the undulator length in Equation (4.18), and instead integrate over η. Changing variables to $z = z'/L_u - 1/2$ and $s = z''/L_u - 1/2$ we find that

$$G = -\frac{j_{C,h}}{2}\int_{-1/2}^{1/2} dz \int_{-1/2}^{1/2} ds\,(z - s)\sin[2x_0(z - s)]e^{-2[2\pi N_u(z-s)\sigma_\eta]^2}. \tag{4.29}$$

In addition to energy exchange, the amplitude equation has an imaginary part proportional to $p(x)$. This term in Equation (4.25) leads to a phase change of the field that accompanies FEL gain. Figure 4.1(a) plots the functions $\bar{g}(x)$ and $\bar{p}(x)$ as a function of the detuning $x_0 = 2\pi N_u(h\eta_0 - \Delta\nu/2)$ when the energy spread $\sigma_\eta = 0.5(2\pi N_u h)$. The functional form of g is very similar to the zero energy spread case shown previously in Figure 3.6, although the peak value has decreased from 0.54 to about 0.46. The value

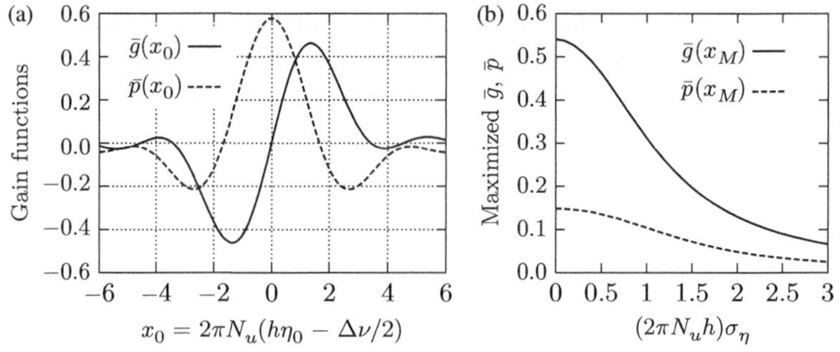

Figure 4.1 (a) Low-gain functions when the energy spread $(2\pi N_u h)\sigma_\eta = 0.5$. The amplitude gain $\bar{g}$ is very similar to its cold counterpart, while the phase change $\bar{p}$ is small at maximum gain. (b) Energy spread dependence of the low-gain functions at the detuning x_M that gives maximal gain.

of p at the detuning that maximizes the gain is small, about 0.13. Therefore, the phase change during FEL amplification is also relatively small.

We are usually free to set the detuning x to maximize the gain, by either choosing the energy or frequency offset (in fact, the FEL itself may pick out the x of maximum gain). In Figure 4.1(b) we plot $\bar{g}$ and $\bar{p}$ at the detuning x_M that maximizes the gain. We see that the phase change $\bar{p}$ is between one-quarter and two-fifths of $\bar{g}$, and is therefore typically rather small.

Furthermore, we see that the gain is not significantly affected by the energy spread provided $(2\pi N_u h)\sigma_\eta \lesssim 0.5$, after which $\bar{g}$ drops quite rapidly. This shows that the FEL gain at higher harmonics is more sensitive to energy spread, which can be attributed to the fact that the normalized undulator radiation bandwidth $\sim 1/hN_u$. On the other hand, the FEL gain for a fixed e-beam energy, undulator period, and undulator length also scales with the constant $j_{C,h} \propto hK^2[\text{JJ}]_h^2$. Hence, under these restrictions one may find that the gain at a certain target wavelength is maximized by operating at a higher FEL harmonic, because in this case $j_{C,h}$ increases as both h and K increase. This will be true provided the energy spread is sufficiently small, $\sigma_\eta \lesssim 1/(4\pi N_u h)$. Such constraints on e-beam energy or undulator length may be imposed by cost and/or size limits of the FEL accelerator facility.

4.3 Solution via Laplace Transformation for Arbitrary FEL Gain

We employed two different perturbation expansions in Section 3.2 and in Section 4.2 to solve the FEL equations in the low-gain limit. Here, we develop the full solution to the linearized 1D FEL equations that will generalize the collective variable approach of Section 3.4.4. We assume emission at the fundamental dominates the dynamics ($\nu \approx 1$), so that the coupled FEL system is described by

$$\left[\frac{\partial}{\partial z} + i\Delta\nu k_u\right] E_\nu(z) = -\kappa_1 n_e \int d\eta\, F_\nu(\eta; z), \tag{4.30}$$

$$\left[\frac{\partial}{\partial z} + 2ik_u\eta\right] F_\nu(\eta; z) = -\chi_1 \frac{dV}{d\eta} E_\nu(z). \tag{4.31}$$

The initial value problem of the coupled Equations (4.30) and (4.31) can be solved by introducing the Laplace transform [2, 3]:

$$\begin{bmatrix} E_{\nu,\mu} \\ F_{\nu,\mu} \end{bmatrix} = \int_0^\infty dz\, e^{i\mu 2\rho k_u z} \begin{bmatrix} E_\nu(z) \\ F_\nu(z) \end{bmatrix}. \tag{4.32}$$

We emphasize that $\nu \approx 1$ represents the frequency that is the Fourier conjugate to the temporal phase θ, while μ gives complex growth as the fields propagate in z. Taking the Laplace transform of the particle and radiation equations lead to the algebraic linear system

$$-2i\mu\rho k_u E_{\nu,\mu} + i\Delta\nu k_u E_{\nu,\mu} = -\kappa_1 n_e \int d\eta\, F_{\nu,\mu}(\eta) + E_\nu(0) \tag{4.33}$$

$$-2i\mu\rho k_u F_{\nu,\mu}(\eta) + 2ik_u\eta F_{\nu,\mu}(\eta) = -\chi_1 E_{\nu,\mu} \frac{dV}{d\eta} + F_\nu(\eta; 0), \tag{4.34}$$

where $E_\nu(0)$ and $F_\nu(\eta; 0)$ are the νth component of the initial radiation field and the initial beam distribution, respectively. Equations (4.33)–(4.34) can be easily solved for $E_{\nu,\mu}$, and the inverse Laplace transformation yields

$$E_\nu(z) = \oint \frac{d\mu}{2\pi i} \frac{e^{-i\mu 2\rho k_u z}}{D(\mu)} \left[E_\nu(0) + \frac{i\kappa_1 n_e}{2\rho k_u N_\lambda} \sum_{j=1}^{N_e} \frac{e^{-i\nu\theta_j(0)}}{\eta_j(0)/\rho - \mu} \right], \tag{4.35}$$

where we have used the definition of F_ν Equation (4.6), and have defined the dispersion function

$$D(\mu) \equiv \mu - \frac{\Delta\nu}{2\rho} - \int d\eta\, \frac{V(\eta)}{(\eta/\rho - \mu)^2}. \tag{4.36}$$

Note that the integration contour in the complex μ plane must be above all singularities/poles of (4.35), so that when $z < 0$ and the contour can be closed at $\Im(\mu) \to +\infty$, it encloses no poles and $E_\nu(z < 0) = 0$.

The first term of Equation (4.35) describes the process of coherent amplification of $E_\nu(0)$, while the second term containing the particle phases $e^{-i\nu\theta_j}$ describes FEL radiation initiated by the electron beam; assuming that the beam is not initially microbunched these phases are random and this term describes the process of self-amplified spontaneous emission.

When $z > 0$ the contour integral of Equation (4.35) encloses all the singularities in the complex μ plane. There are many singularities of kinematic origin, which give the free-streaming solutions for which $\mu = \eta_j(0)/\rho$; however, these vary for each particle according to the initial energy, and we will find that summing over these contributions yield the usual spontaneous emission. Thus, the evolution in a free-electron laser is largely dictated by the poles of $1/D$, which are given by the roots of the dispersion relation:

$$D(\mu) = \mu - \frac{\Delta\nu}{2\rho} - \int d\eta\, \frac{V(\eta)}{(\eta/\rho - \mu)^2} = 0. \tag{4.37}$$

Solutions to $D(\mu) = 0$ for which μ has a positive imaginary part give rise to an exponentially growing electric field amplitude. For e-beams with vanishing energy spread $V(\eta) = \delta(\eta)$, this dispersion relation become $\mu^2(\mu - \Delta\nu/2\rho) = 1$, which in turn reduces to the cubic equation (3.100) when $\Delta\nu = 0$.

Having found the field amplitude, we can compute the power spectral density with some slight adjustments to the formula (1.71). Using the fact that $E_\omega = E_\nu/\omega_1$ and integrating over the area $\mathscr{A}_{\text{tr}}$, we find that

$$\frac{dP}{d\omega} = \frac{\epsilon_0}{\pi c}\frac{\mathscr{A}_{\text{tr}}\lambda_1^2}{T}\left\langle |E_\nu(z)|^2 \right\rangle. \tag{4.38}$$

Here T is the duration of the electron pulse, and $\langle\cdot\rangle$ denotes an ensemble average over the microscopic electron distribution. When calculating $\left\langle\, |E_\nu(z)|^2 \,\right\rangle$ for the SASE term we will use the manipulation

$$\begin{aligned}\left\langle \sum_{j,\ell} e^{-i\nu(\theta_j-\theta_\ell)} G(\eta_j,\eta_\ell)\right\rangle &= \left\langle \sum_j G(\eta_j,\eta_j)\right\rangle + \left\langle \sum_{j\neq\ell} e^{-i\nu(\theta_j-\theta_\ell)} G(\eta_j,\eta_\ell)\right\rangle \\ &\approx N_e \int d\eta\; V(\eta) G(\eta,\eta),\end{aligned} \tag{4.39}$$

where the θ_js are the initial phases, and we drop the sum with $j \neq \ell$ under the assumption that the initial phases are completely random with no correlation. For simplicity, we use the shorthanded notation $\theta_j = \theta_j(0)$ and $\eta_j = \eta_j(0)$ from now on.

4.3.1 Spontaneous Radiation and the Low-Gain Limit

The system dynamics are largely governed by the dispersion function $D(\mu)$ of Equation (4.36). The FEL interaction itself is contained in the third term here, and involves an integral over η of the distribution function V. We can connect the present analysis with our calculation of the spontaneous undulator radiation and the low-gain FEL by assuming this interaction to be weak. Thus, we expand $1/D$ from the integral solution (4.35) as follows:

$$\frac{1}{D(\mu)} = \frac{1}{\mu - \Delta\nu/(2\rho)} + \frac{1}{(\mu - \Delta\nu/(2\rho))^2}\int d\eta\; \frac{V(\eta)}{(\eta/\rho - \mu)^2} + \ldots. \tag{4.40}$$

This expansion is valid mathematically in the limit of vanishing ρ, and hence we also have $\eta/\rho \to \infty$.

We compute the spontaneous radiation amplitude by keeping only the first term in the expansion and applying the residue theorem of contour integration. Then, we get

$$E_\nu(z) = -\frac{i\kappa_1 n_e}{2\rho k_u N_\lambda}\sum_{j=1}^{N_e} e^{-i\nu\theta_j}\frac{e^{-i\Delta\nu k_u z} - e^{-2i\eta_j k_u z}}{\eta_j/\rho - \Delta\nu/2\rho}, \tag{4.41}$$

which is easy to show is the equal to the spontaneous radiation amplitude obtained in the perturbation expansion (4.15). To find the power spectral density, we insert $E_\nu(z)$ into Equation (4.38) and evaluate the ensemble average with the help of Equation (4.39):

$$\left.\frac{dP}{d\omega}\right|_{1D} = \left(\frac{\lambda_1^2}{\mathscr{A}_{\text{tr}}}\right)\frac{I}{I_A}\left(\frac{K[\text{JJ}]}{1+K^2/2}\right)^2\frac{\gamma^2 mc^2 z^2}{\lambda_u^2} \times \int d\eta\, V(\eta)\left\{\frac{\sin[k_u z(\eta - \Delta\nu/2)]}{k_u z(\eta - \Delta\nu/2)}\right\}^2. \tag{4.42}$$

The 1D power spectrum (4.42) is related to undulator radiation power spectral density in the forward direction through (3.114); in particular, we have

$$\left.\frac{dP}{d\omega d\boldsymbol{\phi}}\right|_{\boldsymbol{\phi}=\mathbf{0}} = \left.\frac{dP}{d\omega}\right|_{1D}\frac{\mathscr{A}_{\text{tr}}}{\lambda_1^2}, \tag{4.43}$$

and Equations (4.42) and (4.43) give the well-known formula for undulator radiation, which we derived previously as (2.100).

It is left as an exercise to show that the low-gain FEL theory can be reproduced by keeping the second term of the expansion equation (4.40) into Equation (4.35).

4.3.2 Exponential Growth Regime

In general, the dispersion relation may have a root with positive imaginary part that in turn gives rise to an exponentially growing field amplitude. In keeping with previous notation, we denote this root as μ_3, so that $\Im(\mu_3) > 0$. As $\rho k_u z$ becomes larger than unity the growing solution associated with μ_3 tends to dominate the field dynamics, in which case the field is well described by a single mode. Applying the residue theorem and keeping only the term associated with the growing root μ_3 gives

$$E_\nu(z) = \frac{e^{-2i\rho\mu_3 k_u z}}{D'(\mu_3)}\left[E_\nu(0) + \frac{i\kappa_1 n_e}{2\rho k_u N_\lambda}\sum_{j=1}^{N_e}\frac{e^{-i\nu\theta_j}}{\eta_j/\rho - \mu_3}\right], \tag{4.44}$$

where

$$D'(\mu) = \frac{dD}{d\mu} = 1 - 2\int d\eta\,\frac{V(\eta)}{(\eta/\rho - \mu)^3}. \tag{4.45}$$

The corresponding electron distribution function, which can be obtained from Equation (4.34), is

$$F_{\nu,\mu}(\eta) = \frac{i\chi_1}{2k_u}\frac{dV/d\eta}{(\nu\eta - \mu\rho)}E_{\nu,\mu}. \tag{4.46}$$

The power spectral density of the radiation in the exponential growth regime, which is computed by inserting Equation (4.44) into Equation (4.38), can be written as [2]

$$\frac{dP}{d\omega} = e^{4\mu_I\rho k_u z} g_A\left(\left.\frac{dP}{d\omega}\right|_0 + g_S\frac{\rho\,\gamma_r mc^2}{2\pi}\right), \tag{4.47}$$

where we write the imaginary part $\Im(\mu_3) \equiv \mu_I$, and the initial field and e-beam conditions are given by

$$\left.\frac{dP}{d\omega}\right|_0 \equiv \text{input power spectrum}, \tag{4.48}$$

$$g_A \equiv \frac{1}{|D'(\mu)|^2}, \tag{4.49}$$

$$g_S \equiv \int d\eta \, \frac{V(\eta)}{|\eta/\rho - \mu|^2}. \tag{4.50}$$

The quantities μ_I, g_A and g_S are all functions of the detuning $\Delta\nu$ and the electron beam energy distribution. g_A is a measure of how the initial radiation power and shot noise seed the interaction, while g_S quantifies the relative increase in shot noise seeding as the beam energy spread increases; we will discuss these coupling parameters further near the end of this section. The growth rate μ_I is a very important FEL figure of merit, as it sets the required undulator length to reach saturation. As a simple example, Figure 4.2 plots the growth rate μ_I as a function of $\Delta\nu/2\rho = \Delta\omega/(2\rho\omega_1)$ for various cases of the energy spread, assuming that the beam energy distribution is a flattop of full width $\Delta\eta = \rho\zeta$ (i.e., $V(\eta) = 1/\rho\zeta$ if $|\eta| \leq \rho\zeta/2$ and $V(\eta) = 0$ otherwise). For this distribution, evaluating the integral in the dispersion relation (4.36) leads to

$$\left(\mu - \frac{\Delta\nu}{2\rho}\right)\left(\mu^2 - \frac{\zeta^2}{4}\right) = 1, \tag{4.51}$$

and the roots have closed form expressions. It turns out that this dispersion relation has the same functional form as that obtained including the quantum effects of recoil, and in Appendix C. we write the general solution to the initial value problem for the field, bunching, and collective momentum if μ obeys (4.51). Here, we focus on the growth rate; Figure 4.2 indicates that μ has a positive imaginary part over a frequency width characterized by the high-gain FEL bandwidth $\Delta\nu \sim \rho$. Additionally, the peak growth rate is a decreasing function of the energy width of the distribution function.

For a given electron beam energy spread, it is clearly interesting to compute the maximum value of μ_I, which we take to occur at a frequency detuning $\Delta\nu_m$: $\max_{\Delta\nu} \mu_I(\nu) = \mu_{Im}$ at $\Delta\nu = \Delta\nu_m$. The growth rate at frequencies near its maximum can be approximately determined by expanding μ_I as a Taylor series in $\Delta\nu$:

$$\mu_I(\Delta\nu) = \mu_{Im} - \mu_{I2}(\Delta\nu - \Delta\nu_m)^2 + \dots . \tag{4.52}$$

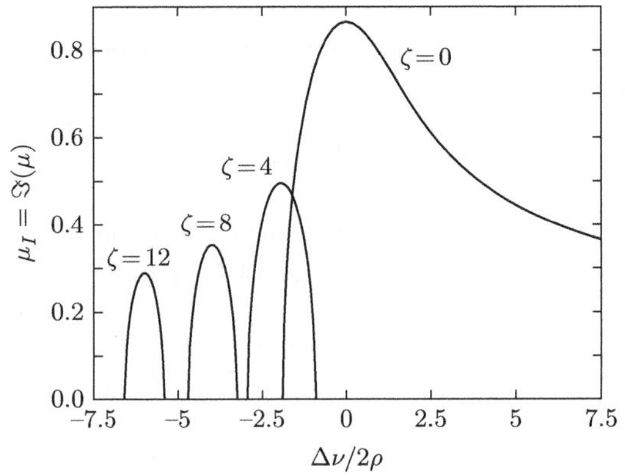

Figure 4.2 The growth rate μ_I as a function of the scaled detuning $\Delta\nu/2\rho$ for a flat-top energy distribution with the full width $\Delta\eta = \zeta\rho$, for various values of ζ. Adapted from Ref. [2].

Thus, by approximating the growth rate as a quadratic function of the frequency difference about its maximum, we may write the exponential gain function as

$$e^{4\mu_I \rho k_u z} \approx e^{z/L_G} \exp\left[-\frac{1}{2}\left(\frac{\omega - \omega_m}{\omega_m \sigma_\nu}\right)^2\right] \tag{4.53}$$

where

$$4\mu_{Im}\rho k_u z \equiv \frac{z}{L_G} \qquad \omega_m \equiv \omega_1(1 + \Delta\nu_m) \qquad \sigma_\nu^2 \equiv \frac{1}{8\mu_{I2}\rho k_u z}. \tag{4.54}$$

Here, we have defined $L_G = (4\mu_{Im}\rho k_u)^{-1}$ to be the power gain length, while $\sigma_\nu \propto (\rho k_u z)^{-1/2}$ is the RMS relative bandwidth of the FEL. Equation (4.53) approximates the frequency dependence of the FEL gain by a Gaussian function. This gain profile describes both the gain curve for coherent amplification of the initial spectral power $dP/d\omega|_0$, and the ensemble averaged spectral profile for the SASE term, since the beam shot noise acts as a white-noise (frequency-independent) seed as shown by Equation (4.47). Note the phrase "ensemble averaged": we will subsequently show that SASE behaves like the chaotic light described in Section 1.2.5, so that any particular instance will be comprised of many longitudinal modes that appear as "spikes" in the single-shot temporal and spectral power profiles.

Now, we apply the Gaussian approximation (4.53) with the definitions (4.54) to express the FEL power spectral density (4.47) as

$$\frac{dP}{d\omega} = g_A e^{z/L_G} \exp\left[-\frac{(\omega - \omega_m)^2}{2(\omega_m \sigma_\nu)^2}\right]\left(\left.\frac{dP}{d\omega}\right|_0 + g_S \frac{\rho\,\gamma_r mc^2}{2\pi}\right). \tag{4.55}$$

Equation (4.55), which is valid in the exponential growth regime $2\rho k_u z \gg 1$, shows how the power grows along the undulator.

For the special case of vanishing energy spread, the various parameters characterizing the 1D FEL have closed form solutions [2, 3]. From (4.36), the dispersion relation for a cold beam is

$$D(\mu) = \mu - \frac{\Delta\nu}{2\rho} - \frac{1}{\mu^2} = 0, \tag{4.56}$$

and the maximal growth rate is at zero detuning $\Delta\nu = 0$. It is then straightforward to show that

$$g_A = \frac{1}{9}, \qquad g_S = 1, \tag{4.57}$$

$$L_G = L_{G0} = \frac{\lambda_u}{4\pi\sqrt{3}\rho}, \qquad \mu_{Im} = \frac{\sqrt{3}}{2}. \tag{4.58}$$

These results are the same as what was predicted by the collective variable model. In addition, we can find the approximate dependence of the growth rate μ on frequency difference $\Delta\nu$ by expanding

$$\mu(\Delta\nu) \approx \mu(0) + \mu_1 \Delta\nu + \mu_2 (\Delta\nu)^2 \tag{4.59}$$

with $\mu(0) = (i\sqrt{3} - 1)/2$ and assuming that $\Delta\nu/\rho \ll 1$. In this case, we solve (4.56) order by order in $\Delta\nu/\rho$, finding that

$$\mu \approx -\frac{1}{2}\left[1 - \frac{\Delta\nu}{3\rho} + \frac{(\Delta\nu)^2}{36\rho^2}\right] + i\frac{\sqrt{3}}{2}\left[1 - \frac{(\Delta\nu)^2}{36\rho^2}\right]. \tag{4.60}$$

Hence, $\mu_{I2} = \sqrt{3}/72\rho^2$, and the RMS bandwidth for a beam with vanishing energy spread is

$$\sigma_\nu = \sigma_{\Delta\omega/\omega} = \sqrt{\frac{3\sqrt{3}\rho}{k_u z}} = \rho\sqrt{\frac{18}{N_G}} \approx \sqrt{\frac{0.83\rho}{z/\lambda_u}}, \tag{4.61}$$

where N_G is the number of power gain lengths of evolution.

In the general case, numerical calculation is necessary to obtain these quantities. Figure 4.3 shows the maximum growth rate μ_{Im} and the corresponding coupling parameters g_S and g_A as a function of the energy spread for a Gaussian energy distribution. As expected, the maximum growth rate decreases (and the growth length increases) as a function of σ_η, since electrons with different energies are resonant with different radiation frequencies. This effect for a Gaussian energy spread can be approximated as affecting the gain length via

$$L_G(\sigma_\eta) = \frac{1}{4\rho k_u \mu_{Im}(\sigma_\eta)} \approx L_{G0}\left[1 + (\sigma_\eta/\rho)^2\right], \tag{4.62}$$

and we see that the gain length is severely lengthened when the spread in energies approaches the FEL bandwidth, i.e., when $\sigma_\eta/\rho \gtrsim 1$. Note that the gain length (4.62) is chosen for an optimal detuning $\Delta\nu$, and that the optimal $\Delta\nu$ becomes more negative as σ_η increases.

Additionally, both g_A and g_S are increasing functions of σ_η. The latter g_S gives the relative strength of the shot noise seeded SASE to that produced by any coherent radiation seed, which is an important quantity if one is interested in generating longitudinally

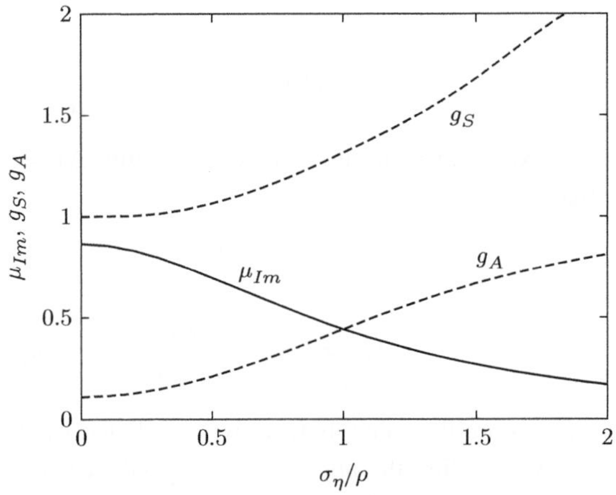

Figure 4.3 Maximum growth rate μ_I, radiation coupling g_A, and shot noise seeding parameter g_S as a function of RMS Gaussian energy spread width σ_η.

coherent FEL light using an external radiation source. As the energy spread increases, one must increase the electromagnetic seed power to overcome the fractional increase in shot noise generated SASE. For small energy spreads $\sigma_\eta \ll \rho$, the quadratic increase in g_S can be ascribed to the energy noise associated with the collective momentum $P(0)$ that we introduced in Section 3.4.4.

In the exponential growth regime, Equation (4.61) implies that the SASE bandwidth decreases as $(\lambda_u/z)^{1/2}$. Hence, during exponential growth the bandwidth decreases more slowly than it does during the spontaneous emission phase where $\Delta\nu \propto \lambda_u/z$. We plot an example illustrating this in Figure 4.4, where we see that the SASE bandwidth follows that of the undulator radiation for the first few gain lengths, after which it decreases as $(\lambda_u/z)^{1/2}$. The radiation force is set to zero for the simulation of the undulator radiation. Note the difference in bandwidth at saturation $\hat{z} \approx 10$ ($z \sim \lambda_u/\rho$) is a factor of order 2–3, whereas after exponential growth the average SASE power is $\sim 10^5$ to 10^7 times larger than that of simple undulator radiation.

Before saturation, the expression (4.55) can be used to compute the characteristics of SASE. For example, the coherence time t_{coh} can be computed from the coherence function

$$\mathcal{C}(\tau) \equiv \frac{\left\langle \int dt\; E(t)E^*(t+\tau)\right\rangle}{\left\langle \int dt\; |E(t)|^2\right\rangle} = \frac{1}{\langle P\rangle}\left\langle \int d\omega\; e^{-i\omega\tau}\frac{dP}{d\omega}\right\rangle, \tag{4.63}$$

where the second line identifies the correlation function with the Fourier transform of the power spectral density using the Weiner–Khinchin theorem. The correlation time $t_{\text{coh}} \equiv \int d\tau\; |\mathcal{C}(\tau)|^2$ is easily evaluated using (4.63) and the power spectral density (4.55); the result is quite similar to that of the chaotic light discussed in Section 1.2.5, with

$$t_{\text{coh}} = \frac{\sqrt{\pi}}{\omega_m \sigma_\nu}. \tag{4.64}$$

Additionally, the quantity $\rho\gamma_r mc^2/2\pi$ in Equation (4.47) [or (4.55)] may be interpreted as the input noise power contained in the electron beam [17]. The noise power is

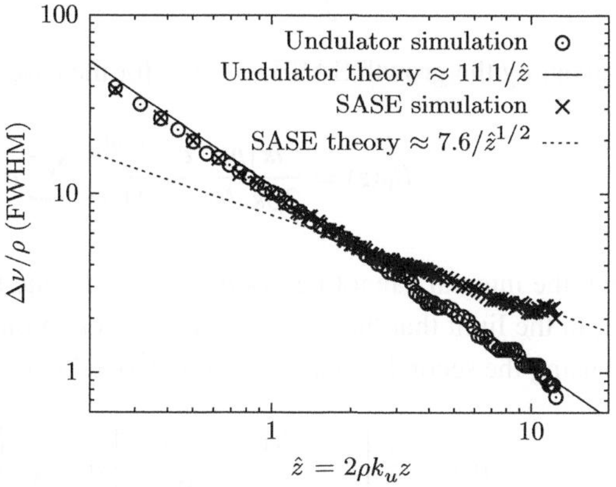

Figure 4.4 Evolution of SASE bandwidth and undulator radiation full-width, half-max (FWHM) bandwidth. The factors 11.1 and 7.6 for the theory lines come from Equations (2.74) and (4.61), respectively, using $\rho = 5 \times 10^{-4}$.

independent of frequency (white noise), and it can be shown to be equal to the power generated from spontaneous undulator radiation in two power gain lengths [5] using manipulations similar to those applied in Equations (3.113) to (3.115). Integrating the SASE term over the frequency, we obtain the electromagnetic power

$$P = \int d\omega \, \frac{dP}{d\omega} = g_S g_A \frac{\rho \gamma_r mc^2}{2\pi} \sqrt{2\pi}\,\omega_1 \sigma_\nu e^{z/L_G}$$
$$= g_S g_A \rho P_{\text{beam}} \frac{e^{z/L_G}}{\sqrt{2} N_{l_{\text{coh}}}}. \tag{4.65}$$

Here $P_{\text{beam}} = (I/e)\gamma_r mc^2$ is the e-beam power and $N_{l_{\text{coh}}} = (I/ec)l_{\text{coh}}$ is the number of electrons in one coherence length $l_{\text{coh}} \equiv ct_{\text{coh}} = \lambda_1/(2\sqrt{\pi}\sigma_\nu)$. Since we expect the saturation power to be about ρP_{beam}, the total amplification factor will be about $N_{l_{\text{coh}}}$, which is a large number whose typical magnitude is 10^5 to 10^7. Furthermore, Equation (4.65) implies that at saturation we have

$$\frac{g_A g_s}{\sqrt{2} N_\lambda} e^{z_{\text{sat}}/L_G} \sim \frac{e^{z_{\text{sat}}/L_G}}{12 N_{l_{\text{coh}}}} \sim 1 \quad \Rightarrow \quad z_{\text{sat}} \sim \frac{\ln(12 N_{l_{\text{coh}}})}{4\pi} \frac{\lambda_u}{\rho}. \tag{4.66}$$

As expected, the saturation length is proportional to λ_u/ρ. However, the fact that $z_{\text{sat}} \approx \lambda_u/\rho$ is somewhat of a numeric coincidence because $\ln(12 N_{l_{\text{coh}}}) \approx 4\pi$ over a wide range of $N_{l_{\text{coh}}}$.

4.3.3 Temporal Fluctuation and Correlation of SASE

The SASE radiation consists of a random collection of a large number of coherent pulses, much like synchrotron radiation. To see this in the time domain, we construct the temporal amplitude by Fourier transforming the field in the frequency representation,

$$E_x(z,t) = \int d\nu \; E_\nu(z) e^{i\Delta\nu[(k_1+k_u)z - \omega_1 t]} e^{i(k_1 z - \omega_1 t)}, \tag{4.67}$$

with E_ν given by the growth SASE solution for the case of vanishing energy spread

$$E_\nu(z) = \frac{i\kappa_1 n_e}{2\rho k_u N_\lambda} \frac{e^{-i\mu 2\rho k_u z}}{\mu D'(\mu)} \sum_{j=1}^{N_e} e^{-i\nu\theta_j(0)}. \tag{4.68}$$

In general, the integral cannot be evaluated exactly due to the dependence of μ on $\Delta\nu$. However, in the limit that the energy spread is negligible, an approximate result can be obtained using the second-order expansion derived in (4.60). Hence, we insert

$$\mu = -\frac{1}{2}\left[1 - \frac{\Delta\nu}{3\rho} + \frac{(\Delta\nu)^2}{36\rho^2}\right] + i\frac{\sqrt{3}}{2}\left[1 - \frac{(\Delta\nu)^2}{36\rho^2}\right] \tag{4.69}$$

into the exponential of μ, and the resulting expression is a Gaussian integral that can be done analytically. We obtain [6]

$$E_x(z,t) \propto \frac{e^{\sqrt{3}\rho k_u z}}{\sqrt{z}} \sum_{j=1}^{N_e} \exp\left\{-i\omega_1\left[t - \frac{z}{c}(1+\rho\Delta\beta) - t_j\right]\right\}$$
$$\times \exp\left\{-\frac{1+i/\sqrt{3}}{4\sigma_\tau^2}\left[t - \frac{z}{c}\left(1+\tfrac{2}{3}\Delta\beta\right) - t_j\right]^2\right\}, \tag{4.70}$$

where the normalized difference of the average electron beam velocity from unity is $\Delta\beta \equiv 1 - \bar{\beta}_z = (1+K^2/2)/2\gamma^2$, and the RMS temporal width

$$\sigma_\tau = \frac{1}{\sqrt{3}\sigma_\omega} \approx \frac{1}{2\omega_1}\sqrt{\frac{z/\lambda_u}{\rho}}. \tag{4.71}$$

The total field profile (4.70) describes a sum of N_e wave packets of RMS pulse length σ_τ that grow exponentially as they propagate. This random collection of modes has the essential properties of chaotic light, although in this case the power grows exponentially with z while its coherence length increases $\sim \sqrt{k_u z}$. Note that the relationship between the RMS temporal and spectral widths of these modes differ from the usual $\sigma_\tau\sigma_\omega = 1/2$ due to the quadratic phase dependence in (4.70). We show an example of such temporal evolution of SASE in Figure 4.5.

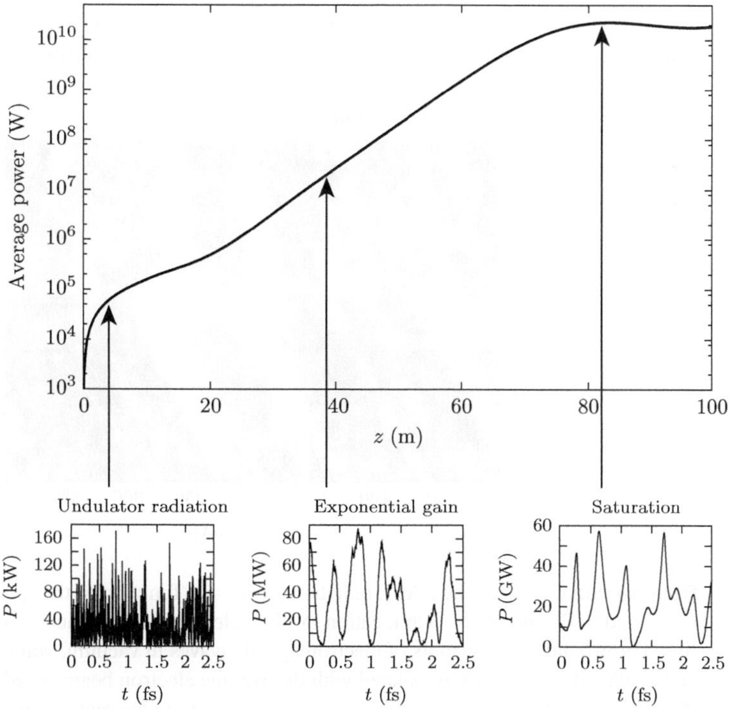

Figure 4.5 Evolution of the LCLS radiation power and temporal structure in a 1 percent time window. Courtesy of H.-D. Nuhn.

The wave packets are distributed randomly in time in a manner quite similar to undulator radiation. The phase velocity is less than the speed of light by the small factor $\rho\Delta\beta$, and it is interesting to note that the group velocity of each wave packet/temporal mode is [7]

$$v_g = \frac{c}{1+2\Delta\beta/3} \approx c\left(1-\tfrac{2}{3}\Delta\beta\right). \tag{4.72}$$

The group velocity is slightly faster than the electrons (by $\Delta\beta/3$), but slower than c because the growing radiation mode is shaped by the FEL gain. Since the FEL gain is tied to the local electron bunching that moves with velocity $\bar{v}_z$, the gain tends to follow the electron beam; the interplay of radiation slippage and FEL gain leads to group velocity (4.72). Figure 4.6 plots simulation results that confirm this interesting property in the exponential growth regime. Additionally, we see that the group velocity of these wave packets is about equal to c after saturation, when the coupling between the radiation and the electron beam is greatly reduced.

As previously mentioned (and shown in Figures 4.5–4.6), the temporal modes of SASE are randomly distributed over the pulse. This is because SASE is initialized by fluctuations in the beam current attributable to the discrete nature of the electron, namely, the shot noise. Thus, SASE is an example of a partially coherent wave, and its temporal fluctuations are those of chaotic light that we discussed in Section 1.2.5. Specifically, there are about $M \sim T/t_{\text{coh}}$ temporal (and spectral) modes or spikes, and the integrated energy fluctuates by a relative amount $1/\sqrt{M}$ from shot to shot. A detailed description of SASE beyond these simple characteristics can be obtained using the techniques of statistical optics described in, e.g., [8]; such methods have been applied to

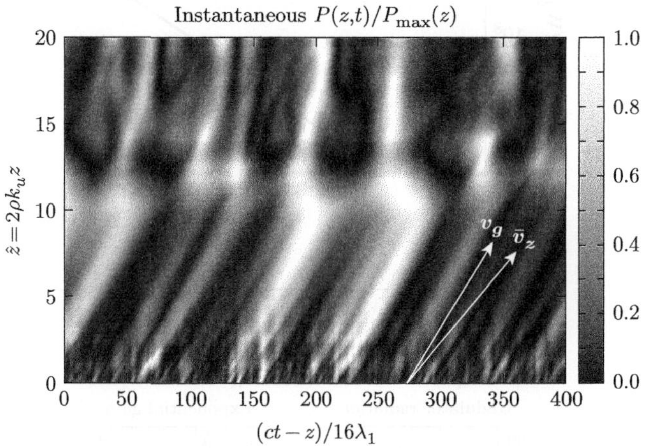

Figure 4.6 Simulation of the SASE radiation power as a function of the speed of light coordinate $ct - z$ and $\hat{z}$. The power at each location in z is scaled by the maximum at that location. Vertical lines correspond to wavefronts of electromagnetic waves in vacuum, while the arrows labeled $\bar{v}_z$ and v_g identify the slopes associated with the average electron beam speed along z and the theoretical SASE group velocity (4.72), respectively. The coherent regions move at the group velocity until saturation near $\hat{z} \sim 10$, after which the radiation becomes nearly uncoupled from the beam and moves approximately at the speed of light. Courtesy of W. Fawley.

SASE light in Refs. [9, 10], among others. Here, we discuss a few such properties in a little more detail.

The characteristic time scale of the temporal fluctuations in SASE power is given by the coherence length which we derived in (4.64):

$$t_{\text{coh}} = \frac{\sqrt{\pi}}{\sigma_\omega} \to \frac{3}{\sqrt{2\pi z/L_{G0}}}\frac{\lambda_1}{c\rho} \sim \frac{\lambda_1}{c\rho}, \tag{4.73}$$

where the last two expressions have inserted the cold, 1D result and then assumed a propagation distance of order a few gain lengths. The field envelope in any given SASE pulse varies over times $\sim t_{\text{coh}}$ as seen in Figure 4.5, while the position and height of the intensity peaks are completely uncorrelated from pulse to pulse. These latter variations lead to shot-to-shot fluctuations in the SASE pulse energy which depend on both the coherence time and the temporal duration of the pulse T. To understand these fluctuations, we first consider a single SASE pulse. Using the fact that the electric field is approximately constant over the temporal duration $T \ll t_{\text{coh}}$, it can be shown [11, 12] that the energy contained within a time interval much less than the coherence time is described by the negative exponential distribution, so that in this limit the probability $p(U)$ to measure the energy U over the time T is

$$T \ll t_{\text{coh}}: \;\; p(U) = \frac{1}{\langle U\rangle}\exp\left(-\frac{U}{\langle U\rangle}\right), \tag{4.74}$$

where $\langle U\rangle$ is the average energy contained in the time T. Equation (4.74) applies if $T \ll t_{\text{coh}}$, for which the field will comprise one longitudinal mode, i.e., $M = 1$. On the other hand, in the limit $M \to \infty$ ($T \to \infty$), the energy will be normally distributed in accordance with the central limit theorem. A probability distribution that interpolates between these two limiting forms is the Gamma probability distribution; this line of reasoning underlies the suggestion in Ref. [9] that the statistics of the energy U in a flat-top SASE pulse with duration T are governed by the Gamma distribution

$$p(U) = \frac{M^M}{\Gamma(M)}\frac{U^{M-1}}{\langle U\rangle^M}\exp\left(-M\frac{U}{\langle U\rangle}\right). \tag{4.75}$$

Here, $\langle U\rangle$ is the ensemble average of the electromagnetic energy for a SASE pulse of duration T, while $\Gamma(M)$ is the Gamma function. Using the properties of the Gamma distribution, M is related to the relative RMS fluctuation in energy σ_U by [6, 9, 13]

$$M = \frac{1}{\sigma_U^2} = \frac{\langle U\rangle^2}{\langle U^2\rangle - \langle U\rangle^2} \approx \begin{cases} T/t_{\text{coh}} & \text{if } T \gg t_{\text{coh}} \\ 1 & \text{if } T \le t_{\text{coh}} \end{cases}. \tag{4.76}$$

Thus, M characterizes the number of longitudinal degrees of freedom (or “modes”) in the pulse, with (4.75)–(4.76) giving its more formal definition. Extensive simulations [9] have shown that (4.75) describes the energy probability distribution very well even up to saturation, while a more thorough statistical analysis has shown that the average number of intensity spikes in the time domain is about $0.7M$ [10].

For hard X-ray wavelengths, the coherence time determined by (4.73) is only of order a few hundred attoseconds, while the SASE pulse duration T is dictated by the length of the electron beam. Typically, the electron beam is between ten and a few hundred femtoseconds in length, so that $M \gg 1$ and the Gamma distribution of shot-to-shot pulse energies approaches a Gaussian distribution with a small relative RMS fluctuation given by $1/\sqrt{M}$. On the other hand, if the electron beam length is comparable to the coherence length so that $T \lesssim t_{\text{coh}}$, than the resulting FEL radiation will be comprised of one longitudinally coherent mode at the expense of shot-to-shot stability: the energy variations from pulse to pulse will approach 100 percent.

The frequency domain exhibits very similar statistical properties. The full SASE bandwidth is about $2\sqrt{\pi}\sigma_\omega$, within which $\sim M$ independent spectral modes of width $2\pi/T$ are randomly distributed. At saturation where $z \sim \lambda_u/\rho$, the full bandwidth for a monoenergetic beam is $\sqrt{6\sqrt{3}}\rho\omega_1 \approx 3.2\rho\omega_1$. Additionally, it is interesting to consider the effect of a monochromator on the radiation statistics. To illustrate the basic physics, here we consider a single SASE pulse that initially comprises many longitudinal modes, so that $M \gg 1$ and $T \gg t_{\text{coh}}$; more complete treatments can be found in Refs. [9, 10]. Passing the SASE pulse through a monochromator selects a certain frequency bandwidth in an analogous manner as the time interval T identifies a temporal region. If we denote the RMS bandwidth of the monochromator by σ_m, after the monochromator the average pulse energy $\langle U\rangle_{\Delta\omega} \approx (\sigma_m/\sigma_\omega)\langle U\rangle$ if $\sigma_m \lesssim \sigma_\omega$, and $\langle U\rangle$ otherwise. Assuming that the monochromator bandwidth is less than that of the SASE, $\sigma_m \lesssim \sigma_\omega$, the monochromator reduces the spectral bandwidth and the number of spectral modes is given by

$$M_F \approx \begin{cases} \dfrac{\sigma_m}{\sigma_\omega}\dfrac{T}{t_{\text{coh}}} & \text{if } \sigma_m \gg 2\pi/T \\ 1 & \text{if } \sigma_m \leq 2\pi/T \end{cases}. \tag{4.77}$$

Thus, the radiation after the monochromator has a narrower bandwidth but larger energy fluctuations from pulse to pulse than the original SASE.

The statistical fluctuations discussed here can be generalized to three dimensions by redefining the total number of modes via $M = M_L M_T^2$, where M_L is the longitudinal mode number (whose limiting values are given by (4.76)), and M_T^2 are the number of transverse modes. Initially, the undulator radiation is also composed of many transverse modes $M_T^2 \gg 1$, so that in 3D the initial fluctuation level is relatively small. Since there is typically one transverse mode with the largest FEL gain, however, the exponential growth tends to preferentially select that single transverse mode, and $M_T \to 1$ after several gain lengths. Thus, the SASE fluctuations near saturation are largely governed by the 1D longitudinal statistics that we described above.

4.4 Quasilinear Theory and Saturation

The exponential field growth cannot continue indefinitely, because one can only extract a finite amount of energy from the electron beam. In the linear theory, the average beam energy is unchanged to first order in $E_\nu(z)$ because the electromagnetic energy is a second-order quantity. Since saturation is a nonlinear phenomenon, it is difficult to obtain analytic results, and simulations are typically required to determine the dynamics of saturation. Nevertheless, there are certain physics we can describe both qualitatively and quantitatively.

We begin this section by discussing the quasilinear treatment of saturation [2, 14], which extends the linear analysis of Section 4.3 by allowing the background distribution function $V(\eta)$ to become a slow function of the electromagnetic power. Since V now varies in z, the energy exchange from the electrons to the field slows and eventually stops, and quasilinear theory provides a useful picture of the dynamics up to saturation. After showing some numerical examples of this model, we then briefly compare the quasilinear theory to other theories, and finally discuss other general qualitative features of saturation physics by referencing simulation results.

The quasilinear theory of saturation obtains by relaxing our assumption that the background distribution function V is independent of the evolution coordinate z, i.e., by taking $V(\eta) \to V(\eta; z)$. We derive the quasilinear equations in a rather general manner which can also be used to study a temporally varying electron beam. It is convenient to introduce the dimensionless coordinates

$$\hat{z} \equiv 2k_u\rho z \qquad \hat{\eta} \equiv \frac{\eta}{\rho} \tag{4.78}$$

which, as indicated in Section 3.4.3, will typically be $O(1)$ for high-gain FELs. These variables will also be useful when we study the 3D theory. Additionally, we define the scaled field variables

$$a_\nu \equiv \frac{eK[\mathrm{JJ}]}{4\gamma_r^2 mc^2 k_u \rho^2} E_\nu = \frac{\chi_1}{2k_u\rho^2} E_\nu \qquad f \equiv \rho F. \tag{4.79}$$

The electromagnetic scaling was first introduced in Section 3.4.3, while the second equation for f insures that the normalization $\int d\eta\; F = \int d\hat{\eta}\; f$. In terms of these variables, the Klimontovich equation for the distribution function is

$$\left\{ \frac{\partial}{\partial \hat{z}} + \hat{\eta}\frac{\partial}{\partial \theta} + \int d\nu\; \left[a_\nu(z) e^{i\nu\theta} + c.c. \right] \frac{\partial}{\partial \hat{\eta}} \right\} f(\theta, \hat{\eta}; \hat{z}) = 0. \tag{4.80}$$

Using the Fourier representation of f and applying $\int d\theta\; e^{-i\nu\theta}$ to (4.80), we find

$$\begin{aligned} \left[\frac{\partial}{\partial \hat{z}} + i\nu\hat{\eta} \right] f_\nu(\eta; z) &+ \int d\nu'\; a_{\nu'}(z) \frac{\partial}{\partial \hat{\eta}} f_{\nu-\nu'}(\hat{\eta}; \hat{z}) \\ &+ \int d\nu'\; a^*_{\nu'}(z) \frac{\partial}{\partial \hat{\eta}} f_{\nu+\nu'}(\hat{\eta}; \hat{z}) = 0. \end{aligned} \tag{4.81}$$

We know that the electromagnetic field is resonantly driven near the fundamental frequency, implying that $a_{\nu'}$ is large for only a small region about $\nu' = 1$; typically $|\nu' - 1| \lesssim \rho \ll 1$, while the power in the harmonics is at most a few percent. In this case, the second line in (4.81) predominantly describes the coupling of the radiation to the higher-order harmonics of the distribution function when $\nu \approx 1$. Neglecting these harmonic contributions, which are typically unimportant before saturation, we have

$$\left[\frac{\partial}{\partial \hat{z}} + i\hat{\eta} \right] f_\nu(\hat{\eta}; \hat{z}) + \int d\nu'\; a_{\nu'}(z) \frac{\partial}{\partial \hat{\eta}} f_{\nu-\nu'}(\hat{\eta}; \hat{z}) = 0. \tag{4.82}$$

Equation (4.82) applies if $\nu \approx 1$ and $a_{\nu'}$ is localized about the fundamental $\nu' \approx 1$, so that $\nu - \nu' \approx 0$ and the convolution couples only to the low-frequency (long-wavelength) components of the distribution function. In other words, (4.82) is the generalization of Equation (4.47) that allows the background distribution function V to be a function of θ (giving a finite shape to the electron beam) and z (implying that V can now dynamically evolve). Note that the approximation leading from (4.81) to (4.82) breaks down at and after saturation, when the other harmonic components f_ν become significant. For this reason, the quasilinear approximation cannot be used after saturation.

To simplify our description of saturation physics, we will again assume that the background is independent of phase. Thus, the quasilinear equations describing saturation obtain by taking the smooth background distribution $f_{\nu\approx 0}$ to be uniform in θ but now dependent on z:

$$f_{\nu\approx 0}(\hat{\eta};\hat{z}) = \delta(\nu)V(\hat{\eta};\hat{z}). \tag{4.83}$$

Inserting the uniform background (4.83) into the expression (4.82), the micro-distribution function for frequencies near the fundamental is governed by

$$\left[\frac{\partial}{\partial \hat{z}} + i\nu\hat{\eta}\right] f_\nu(\eta;z) + a_\nu(z)\frac{\partial}{\partial \hat{\eta}}V(\hat{\eta};\hat{z}) = 0. \tag{4.84}$$

While the evolution of V can be obtained from the $\nu = 0$ component of (4.82)[1], it is more straightforward to integrate (4.80) over θ. Using the facts that $\int d\theta f = (2\pi cT/\lambda_1)V$ and that $f_\nu = f^*_{-\nu}$ (since f is real), averaging Equation (4.80) over phase gives

$$\frac{\partial}{\partial \hat{z}}V(\hat{\eta};\hat{z}) + \frac{\lambda_1}{cT}2\Re \int d\nu\; a^*_\nu \frac{\partial f_\nu}{\partial \hat{\eta}} = 0. \tag{4.85}$$

Finally, the Maxwell equation (4.12) in scaled variables is

$$\left[\frac{\partial}{\partial \hat{z}} + \frac{i(\nu-1)}{2\rho}\right] a_\nu(\hat{z}) = -\int d\hat{\eta}\, f_\nu(\hat{\eta};\hat{z}). \tag{4.86}$$

The Equations (4.84) and (4.86) recover the linearized Maxwell–Klimontovich equations of Section 4.1 when the field a_ν is sufficiently small, in which case (4.85) implies that V is constant. As the field grows, the background distribution changes with $\hat{z}$ such that the average beam energy decreases while its energy spread increases. The latter of these reduces the effective growth rate of the field and ultimately limits the FEL power such that $(\lambda_1/cT)\int d\nu\; |a_\nu|^2 \sim 1$, which in physical units implies a saturation power $P \sim \rho P_{\text{beam}}$.

An important characteristic of the quasilinear equations (4.84)–(4.86) is that they conserve the total (field + particle) energy, which is in contrast to the linear system (4.12) and (4.11). This can be shown by calculating the change in kinetic energy

$$\begin{aligned}\frac{d}{d\hat{z}}\int d\eta\eta V = \int d\hat{\eta}\, \frac{\partial V}{\partial \hat{z}}\hat{\eta} &= -\frac{\lambda_1}{cT}\int d\hat{\eta}\; 2\Re \int d\nu\; a^*_\nu \frac{\partial f_\nu}{\partial \eta}\hat{\eta} \\ &= -\frac{\lambda_1}{cT}\frac{d}{d\hat{z}}\int d\nu\; |a_\nu|^2 .\end{aligned} \tag{4.87}$$

[1] Multiply (4.82) by $\delta(\nu)$, integrate over ν and then identify the resulting $\delta(0)$ with cT/λ_1.

Hence, the sum of the kinetic and field energies is constant, and total energy is conserved.

One can calculate higher moments of the distribution (such as the energy spread), or, better, determine the background energy distribution $V(\hat{\eta}, \hat{z})$ by solving the quasilinear equations numerically. To simplify the equations, we assume that the evolution of V occurs over a much longer time scale than that given by the field growth rate, so that the field variables are well approximated by the growing mode:

$$a_\nu(z) = a_\nu(0)\exp\left[-i\int_0^{\hat{z}} d\hat{z}'\, \mu_\nu(\hat{z}')\right] \tag{4.88}$$

$$f_\nu(z) = f_\nu(0)\exp\left[-i\int_0^{\hat{z}} d\hat{z}'\, \mu_\nu(\hat{z}')\right], \tag{4.89}$$

where $\mu_\nu(\hat{z})$ has a positive imaginary part (growth rate) that we will find satisfies the 1D dispersion relation (4.36) in the linear limit. By using the *ansatz* (4.88)–(4.89), the two equations (4.84) and (4.86) can be combined into a single equation for $\mu_\nu(\hat{z})$, while (4.85) is simplified:

$$\mu_\nu(\hat{z}) - \frac{\nu - 1}{2\rho} = \int d\hat{\eta}\, \frac{V(\hat{\eta};\hat{z})}{[\hat{\eta} - \mu_\nu(\hat{z})]^2} \tag{4.90}$$

$$\frac{\partial}{\partial \hat{z}} V(\hat{\eta};\hat{z}) = -\frac{2\lambda_1}{cT}\int d\nu\, \left|a_\nu(\hat{z})\right|^2 \frac{\partial}{\partial \hat{\eta}}\left\{\Im\left[\frac{\partial V/\partial \hat{\eta}}{\mu_\nu(\hat{z}) - \hat{\eta}}\right]\right\}. \tag{4.91}$$

As advertised, Equation (4.90) looks just like the usual dispersion relation for μ, although now it is a function of $\hat{z}$ determined by the slowly varying distribution V. The second equation (4.91) describes the evolution of V due to the FEL interaction up to saturation.

The simplest example illustrating the dynamics of (4.90) and (4.91) is when the radiation is monochromatic, so that $a_\nu = A(\hat{z})\delta(\nu - 1)$. Then, we have $|a_\nu|^2 = A^2(cT/\lambda_1)\delta(\nu - 1)$, and the quasilinear equations become

$$\mu(\hat{z}) - \frac{\nu - 1}{2\rho} = \int d\hat{\eta}\, \frac{V(\hat{\eta};\hat{z})}{[\hat{\eta} - \mu(\hat{z})]^2} \tag{4.92}$$

$$\frac{\partial}{\partial \hat{z}} V(\hat{\eta};\hat{z}) = -2\left|A(\hat{z})\right|^2 \frac{\partial}{\partial \hat{\eta}}\left\{\Im\left[\frac{\partial V/\partial \hat{\eta}}{\mu(\hat{z}) - \hat{\eta}}\right]\right\}. \tag{4.93}$$

Saturation is defined when the V evolves such that the growth rate $\Im(\mu) = 0$, which from (4.91) implies that $\partial V/\partial \hat{z} = 0$ and we reach an equilibrium with constant power.

The set of Equations (4.92)–(4.93) can be easily solved numerically given the initial conditions. In the following example, we present the single-frequency quasilinear evolution using the initial condition $A(0) = 2 \times 10^{-3}$ and assuming that the field is on-resonance with $\nu = 1$. Furthermore, we initialize V using a Gaussian energy distribution with RMS energy spread equal to 0.1ρ, so that $\langle\hat{\eta}^2\rangle^{1/2} = 0.1$. In Figure 4.7 we show that the power saturates when $\Im(\mu) = 0$, at which point the energy spread has increased to about ρ while the radiation power $|A|^2 \sim 1$. Furthermore, we plot the initial and final background distribution V in Figure 4.7(c). One can see that the final energy distribution

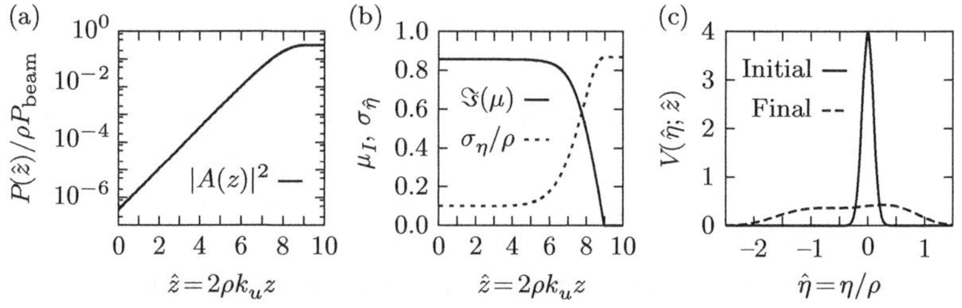

Figure 4.7 Dynamics of the quasilinear equations for monochromatic radiation. Plot (a) shows that the power initially grows exponentially, but slows and then stops as $P \to \rho P_{\text{beam}}$. At this point the growth rate has decreased from its linear value $\Im(\mu) \approx \sqrt{3}/2$ to zero, while the normalized energy spread increases toward unity as shown in (b). Panel (c) shows the background distribution function initially and at saturation where $\hat{z} \approx 9$.

is not only broader (with a larger energy spread), but also becomes flat in the resonant region $\hat{\eta} \approx \Re[\mu(0)] \approx -0.5$. This is indicative of particle trapping at saturation.

In addition to describing the growth of a single frequency component, the quasilinear equations can also be used to illustrate the ensemble averaged growth and saturation of self-amplified spontaneous emission. To study the saturation of SASE, we assume that $V(\hat{\eta};\hat{z}=0)$ is given, and that V is statistically stationary at all $\hat{z}$. This is a good approximation for the long and uniform electron beams that we study here, because the evolution of V is determined by the total radiation power whose relative fluctuation is of order $1/\sqrt{M_L}$, where M_L is the number of longitudinal modes. For long beams the radiation is characterized by many longitudinal modes, and $\langle V(\hat{\eta};\hat{z})\rangle \approx V(\hat{\eta};\hat{z})$. Thus, the ensemble average of (4.90) is trivial, while the only statistically varying part of (4.91) comes from the fluctuating radiation field. To evaluate this, we use the definition of a_ν to find that

$$\frac{\lambda_1}{cT}\left\langle |a_\nu(\hat{z})|^2\right\rangle = \frac{\lambda_1}{cT}\left\langle |a_\nu(0)|^2\right\rangle e^{\int d\hat{z}'\,\Im[2\mu_\nu(\hat{z}')]} = \frac{1}{\rho P_{\text{beam}}}\frac{dP(\hat{z})}{d\nu}. \tag{4.94}$$

Comparing (4.94) to our result (4.47) for the ensemble averaged initial SASE seeding, we find that the equation for V is given by

$$\frac{\partial}{\partial \hat{z}}V(\hat{\eta};\hat{z}) = \frac{2g_A g_s}{N_\lambda}\int d\nu\; e^{\int d\hat{z}'\,\Im[2\mu_\nu(\hat{z}')]}\frac{\partial}{\partial \hat{\eta}}\left[\Im\left(\frac{\partial V/\partial \hat{\eta}}{\mu-\hat{\eta}}\right)\right], \tag{4.95}$$

where g_s and g_A are $\sim O(1)$ constants that were defined with respect to $V(\hat{\eta};0)$ in Equations (4.49) and (4.50), and plotted in Figure 4.3. The integrand is nonzero for those frequencies which have positive growth rates, which are typically given by those within the FEL bandwidth $|\nu - 1| \lesssim \sigma_\nu$. Furthermore, the right-hand side of (4.95) becomes of order one when

$$\frac{g_A g_s}{N_\lambda}\sigma_\nu e^{2\mu\hat{z}} \sim \frac{1}{10N_{l_{\text{coh}}}}e^{2\mu\hat{z}} \sim 1 \quad \Rightarrow \quad \frac{z}{L_G} \sim \ln(10N_{l_{\text{coh}}}), \tag{4.96}$$

where $N_{l_{\text{coh}}}$ is number of electrons in one coherence length. Thus, the saturation distance is a numerical factor times the gain length. The numerical factor depends only

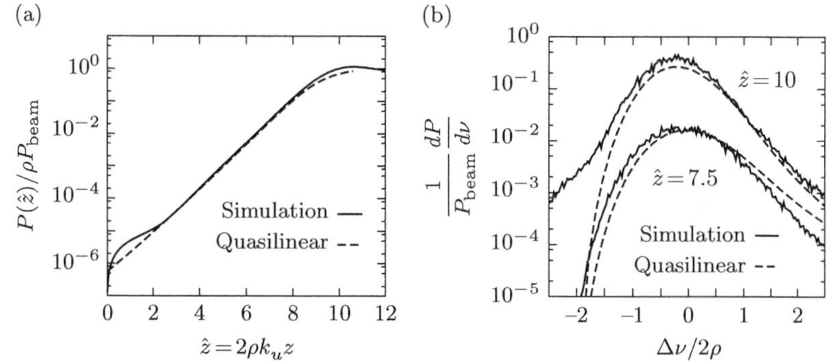

Figure 4.8 Power evolution in SASE. Panel (a) compares the total (integrated) ensemble averaged power predicted by quasilinear theory to that obtained via 1D simulations, showing nearly identical linear evolution, with reasonable agreement for the saturation length and power. Panel (b) shows the ensemble averaged power spectral density near the end of the linear stage $\hat{z} = 7.5$ and near saturation $\hat{z} = 10$. In the linear stage the dynamics are very well predicted, while the qualitative features of saturation are also reproduced. At $\hat{z} = 10$ the integrated power from quasilinear theory is a little more than one-third that obtained by the 1D simulation.

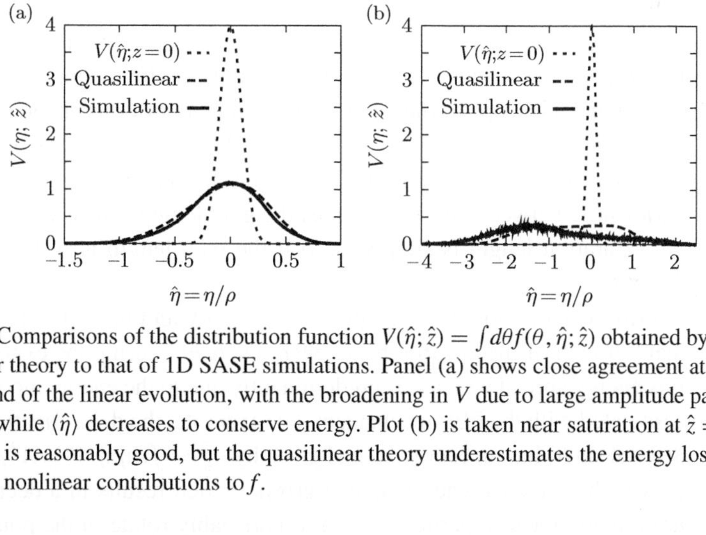

Figure 4.9 Comparisons of the distribution function $V(\hat{\eta}; \hat{z}) = \int d\theta f(\theta, \hat{\eta}; \hat{z})$ obtained by quasilinear theory to that of 1D SASE simulations. Panel (a) shows close agreement at $\hat{z} = 7.5$ near the end of the linear evolution, with the broadening in V due to large amplitude particle bunching while $\langle\hat{\eta}\rangle$ decreases to conserve energy. Plot (b) is taken near saturation at $\hat{z} = 10$. The agreement is reasonably good, but the quasilinear theory underestimates the energy loss by neglecting nonlinear contributions to f.

logarithmically on the number of electrons in one coherence length, and typically $\ln(10N_{l_{coh}}) \sim 16$ to 20.

We show the ensemble-averaged dynamics of SASE predicted by the quasilinear theory in Figures 4.8 and 4.9, along with results obtained by averaging 100 runs in a 1D FEL code. Here, we again use an initial energy spread $\sigma_{\hat{\eta}} = 0.1$, and choose $N_\lambda = 1250$ and $\rho = 5 \times 10^{-4}$. Figure 4.8 compares the power evolution, with (a) showing the total integrated power, while (b) plots the ensemble averaged power spectral density at two undulator positions. The power evolution predicted by the quasilinear theory is quite similar to that shown in the simulation up to saturation, with both yielding a saturation length of $\hat{z} \approx 10$. After exponentiating by $\sim 10^6$, the quasilinear approach well predicts the saturation length, while differing in total power by a factor of 3. In Figure 4.8(b) we plot the power spectral density at the end of the linear evolution $\hat{z} = 7.5$ and near saturation $\hat{z} = 10$. Again, the linear agreement is very good, with the ensemble averaged power spectral density growing according to the growth rate $\Im[\mu(\nu)]$. While the qualitative features at saturation are described reasonably well, the saturated SASE spectrum is broader due to nonlinear effects.

To illustrate the quasilinear saturation from another perspective, we plot the background distribution $V(\hat{\eta}; \hat{z})$ at two locations in Figure 4.9. Again, panel (a) shows that evolution up to the end of the linear regime at $\hat{z} = 7.5$ is very well described by the quasilinear theory. Note that while the growth rate has changed little, V has undergone significant evolution, having both broadened and decreased in mean energy. Panel (b) shows V near saturation $\hat{z} = 10$. While the agreement is still rather good, the simulation indicates a wider V with more energy loss. The discrepancy can largely be attributed to the harmonic components of the highly nonlinear distribution function.

As we have shown, the quasilinear theory gives a reasonably good picture of the evolution up to saturation. However, a complete description of the nonlinear physics after saturation would also include the higher-order harmonics of the distribution function. Other analytic results regarding saturation have been derived from the 1D steady-state equations in the time domain (see, e.g., [15, 16, 17, 18, 19]); these approximate models typically join some form of the linear, collective description of the FEL to certain nonlinear dynamics associated with saturation (like particle trapping). While these analyses are beyond the scope of this book, in the remainder of this chapter we will discuss some of the important general physics regarding FEL saturation. As we mentioned in (4.96) and the discussion thereafter, the saturation distance is a numerical factor whose value varies little between 18 and 20 times the gain length. Since we also have $4\pi\sqrt{3} \approx 20$, the saturation length is simply λ_u/ρ if the gain length is that of cold 1D theory, i.e., $L_G \approx L_{G0}$.

Prior to saturation both the quasilinear theory and the 1D simulations display significant growth of the electron beam energy spread. During the exponential growth phase this energy spread increase is predominantly due to the energy modulation and bunching associated with the FEL mechanism. For example, the delta-function energy modulation $\delta(\hat{\eta} - A\sin\theta)$ has an energy spread given by $\sigma_{\hat{\eta}}^2 = \int d\hat{\eta}\, \hat{\eta}^2 V = A^2/2$. As the modulation grows, the effective energy spread grows, which results in a decrease in gain. As saturation approaches, particles begin to noticeably rotate in the ponderomotive potential formed by the radiation and undulator fields. At this point, higher harmonic components of the distribution function become more important as particles begin to rotate in phase space, executing the so-called synchrotron oscillations that we show in Figure 4.10.

After saturation, the periodic gain and loss of particle energy associated with the synchrotron motion results in oscillations of the radiation energy over many undulator periods. This can in turn resonantly couple the FEL radiation at frequency ω to fields at frequency $\omega \pm c\Omega_s$, which can continuously extract energy from the mode of interest. This is known as the sideband instability, which is further explained in, e.g., [20, 21, 22]. Further analytic progress in the saturation regime is generally not possible, and simulations are typically used to help describe and predict the dynamics.

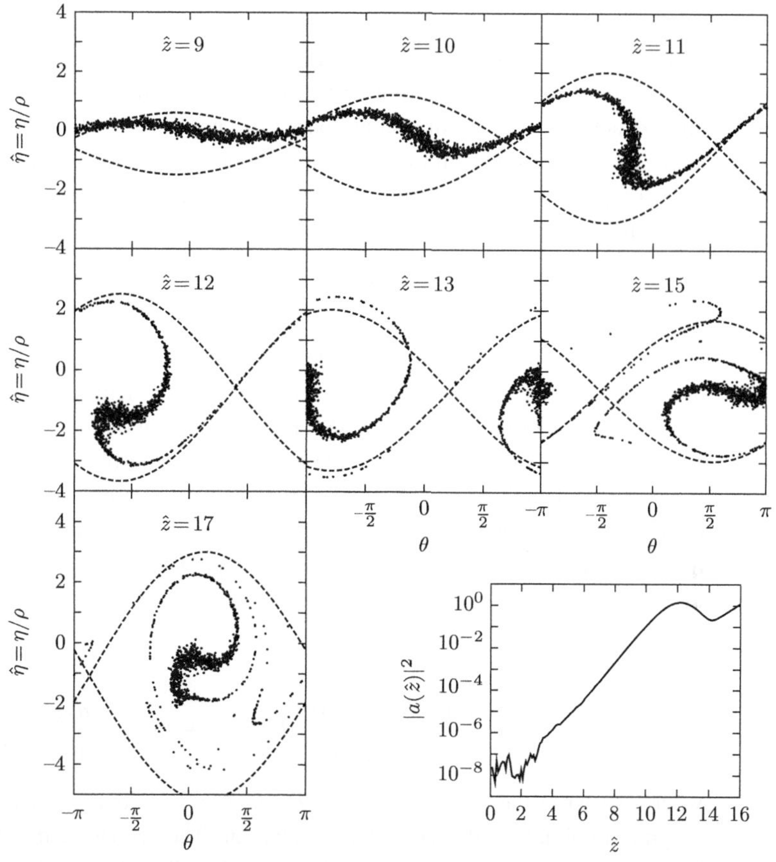

Figure 4.10 Evolution of the particle distribution during FEL gain and saturation. The lower-right panel shows the power evolution as a function of the scaled distance $\hat{z}$. The remaining panels show particle phase space portraits at various location along the undulator, for an initial energy spread $\sigma_{\hat{\eta}} = \sigma_\eta/\rho = 0.1$. The electron positions are plotted as the black dots, while the dashed lines denote the separatrices of the ponderomotive potential formed by the instantaneous value of the complex electric field. The separatrices partition phase space into three regions: the two passing regions which lie above and below the separatrices and the trapped region that lies in between. In this example, saturation occurs at $\hat{z} \approx 12$ ($z \approx 21L_G$), after which time the field amplitude oscillates in z as the trapped particles rotate in the ponderomotive potential.

4.5 Undulator Tapering after Gain Saturation

At first glance, particle trapping in the saturation regime appears to prevent further energy extraction from the electron beam. Electrons alternately lose and then gain energy as they make synchrotron oscillations in the ponderomotive bucket, so that on average the field energy stays approximately constant. However, Kroll, Morton and Rosenbluth (hereafter referred to as KMR) showed in Ref. [23] that it is possible to extract more energy from the e-beam by varying the undulator parameter K after saturation. This extra field energy is generated by effectively adjusting the undulator parameter K such that the FEL resonance condition is maintained even as electrons lose energy.

In order to extract more energy one must decrease K, so that this process is referred to as undulator tapering. FEL experiments from microwave frequencies to hard X-rays have demonstrated that tapering an undulator substantially increases the output power and extraction efficiency of a single-pass FEL amplifier [24, 25, 26]. In this section, we apply the KMR formalism to analyze undulator tapering for a high-gain FEL.

In a tapered undulator where the undulator parameter K is a function of z, we can define the resonant energy $\gamma_r mc^2$ via

$$\frac{1 + K(z)^2/2}{2\gamma_r(z)^2} = \frac{\lambda_1}{\lambda_u} = \text{constant}. \tag{4.97}$$

We emphasize that the right-hand side of the above equation is a constant, since λ_1 is determined in the exponential growth regime, and is defined to be equal to the central wavelength of the seeded or SASE FEL. Following our derivation of the pendulum equations in Section 3.3.1, we define θ as the electron phase relative to the resonant wave of wavelength λ_1, and $\eta = (\gamma - \gamma_r)/\gamma_r$ as the relative energy deviation from the resonant energy. We also assume the total energy change is still a small fraction of the initial beam energy, so that $\gamma_0 = \gamma_r(z = 0) \approx \gamma_r$. The electron's longitudinal motion is then given by

$$\frac{d\theta}{dz} = 2k_u\eta\,, \tag{4.98}$$

$$\frac{d\eta}{dz} = -\frac{eK[\text{JJ}]}{2\gamma_0^2 mc^2}|E|\sin(\theta + \phi) - \frac{1}{\gamma_0}\frac{d\gamma_r}{dz}\,, \tag{4.99}$$

where we write the complex field as $E = |E|\,e^{i\phi}$, so that $|E|$ and ϕ are the slowly-varying amplitude and phase of the radiation. The second term on the right side of Equation (4.99) is a prescribed function that describes the undulator tapering in terms of its effect on the resonant energy.

Following KMR [23], we define the synchronous phase ψ_r via

$$-\frac{1}{\gamma_0}\frac{d\gamma_r}{dz} = \frac{eK[\text{JJ}]}{2\gamma_0^2 mc^2}|E|\sin\psi_r. \tag{4.100}$$

Note that when $\theta = \psi_r - \phi$, the right-hand side of (4.99) vanishes, η is a constant, and $d\gamma/dz = d\gamma_r/dz$. A solution that is consistent with this and Equation (4.98) has $\gamma = \gamma_r(z)$, in which case the electron energy loss follows the externally prescribed undulator taper. Hence, the synchronous phase plays an important role in the physics of post-saturation undulator tapering.

In terms of the synchronous phase (4.100), the dynamical equations (4.98)–(4.99) are governed by the Hamiltonian

$$H = k_u\eta^2 - \frac{eK[\text{JJ}]}{2\gamma_0^2 mc^2}|E|\left[\cos(\theta + \phi) + \theta\sin\psi_r\right]. \tag{4.101}$$

If we compare this H to the Hamiltonian in Equation (3.23), we see (4.101) has the additional term $\theta\sin\psi_r$ due to the change of the synchronous phase with K. The Hamiltonian (4.101) is formally the same as that which describes the longitudinal/synchrotron motion in an rf accelerator [27].

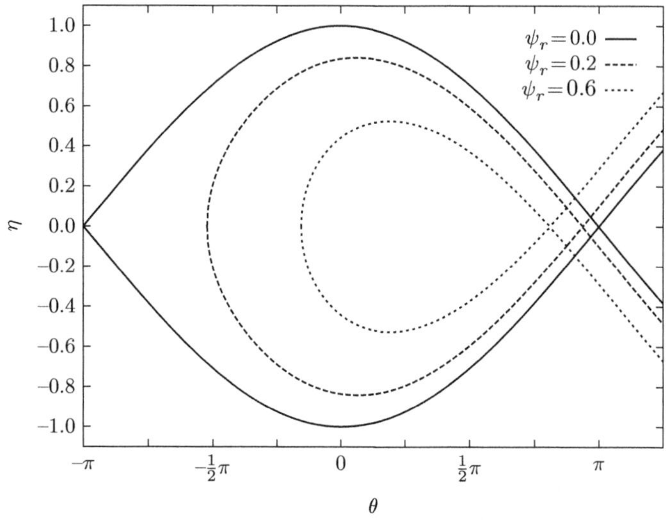

Figure 4.11 FEL ponderomotive bucket for trapping electrons as a function of the synchronous phase ψ_r for $\psi_r = 0$ (solid), $\psi_r = 0.2$ (dashed) and $\psi_r = 0.6$ (dotted).

In a tapered FEL a "synchronous" electron is one whose ponderomotive phase $\psi \equiv \theta + \phi$ equals the synchronous phase ψ_r, and the energy of a synchronous electron follows the resonant energy $\gamma_r mc^2$ throughout the undulator. Electrons with small energy deviations $\eta \neq 0$ perform synchrotron oscillations about the synchronous particle in the ponderomotive bucket. Figure 4.11 shows this ponderomotive bucket for various synchronous phases ψ_r when $\phi = 0$. When $\sin\psi_r > 0$ the bucket decelerates as the resonant energy decreases, and electrons that remain in the bucket radiate the energy difference into the field. Electrons will remain trapped in the bucket provided $|E|$ and ϕ change adiabatically with z, in which case the radiation power can be significantly increased over its nominal "saturation" value.

The electron beam bunching remains approximately constant once electrons are trapped in the ponderomotive bucket. Thus, the electric field E grows linearly with propagation distance in the tapered/trapped regime. If $E \propto z$, Equation (4.100) indicates that the kinetic energy loss $\Delta\gamma_r \propto z^2$, which is consistent with energy conservation since the radiation power also scales as z^2. The associated undulator taper can be obtained from Equation (4.97),

$$\frac{\Delta K}{K} = \frac{4\lambda_1}{\lambda_u}\frac{\Delta\gamma_r}{\gamma_r} \propto z^2 , \tag{4.102}$$

so that a quadratic taper profile maximizes energy extraction. Typically, one starts to taper the undulator strength about two gain lengths before FEL saturation when the e-beam energy loss becomes a significant fraction of ρ, while the precise profile of the quadratic taper is optimized either experimentally or numerically.

Figure 4.12 compares the performance of a seeded FEL amplifier in the absence of undulator tapering (dashed) to that with a numerically optimized quadratic taper (solid). The initial energy spread is chosen to be relatively small, $\sigma_\eta = 0.2\rho$, and the quadratic

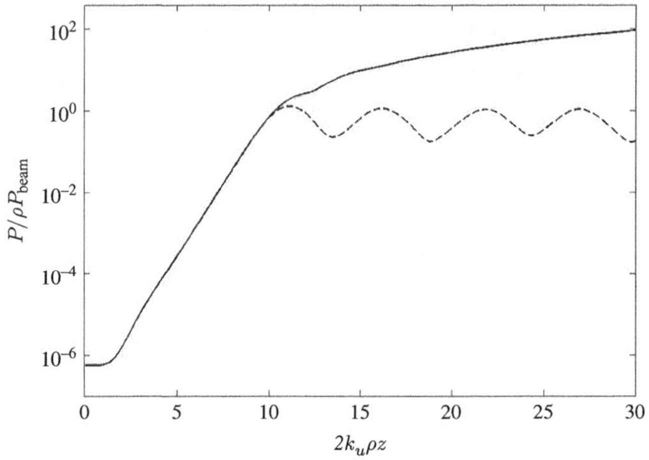

Figure 4.12 FEL power as a function of the undulator distance for a tapered undulator (solid) and a constant undulator (dashed).

taper begins at $\hat{z} = 2k_u\rho z = 9$. The tapered undulator can enhance the output FEL power by two orders of magnitude over that with a constant K. The reason for this large difference is shown in Figure 4.13, where we plot the macro-particle phase-space at several locations along the tapered undulator. Notice how a significant fraction of the electrons become trapped in the wave and are subsequently moved to lower energy as the undulator strength decreases.

While Figures 4.12 and 4.13 illustrates the basic physics, it is an idealized situation starting from a longitudinally coherent (seeded) FEL in 1D. If we instead consider SASE, the radiation field will exhibit large variations over a coherence length $\sim \lambda_1/\rho$. This in turn can lead to de-trapping of electrons as the changing electric field slips over the electrons, so that the efficiency of undulator tapering is somewhat limited for SASE; typical tapering experiments have shown that the SASE power can be increased by a factor between two and four [26]. In addition, 3D effects such as diffraction can also degrade the taper efficiency even when the initial E-field is longitudinally coherent. For example, numerical simulations indicate that the initial quadratic taper law eventually becomes linear with z for sufficiently long undulators. Eventually, all the particles become de-trapped from the bucket due to these deleterious effects, and the radiation power comes to a hard saturation. Further discussion of tapering is beyond the scope of the book, and can be found in the Refs. [28, 29, 30, 31].

4.6 Superradiance

The spontaneous emission from a collection of particles can often be computed by assuming that each particle radiates independently of all others, in which case the total field energy scales linearly with the number of radiating particles. However, this picture is incomplete and gives the wrong answer if the particles are initially correlated or if the

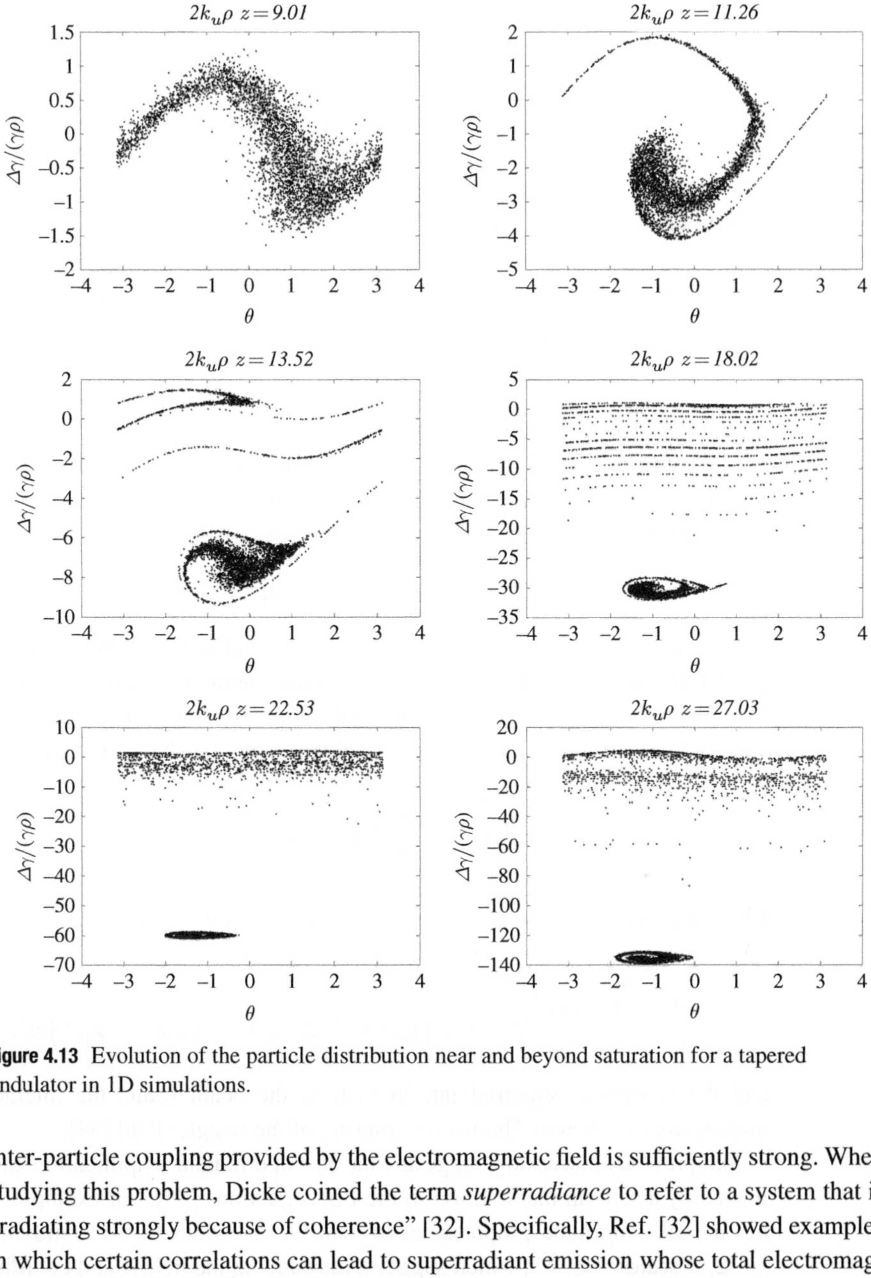

Figure 4.13 Evolution of the particle distribution near and beyond saturation for a tapered undulator in 1D simulations.

inter-particle coupling provided by the electromagnetic field is sufficiently strong. When studying this problem, Dicke coined the term *superradiance* to refer to a system that is "radiating strongly because of coherence" [32]. Specifically, Ref. [32] showed examples in which certain correlations can lead to superradiant emission whose total electromagnetic energy scales quadratically with the number of emitters. Since that time, the word "superradiance" has been applied to many electromagnetic processes whose radiated energy from N_e particles scales as N_e^2; equivalently, it can refer to situations in which the output field intensity is proportional to the square of the particle density. Here, we briefly discuss one specific solution of the FEL that exhibits superradiance.

In normal steady-state operation, the FEL radiation intensity at saturation scales $\propto \rho(I/\sigma_r^2) \propto n_e^{4/3}$, while the energy scales $\propto \rho N_e \propto n_e^{1/3} N_e$. As we explained in

Section 3.4.5, the latter scaling can be understood as a coherent enhancement of the spontaneous undulator radiation. In addition to this steady-state operation, Bonifacio and collaborators pointed out in Ref. [33] that the FEL can also exhibit superradiant emission with a field intensity that scales as n_e^2 (see also [34] and the references therein). In this section we discuss one particular superradiant example that arises as a self-consistent solution to the FEL equations in the particle trapping regime. We reserved presenting this solution for this section because it relies on the physics of particle trapping and can lead to powers in excess of the steady-state saturation power ρP_{beam}. To understand the basic physics and scalings, consider an FEL radiation pulse that slips ahead into an initially unperturbed part of the electron beam. If the field is sufficiently intense, electrons are trapped in the ponderomotive potential and begin to rotate in phase space. During the first half-period the particles lose energy to the wave, while the energy transfer reverses directions in the second half. Hence, only that part of the field that slips over the beam during one-half of a synchrotron period is amplified. A pulse with peak power $|a|^2$ grows such that its width is proportional to the synchrotron period $\sim |a|^{-1/2}$, so that the total electromagnetic energy scales as

$$U_{\text{EM}} = P\Delta t \sim |a|^2 \times |a|^{-1/2} = |a|^{3/2}. \tag{4.103}$$

Meanwhile, the energy of an average electron changes by an amount of order the ponderomotive bucket height $|a|^{1/2}$ due to the interaction, while the number of electrons that have lost energy is proportional to the propagation distance (slippage) of the superradiant pulse over the e-beam $\propto \hat{z}$. Hence, the total change in kinetic energy scales as

$$U_{\text{KE}} \propto \sum_j \Delta\eta_j \sim \hat{z}\Delta\eta_j \sim -\hat{z}\,|a|^{1/2}. \tag{4.104}$$

Evidently, energy is conserved if $|a| \sim \hat{z}$. In this case, the power $|a|^2 \sim \hat{z}^2$, while the intensity

$$\frac{P}{\mathscr{A}_{\text{tr}}} \approx \frac{I}{e}\frac{\rho\gamma mc^2}{\mathscr{A}_{\text{tr}}}\hat{z}^2 = cn_e\left(\rho\gamma mc^2\right)(2\rho k_u z)^2 = \left(\gamma mc^3\kappa_1\chi_1 z^2\right)n_e^2, \tag{4.105}$$

and the process is superradiant. In fact, in the beam frame the interaction can be understood as coherent Thomson scattering of the wiggler field [34].

The previous discussion suggests that the superradiant amplification can be approximated by a self-similar solution to the FEL equations [35]. For example, the radiation has a peak amplitude proportional to $\hat{z}$, while its width in the comoving coordinate $\hat{z} - 2\rho\theta$ scales $\sim \hat{z}^{-1/2}$, which suggests introducing the field A and coordinate y via

$$A \equiv a/\hat{z} \qquad y \equiv \sqrt{\hat{z}}(\hat{z} - 2\rho\theta) \tag{4.106}$$

where A is approximately a function of y only. Then, the left-hand side of the Maxwell equation (3.83) becomes

$$\left(\frac{\partial}{\partial\hat{z}} + \frac{1}{2\rho}\frac{\partial}{\partial\theta}\right)a = \left(\frac{y}{2}\frac{\partial}{\partial y} + \hat{z}\frac{\partial}{\partial\hat{z}} + 1\right)A \approx \frac{y}{2}\frac{dA}{dy} + A. \tag{4.107}$$

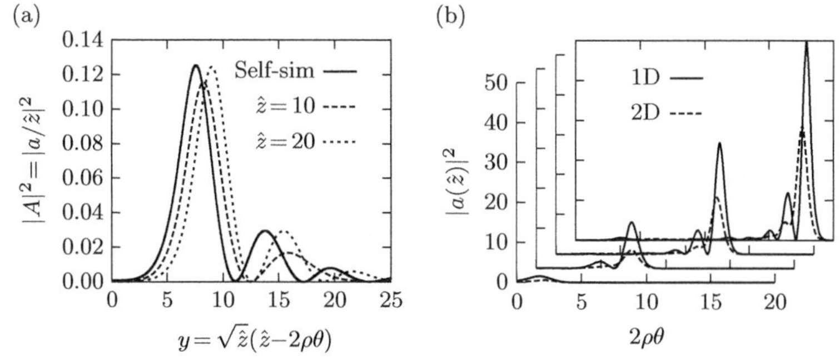

Figure 4.14 (a) Comparison of the self-similar, superradiant field profile with that obtained from solving the full 1D FEL equations at several propagation distances. (b) Comparison of the 1D and 2D superradiant power profiles.

As we mentioned earlier, a typical electron changes its energy by an amount of order the ponderomotive bucket height $\sim |a|^{1/2} \sim \hat{z}^{1/2}$. Hence, we define the self-similar momentum $p_j \equiv \hat{\eta}/\sqrt{\hat{z}}$, which leads to

$$\frac{1}{\hat{z}}\frac{d\hat{\eta}_j}{d\hat{z}} = \left(\frac{\partial}{\partial y} + \frac{y}{2\hat{z}^{3/2}}\frac{\partial}{\partial y} + \frac{1}{\sqrt{\hat{z}}}\frac{\partial}{\partial \hat{z}} + \frac{1}{2\hat{z}^{3/2}}\right) p_j \approx \frac{dp_j}{dy} \tag{4.108}$$

for $\hat{z} \gg 1$, so that the self-similar assumptions are self-consistent after long propagation lengths. We append the right-hand sides of Equations (4.107) and (4.108), and include the phase equation to obtain the following set of ODEs describing the self-similar superradiant FEL pulse [35]:

$$\frac{d\theta_j}{dy} = p_j(y). \tag{4.109}$$

$$\frac{dp_j}{dy} = A(y)e^{i\theta_j(y)} + A(y)^* e^{-i\theta_j(y)} \tag{4.110}$$

$$\frac{y}{2}\frac{dA}{dy} + A = -\frac{1}{N_\Delta}\sum_j e^{-i\theta_j(y)}. \tag{4.111}$$

Note that Equation (4.111) implies that at $y = 0$ the sum of the bunching and the field A vanishes, i.e., $A(0) = -\sum_j e^{-i\theta_j(0)}/N_\Delta \equiv -b(0)$. Thus, the truly self-similar solution requires imposing an initial coherent bunching that is a step-like function along the beam. While possible in principle, a more realistic means of seeding the FEL uses an initially sharp external radiation source. We compare the self-similar solution using initial conditions $|A(0)| = |b(0)| = 0.03$ to results that solve the full 1D FEL equations assuming $a(0) = 0.1$ and $b(0) = 0$ in Figure 4.14(a). We plot the self-similar field power $|A|^2 = \left|a/\hat{z}\right|^2$ as a function of the self-similar coordinate $y = \sqrt{\hat{z}}(\hat{z} - 2\rho\theta)$ for three different propagation lengths. Even with the different boundary conditions the field is reasonably well-described by the self-similar solution. More importantly, the field power grows quadratically with $\hat{z}$, and the emission is superradiant.

Simulations also indicate that the self-similar, superradiant behavior derived in 1D survives when the 3D effect of diffraction is included. This is largely because the FEL bunching provides some optical guiding of the field in a manner similar to that of optical fibers: the bunching modifies the index of refraction such that the field is partially guided along the electron beam. The largely refractive guiding mentioned here is somewhat different from the gain guiding associated with high-gain FELs that we will discuss in the next chapter. We compare 2D, cylindrically symmetric simulation results to those in 1D in Figure 4.14(b), where we use the same parameters as Figure 4.14(a), and set the scaled e-beam size to be $\hat{\sigma}_x = 4$ and the initial Gaussian field profile to have $\sigma_r = 2/\sqrt{\rho k_1 k_u}$. While the peak power is reduced by a factor ~ 2, we see that it continues to scale quadratically with distance, and the superradiant, self-similar pulse structure is largely preserved.

The simulations in Figure 4.14 use a beam whose electrons are initially equally spaced in phase, so that the fluctuations due to shot noise are suppressed and SASE is eliminated. When shot noise fluctuations are included, however, the resulting SASE eventually disturbs the electron beam and hence the self-similar amplification. For this reason we expect that the superradiant amplification demonstrated here applies over propagation lengths of order the SASE saturation length, which typically restricts $\hat{z} \lesssim 4\pi$.

Finally, there has been some experimental success in measuring superradiant behavior in FELs. The first observation was reported in Ref. [36], while more recent experimental results include Ref. [37]

References

[1] W. Colson, "The nonlinear wave equation for higher harmonics in free-electron lasers," *IEEE J. Quantum Electron.*, vol. 17, p. 1417, 1981.

[2] K.-J. Kim, "An analysis of self-amplified spontaneous emission," *Nucl. Instrum. Methods Phys. Res., Sect. A*, vol. 250, p. 396, 1986.

[3] J.-M. Wang and L.-H. Yu, "A transient analysis of a bunched beam free electron laser," *Nucl. Instrum. Methods Phys. Res., Sect. A*, vol. 250, p. 484, 1986.

[4] K.-J. Kim, "Three-dimensional analysis of coherent amplification and self-amplified spontaneous emission in free electron lasers," *Phys. Rev. Lett.*, vol. 57, p. 1871, 1986.

[5] L.-H. Yu and S. Krinsky, "Amplified spontaneous emission in a single pass free electron laser," *Nucl. Instrum. Methods Phys. Res., Sect. A*, vol. 285, p. 119, 1989.

[6] K.-J. Kim, "Temporal and transverse coherence of SASE," in *Towards X-ray Free Electron Lasers*, ser. AIP Conference Proceedings 413, R. Bonifacio and W. Barletta, Eds. New York: AIP, 1997.

[7] R. Bonifacio, L. D. Salvo, P. Pierini, N. Piovella, and C. Pellegrini, "Spectrum, temporal structure, and fluctuations in a high-gain free-electron laser starting from noise," *Phys. Rev. Lett.*, vol. 73, p. 70, 1994.

[8] J. Goodman, *Statistical Optics*. New York: John Wiley & Sons, Inc., 2000.

[9] E. L. Saldin, E. A. Schneidmiller, and M. V. Yurkov, "Statistical properties of radiation from VUV and X-ray free electron laser," *Opt. Commun.*, vol. 148, p. 383, 1998.

[10] S. Krinsky and R. L. Gluckstern, "Analysis of statistical correlations and intensity spiking in the self-amplified spontaneous-emission free-electron laser," *Phys. Rev. ST Accel. Beams*, vol. 6, p. 050701, 2003.

[11] S. O. Rice, "Mathematical analysis of random noise," *Bell Systems Tech. J.*, vol. 23, p. 282, 1945.
[12] —, "Mathematical analysis of random noise," *Bell Systems Tech. J.*, vol. 24, p. 46, 1945.
[13] L.-H. Yu and S. Krinsky, "Analytical theory of intensity fluctuations in SASE," *Nucl. Instrum. Methods Phys. Res., Sect. A*, vol. 407, p. 261, 1998.
[14] N. A. Vinokurov, Z. Huang, O. A. Shevchenko, and K.-J. Kim, "Quasilinear theory of high-gain FEL saturation," *Nucl. Instrum. Methods Phys. Res., Sect. A*, vol. 475, p. 74, 2001.
[15] R. Bonifacio, F. Casagrande, and L. D. S. Souza, "Collective variable description of a free-electron laser," *Phys. Rev. A*, vol. 33, p. 2836, 1986.
[16] C. Marnoli, N. Sterpi, M. Vasconi, and R. Bonifacio, "Three-mode treatment of a high-gain steady-state free-electron laser," *Phys. Rev. A*, vol. 44, p. 5206, 1991.
[17] R. L. Gluckstern, S. Krinsky, and H. Okamoto, "Analysis of the saturation of a high-gain free-electron laser," *Phys. Rev. E*, vol. 47, p. 4412, 1993.
[18] G. Dattoli and P. Ottaviani, "Semi-analytical models of free electron laser saturation," *Opt. Commun.*, vol. 204, p. 283, 2002.
[19] S. Krinsky, "Saturation of a high-gain single-pass FEL," *Nucl. Instrum. Methods Phys. Res., Sect. A*, vol. 528, p. 52, 2004.
[20] W. B. Colson and R. A. Freedman, "Synchrotron instability for long pulses in free electron laser oscillators," *Opt. Commun.*, vol. 46, p. 37, 1983.
[21] J. C. Goldstein, "Theory of the sideband instability in free electron lasers," *Nucl. Instrum. Methods Phys. Res., Sect. A*, vol. 237, p. 27, 1985.
[22] W. B. Colson, "The trapped-particle instability in free electron laser oscillators and amplifiers," *Nucl. Instrum. Methods Phys. Res., Sect. A*, vol. 250, p. 168, 1986.
[23] N. M. Kroll, P. L. Morton, and M. N. Rosenbluth, "Free-electron lasers with variable parameter wigglers," *IEEE J. Quantum Electron.*, vol. 17, p. 1436, 1981.
[24] T. J. Orzechowski, B. R. Anderson, J. C. Clark, W. M. Fawley, A. C. Paul, D. Prosnitz, E. T. Scharlemann, S. M. Yarema, D. B. Hopkins, A. M. Sessler, and J. S. Wurtele, "High-efficiency extraction of microwave radiation from a tapered-wiggler free-electron laser," *Phys. Rev. Lett.*, vol. 57, p. 2172, 1986.
[25] X. J. Wang, H. P. Freund, D. Harder, W. H. Miner, J. B. Murphy, H. Qian, Y. Shen, and X. Yang, "Efficiency and spectrum enhancement in a tapered free-electron laser amplifier," *Phys. Rev. Lett.*, vol. 103, p. 154801, 2009.
[26] D. Ratner, A. Brachmann, F. J., D. Y. Ding, D. Dowell, P. Emma, J. Frisch, S. Gilevich, G. Hays, P. Hering, Z. Huang, R. Iverson, H. Loos, A. Miahnahri, H.-D. Nuhn, J. Turner, J. Welch, W. White, J. Wu, D. Xiang, G. Yocky, and W. M. Fawley, "FEL gain length and taper measurements at LCLS," in *Proceedings of the 2009 FEL Conference*, 2009.
[27] H. Wiedemann, *Particle Accelerator Physics I and II*, 2nd ed. Berlin: Springer-Verlag, 1999.
[28] W. M. Fawley, "'Optical guiding' limits on extraction efficiencies of single pass, tapered wiggler amplifiers," *Nucl. Instrum. Methods Phys. Res., Sect. A*, vol. 375, p. 550, 1996.
[29] W. M. Fawley, Z. Huang, K.-J. Kim, and N. A. Vinokurov, "Tapered undulators for SASE FELs," *Nucl. Instrum. Methods Phys. Res., Sect. A*, vol. 483, no. 12, p. 537, 2002.
[30] Y. Jiao, J. Wu, Y. Cai, A. W. Chao, W. M. Fawley, J. Frisch, Z. Huang, H.-D. Nuhn, C. P. S., and Reiche, "Modeling and multidimensional optimization of a tapered free electron laser," *Phys. Rev. ST Accel. Beams*, vol. 15, p. 050704, 2012.

[31] E. A. Schneidmiller and M. V. Yurkov, "Optimization of a high efficiency free electron laser amplifier," *Phys. Rev. ST Accel. Beams*, vol. 18, p. 030705, 2015.

[32] R. H. Dicke, "Coherence in spontaneous radiation processes," *Phys. Rev.*, vol. 93, p. 99, 1953.

[33] R. Bonifacio and F. Casagrande, "The superradiant regime of a free electron laser," *Nucl. Instrum. Methods Phys. Res., Sect. A*, vol. 239, p. 36, 1985.

[34] R. Bonifacio, F. Casagrande, L. D. Salvo, P. Pierini, and N. Piovella, "Physics of the high-gain FEL and superradiance," *Riv. Nuovo Cimento*, vol. 13, p. 1, 1990.

[35] R. Bonifacio, L. D. Salvo, P. Pierini, P. Pierini, and N. Piovella, "The superradiant regime of a FEL: analytical and numerical results," *Nucl. Instrum. Methods Phys. Res., Sect. A*, vol. 296, p. 358, 1990.

[36] T. Watanabe, X. J. Wang, J. B. Murphy, J. Rose, Y. Shen, T. Tsang, L. Giannessi, P. Musumeci, and S. Reiche, "Experimental characterization of superradiance in a single-pass high-gain laser-seeded free-electron laser amplifier," *Phys. Rev. Lett.*, vol. 98, p. 034802, 2007.

[37] L. Giannessi, M. Bellaveglia, E. Chiadroni, A. Cianchi, M. E. Couprie, M. Del Franco, G. Di Pirro, M. Ferrario, G. Gatti, M. Labat, G. Marcus, A. Mostacci, A. Petralia, V. Petrillo, M. Quattromini, J. V. Rau, S. Spampinati, and V. Surrenti, "Superradiant cascade in a seeded free-electron laser," *Phys. Rev. Lett.*, vol. 110, p. 044801, 2013.

5 3D FEL Analysis

Chapter 5 extends the one-dimensional (1D) theoretical analysis of FELs to the three-dimensional (3D) regime. Although the FEL interaction is predominantly longitudinal in nature, transverse physics cannot be neglected if one wants to have a complete picture of the FEL. Specifically, we must understand the roles of radiation diffraction and guiding, and how the electron's betatron motion in the undulator affects performance. We first describe these effects qualitatively in Section 5.1, where we emphasize the underlying physical picture. In Section 5.2 we revisit the electron trajectory in the undulator, taking into account the 3D undulator magnetic field and the coupling of the transverse degrees of freedom to the longitudinal motion. Section 5.3 generalizes the FEL pendulum equations and the 1D field equation to 3D by including these transverse effects. The low-gain solution will be presented in Section 5.4 including a generalized Madey theorem. To solve the coupled Maxwell–Klimotovich equations in the high-gain regime, Van Kampen's normal mode expansion is introduced in Section 5.5, and a 3D dispersion relation is derived for the radiation growth rate in terms of four universal scaled parameters. Finally, a simple variational solution is discussed and a handy fitting formula for the FEL gain length is presented near the end of this chapter.

5.1 Qualitative Discussion

5.1.1 Diffraction and Guiding

A remarkable feature of a SASE FEL is its transverse coherence. As we have discussed, the spontaneous undulator radiation has a transverse phase space area that is determined by the electron beam emittance $(2\pi\varepsilon_x)^2$. This area is typically much larger than the diffraction-limited phase space area $(\lambda/2)^2$, especially at X-ray wavelengths, so that undulator radiation is composed of many transverse modes. Thus, in a SASE FEL, the initial transverse phase space of the spontaneous emission also consists of an incoherent sum of many spatial modes. However, since the FEL interaction is localized within the electron beam near the peak electron density, there is one "dominant" mode whose transverse size σ_r is dictated by the beam area, and whose natural divergence satisfies $\sigma_r\sigma_{r'} = \lambda/4\pi$. Higher-order spatial modes either diffract more, which results in greater effective losses, or are of larger spatial extent and couple less efficiently to the particles. Thus, the fundamental mode has the highest effective gain, so that it eventually

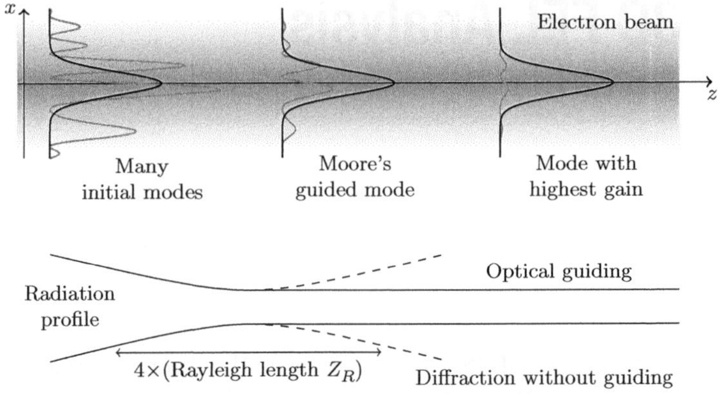

Figure 5.1 Illustration of Moore's guided mode. In the top panel the preferentially guided mode is plotted in black, while the higher-order modes are in gray. The intensity at each z location is scaled to keep the height of the guided (black) mode invariant, so that what appears to be a decrease in the power in the higher-order (gray) modes is actually the larger gain of the Gaussian guided profile outstripping the smaller gain associated with all other modes. The bottom panel compares the natural diffraction of the radiation with that of the guided mode generated by FEL gain.

becomes the preferred spatial distribution for the SASE radiation. This surviving fundamental mode appears to be guided after a sufficient undulator distance, a phenomenon commonly referred to as "optical guiding" or "gain guiding" [1, 2].

We illustrate the general idea of gain guiding schematically in Figure 5.1. Since gain is only effective within the central area, one "matched" transverse mode shape is selected over all others, and this mode then appears to be guided over many vacuum Rayleigh lengths due to the gain. The transverse mode selection is also clearly evident in Figure 5.2, which was obtained from a 3D GENESIS simulation of SASE. Initially, the radiation power is randomly distributed in the transverse plane, but after a sufficient propagation length only one localized coherent mode survives. For one Gaussian-like transverse mode to completely dominate in this way, there must be enough propagation distance for the competing modes to communicate transversely via diffraction.

In the 1D analysis, we introduced the important FEL scaling or Pierce parameter ρ, defined through the relation $n_e\kappa_1\chi_1 = 4k_u^2\rho^3$ which is equivalent to

$$\rho = \left(\frac{e^2K^2[\mathrm{JJ}]^2 n_e}{32\epsilon_0\gamma_r^3 mc^2k_u^2}\right)^{1/3} = \left[\frac{1}{8\pi}\frac{I}{I_A}\left(\frac{K[\mathrm{JJ}]}{1+K^2/2}\right)^2\frac{\gamma_r\lambda_1^2}{2\pi\sigma_x^2}\right]^{1/3}, \tag{5.1}$$

where $I_A = ec/r_e \approx 17045$ A is the Alfvén current. Many important characteristics of the FEL scale with ρ: the gain length and saturation length scale inversely with ρ, while the bandwidth is proportional to ρ. As shown in the previous chapter, for vanishing e-beam energy spread the ideal gain length is given by

$$L_{G0} = \frac{\lambda_u}{4\sqrt{3}\pi\rho}. \tag{5.2}$$

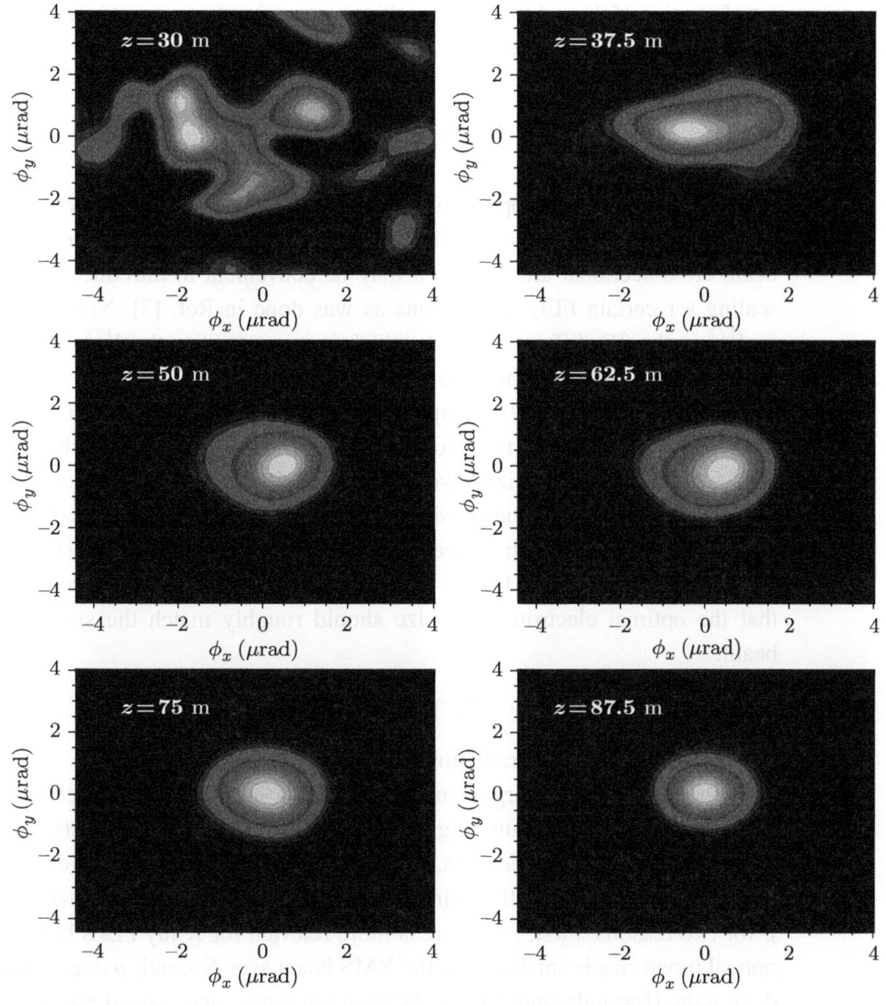

Figure 5.2 Evolution of the LCLS radiation angular distribution at different z location. Courtesy of S. Reiche.

When 3D effects are included, a different dimensionless combination of parameters may govern the gain characteristics of the FEL. To see this, consider the extreme case where the effect of diffraction is "large," meaning that the radiation mode size is significantly larger than the electron beam size. To better describe the interaction between the electrons and the radiation in this 3D limit, the beam area $\mathscr{A}_{\text{tr}} = 2\pi\sigma_x^2$ in Equation (5.1) should be replaced by the diffraction-limited cross-section introduced in Chapter 1.2.1, i.e.,

$$2\pi\sigma_x^2 \to 2\pi\frac{\lambda_1}{4\pi}Z_R. \tag{5.3}$$

Here Z_R is the Rayleigh length of the radiation, which from our discussion on gain guiding ought to be of order a few gain lengths. Thus, by inserting $2\pi\sigma_x^2 \to \lambda_1 L_G$

into Equation (5.1) and then the resulting expression for ρ into (5.2), one can solve the resulting algebraic equation for the gain length L_G to find

$$L_G^{-1} = \frac{4\pi}{\lambda_u}\frac{3^{3/4}}{2}\sqrt{\frac{I}{\gamma I_A}\frac{K^2[\mathrm{JJ}]^2}{(1+K^2/2)}}. \tag{5.4}$$

This equation gives an approximate formula for the growth rate when the 3D effect of diffraction dominates, specifically, when the optical mode is larger than the electron beam cross-sectional area. Thus, it may be convenient to introduce the diffraction D-scaling for certain FEL applications as was done in Ref. [3]. Notice that L_G^{-1} scales as $I^{1/2}$ in the 3D diffractive limit, which is in contrast to the $I^{1/3}$ behavior that characterizes the 1D limit when the electron beam size is larger than that of the optical mode. Additionally, the D-scaling shows that shrinking the electron beam cross-section much below that of the radiation mode does not further reduce the gain length. In fact, reducing the beam size beyond a certain point actually tends to increase the gain length, since decreasing the physical beam size necessarily increases the angular spread of an electron beam with nonzero emittance. While we will investigate the e-beam and radiation divergence further in the next section, it is evident from this discussion that the optimal electron beam size should roughly match the size of the radiation beam:

$$\sigma_x \sim \sigma_r = \sqrt{\varepsilon_r Z_R} \sim \sqrt{\varepsilon_r L_G}, \tag{5.5}$$

where $\varepsilon_r = \lambda_1/4\pi$ is the radiation emittance.

The above qualitative arguments are useful for understanding the effect of diffraction and for estimating the gain length of certain high-gain FEL projects operating in the infrared and visible wavelengths, where the optical mode size is larger than the e-beam size. Nevertheless, we will continue to scale quantities by the dimensionless parameter ρ for two reasons. First, ρ-scaling is more relevant for X-ray FELs because the typical optical mode size is smaller than the RMS beam size. Second, ρ does not require introducing the (formally undetermined) Rayleigh range, and instead relies on the electron beam cross-sectional area as shown in Equation (5.1).

5.1.2 Beam Emittance and Focusing

An electron beam with finite emittance ε_x has a RMS angular spread $\sigma_{x'} = \varepsilon_x/\sigma_x$, so that its size will expand in free space as we discussed in Chapter 1.1.4 and expressed in Equation (1.27). Hence, to keep a nearly constant e-beam size and maximize the FEL interaction in a long undulator channel requires proper electron focusing. As shown in the next section, the undulator magnetic field does provide a "natural" focusing effect. The natural focusing strength, however, is typically too weak, so that external focusing by quadrupole magnets is often required. This focusing is used to decrease the beam size, thereby increasing the ρ parameter and decreasing the gain length. As mentioned in the previous section, decreasing the beam size below that of the optical mode may actually degrade the FEL performance, because the increasing angular spread introduces a spread in the resonant wavelength. This effect is similar to that of energy spread, and can be understood by considering the FEL resonance condition

$$\lambda_1(\psi) = \frac{\lambda_u}{2\gamma^2}\left(1 + \frac{K^2}{2} + \gamma^2\psi^2\right), \tag{5.6}$$

where ψ is the angle the particle trajectory[1] makes with respect to the z axis. From Equation (5.6), we see that the spread in particle angles given by $\Delta\psi = \sigma_{x'}$ causes a spread in the resonant wavelength

$$\frac{\Delta\lambda}{\lambda_1} = \sigma_{x'}^2 \frac{\lambda_u}{\lambda_1} = \frac{\varepsilon_x}{\beta_x}\frac{\lambda_u}{\lambda_1}. \tag{5.7}$$

To not adversely affect the FEL gain, we demand that the induced wavelength variation due to the angular spread be less than the FEL bandwidth $\sim \rho$, namely that

$$\frac{\Delta\lambda}{\lambda_1} = \sigma_{x'}^2 \frac{\lambda_u}{\lambda_1} \lesssim \rho \approx \frac{\lambda_u}{4\pi L_G}. \tag{5.8}$$

Due to optical guiding, the radiation Rayleigh range is of order the gain length, $Z_R \sim L_G$, so that (5.8) implies that the electron beam angular divergence should be no more than that of the radiation:

$$\sigma_{x'} = \sqrt{\frac{\varepsilon_x}{\beta_x}} \leq \sqrt{\frac{\varepsilon_r}{L_G}} \sim \sigma_{r'}. \tag{5.9}$$

The inequalities regarding the beam size (5.5) and angular divergence (5.9) together require

$$\varepsilon_{x,y} \lesssim \varepsilon_r = \frac{\lambda_1}{4\pi}, \tag{5.10}$$

while the optimal focusing beta function for a given emittance saturates the inequality (5.9):

$$\beta_x \sim L_G \frac{\varepsilon_x}{\varepsilon_r}. \tag{5.11}$$

A smaller beam emittance allows for a tighter focused beam size and hence a smaller gain length. In the following sections, we expand upon these qualitative arguments by quantitatively studying the effects of diffraction, guiding, beam emittance, and betatron motion on the FEL gain.

5.2 Electron Trajectory

In Chapters 3 and 4 we considered 1D motion in the undulator field

$$\boldsymbol{B}(0, 0, z) = -B_0 \sin(k_u z)\hat{y}, \tag{5.12}$$

[1] We found essentially the same formula in Equation (2.52) in terms of the optical angle ϕ, since one can exchange $\phi \leftrightarrow \psi$ by redefining the optical axis.

and showed that the electron trajectory in the transverse plane was given by the "wiggle motion"

$$x_w(z) = \frac{K}{\gamma k_u}\sin(k_u z) \qquad\qquad y_w(z) = 0, \tag{5.13}$$

where $K \equiv eB_0/mck_u$ is the dimensionless deflection (or undulator) parameter. Equation (5.13) represents the ideal trajectory of an electron injected along the optical axis of the undulator. Our description of the full transverse dynamics associated with arbitrary initial conditions will average over the fast oscillations in the undulator field in a manner similar to that of the previous chapter. There, the averaged equations described the slowly evolving longitudinal coordinates (θ, η); in this chapter, we will include the transverse degrees of freedom, paying particular attention to the variation in phase θ caused by nonzero $(\boldsymbol{x}, \boldsymbol{p})$. We write the transverse coordinates as a sum of the wiggle motion and the slow (betatron) evolution:

$$x(z) = x_w(z) + x_\beta(z), \tag{5.14}$$

$$y(z) = y_\beta(z). \tag{5.15}$$

In Equations (5.14)–(5.15), $\boldsymbol{x}_\beta(z)$ is the slowly evolving part of $\boldsymbol{x}$ that represents the transverse beam envelope. If the magnetic field (5.12) described an actual undulator, for example, then $\boldsymbol{x}_\beta(z) = \boldsymbol{x}(0) + \boldsymbol{x}'(0)z$, and the transverse dynamics would be simple rectilinear motion[2]. Realizable undulator fields, however, focus the electrons transversely which results in a slow oscillatory motion whose period is much longer than the undulator wavelength λ_u. The combined fast (wiggle) and slow (betatron) motion is shown in Figure 5.3.

In general, the amplitude of the betatron oscillation is larger than that of the undulator wiggle motion. To determine the betatron motion more precisely, we must investigate the magnetic field at points other than those along the z axis.

5.2.1 Natural Focusing in an Undulator

The undulator field equation (5.12) is only valid very near the $y = 0$ plane because it does not satisfy the vacuum Maxwell equations in 3D (i.e., the curl of Equation (5.12)

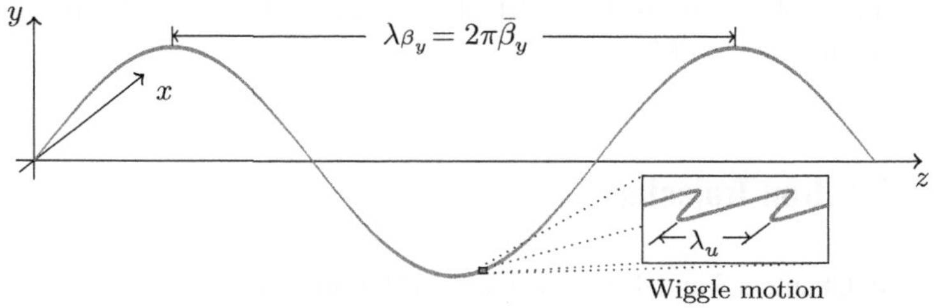

Figure 5.3 Combined motion of the wiggler and betatron oscillations in a planar undulator. The wiggle motion is along $\hat{x}$, while the betatron oscillation has a much longer period along $\hat{y}$.

[2] This is the transverse motion we used in Chapter 2.4 to study undulator radiation. It is an appropriate approximation if the undulator length L_u is much shorter than the natural undulator focusing length $\sim \gamma\lambda_u/K$.

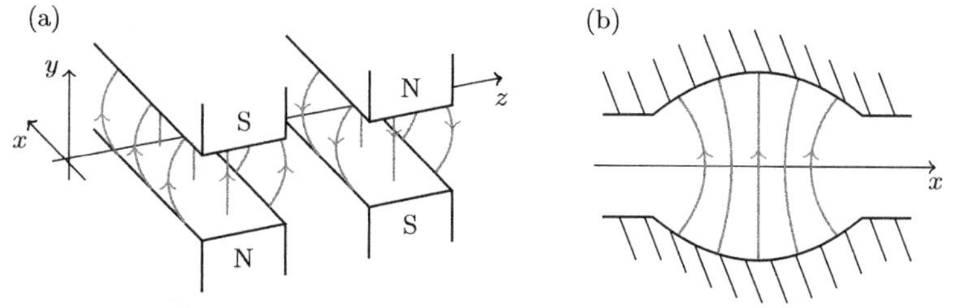

Figure 5.4 (a) The fringe magnetic field of a flat face planar undulator that gives rise to natural focusing. (b) Illustration of a parabolic-pole faced undulator that focuses the beam in both directions.

does not vanish). An exact solution of Maxwell's equations describing a planar undulator with flat poles is

$$\boldsymbol{B}(\boldsymbol{x};z) = -B_0 \cosh(k_u y)\sin(k_u z)\hat{y} - B_0 \sinh(k_u y)\cos(k_u z)\hat{z}. \tag{5.16}$$

This reduces to Equation (5.12) when $y = 0$, while the $\hat{z}$-component of $\boldsymbol{B}$ accounts for the fringe field shown in Figure 5.4(a). The Lorentz force on an electron due to the 3D undulator field is given by

$$-e[\boldsymbol{v} \times \boldsymbol{B}] = -e \begin{bmatrix} B_z \dfrac{dy}{dt} - B_y \dfrac{dz}{dt} \\ -B_z \dfrac{dx}{dt} \\ B_y \dfrac{dx}{dt} \end{bmatrix} = \frac{d}{dt} \begin{bmatrix} \gamma m \dfrac{dx}{dt} \\ \gamma m \dfrac{dy}{dt} \\ \gamma m \dfrac{dz}{dt} \end{bmatrix}. \tag{5.17}$$

The velocity along x can be determined if we notice from (5.16) that

$$-e\left(B_z \frac{dy}{dt} - B_y \frac{dz}{dt}\right) = \frac{eB_0}{k_u}\frac{d}{dt}\left[\cosh(k_u y)\cos(k_u z)\right]. \tag{5.18}$$

Thus, the $\hat{x}$-component of (5.17) can be directly integrated, and we find that

$$x' \equiv \frac{dx}{dz} = \frac{dx/dt}{dz/dt} \approx \frac{K}{\gamma}\cosh(k_u y)\cos(k_u z) + x'(0), \tag{5.19}$$

where we approximate $dz/dt \approx c$. This result follows simply from the Hamiltonian formalism of Appendix A, for which (5.19) is a consequence of the conservation of canonical momentum along $\hat{x}$. Furthermore, we see that (5.19) yields similar wiggle motion to the 1D result (5.13), only now the oscillation amplitude depends slowly on the y coordinate. Inserting the velocity (5.19) into the $\hat{y}$-component of the Lorentz force (5.17) and neglecting the slow and small time dependence of γ, we find that the motion in the vertical plane is governed by

$$\begin{aligned} y'' &\approx -\frac{K^2 k_u}{\gamma_r^2}\cos^2(k_u z)\sinh(k_u y)\cosh(k_u y) \\ &\approx -\left(\frac{Kk_u}{\gamma_r}\right)^2 \cos^2(k_u z) y \end{aligned} \tag{5.20}$$

to first order in $k_u y$. After averaging over an undulator period, this leads to the harmonic oscillator equation

$$y'' = -k_{n0}^2 y, \quad \text{with} \quad k_{n0} = \frac{Kk_u}{\sqrt{2}\gamma} \equiv \frac{1}{\beta_n}. \tag{5.21}$$

We see that there is a restorative force along y, so that in the vertical plane the beam's natural beta function β_n is given by the inverse of the oscillation frequency k_{n0}. The natural focusing action in y (but not x) could have been anticipated by realizing that the fast oscillations lead to an average (ponderomotive) force that pushes particles toward regions of lower field strength.

Natural focusing in both planes can be achieved by designing the magnitude of the magnetic field to be a minimum on axis. For example, one can shape the undulator pole faces to have a parabolic profile [4] as shown in Figure 5.4(b), in which case the magnetic field is

$$\begin{aligned} \boldsymbol{B} = -\frac{B_0}{k_y}\big[&k_x \sinh(k_x x)\sinh(k_y y)\sin(k_u z)\hat{x} \\ &+ k_y \cosh(k_x x)\cosh(k_y y)\sin(k_u z)\hat{y} \\ &+ k_u \cosh(k_x x)\sinh(k_y y)\cos(k_u z)\hat{z}\big], \end{aligned} \tag{5.22}$$

with $k_x^2 + k_y^2 = k_u^2$ to satisfy Maxwell's equations in vacuum. The field (5.22) leads to natural focusing in both directions, and it can be shown that the natural focusing strength in the x and y planes satisfy $k_{nx}^2 + k_{ny}^2 = k_{n0}^2$.

To summarize, the transverse motion in a planar undulator is given by

$$x = x_w + x_\beta \qquad\qquad y = y_\beta \tag{5.23}$$

with the wiggle motion x_w given by integrating the velocity equations in the wiggler (5.19), and the betatron motion $\boldsymbol{x}_\beta$ is given by

$$x_\beta = x_0\cos(k_{nx}z) + \frac{x_0'}{k_{nx}}\sin(k_{nx}z) \tag{5.24a}$$

$$y_\beta = y_0\cos(k_{ny}z) + \frac{y_0'}{k_{ny}}\sin(k_{ny}z). \tag{5.24b}$$

Here, the value of the natural undulator focusing k_n depends on the pole shape, with $k_{nx}^2 + k_{ny}^2 = k_{n0}^2$. The simple harmonic motion in the transverse direction conserves the energy $H_y = (p_y^2 + k_{ny}^2 y^2)/2$ as the electron follows an ellipse in the phase space $(y, p_y) = (y, y')$. If the poles are shaped the invariant H_x can be constructed in an analogous manner; in either case the motion conserves the energy H_x, whose limit is $p_x^2/2$ for $k_{nx} \to 0$.

While the evolution of the transverse degrees of freedom is largely independent of the longitudinal ones, the equation of motion for the ponderomotive phase does depend upon $\boldsymbol{p}$. This coupling arises because particles of equal energies but different transverse momenta have different longitudinal velocities, which results in a spread of θ due to the variation in the resonance condition. To be explicit, we consider the average longitudinal velocity scaled by c:

$$\frac{\bar{v}_z}{c} = \sqrt{1 - \overline{\boldsymbol{x}'^2} - \frac{1}{\gamma^2}} \approx 1 - \frac{1}{2}\left(\overline{x'^2 + y'^2} + \frac{1}{\gamma^2}\right). \tag{5.25}$$

For a planar undulator with flat poles, (5.19) expresses both the wiggle motion and slow drift along $\hat{x}$, while (5.21) describes the simple harmonic motion in $\hat{y}$. Averaging over an undulator period, we have

$$\begin{aligned}\overline{x'^2 + y'^2} &= \frac{K^2}{2\gamma^2}\cosh^2(k_u y) + x'(0)^2 + y'(z)^2 \\ &\approx \frac{K^2}{2\gamma^2}(1 + k_u^2 y^2) + x'(0)^2 + y'(z)^2 = \frac{K^2}{2\gamma^2} + 2H_\perp,\end{aligned} \tag{5.26}$$

where we have expressed the angular spread in terms of the energy $H_\perp = H_x + H_y$, so that (5.26) is valid for arbitrary pole shape. Thus, natural undulator focusing preserves the average longitudinal velocity for each electron. To see why this is so, consider the betatron motion along y shown in Figure 5.3: y' is maximum on axis while zero at maximum y. The magnetic field at this turning point is larger than that on axis, resulting in wiggle oscillations along x that have a larger amplitude. As shown in (5.26), these two effects compensate each other, so that $\overline{x'^2 + y'^2}$ and, hence, $\bar{v}_z$, is constant. Thus, we have

$$\frac{\bar{v}_z}{c} = 1 - \frac{1 + K^2/2}{2\gamma^2} - \frac{\boldsymbol{p}^2 + k_\beta^2 \boldsymbol{x}^2}{2} = 1 - \frac{1 + K^2/2}{2\gamma^2} - H_\perp. \tag{5.27}$$

The values of H_x and H_y depend on the initial conditions $(\boldsymbol{x}(0), \boldsymbol{p}(0))$, which in turn vary from electron to electron. Along any particular particle trajectory, however, both H_x and H_y are conserved. If we initialize the transverse electron distribution function to be constant along lines of fixed H_x and H_y, than the resulting matched beam envelope is invariant along the undulator as we show in Figure (5.5). For a beam of emittance ε_x, the matched RMS beam size is

$$\sigma_y \sim \sqrt{\frac{\varepsilon_x \gamma_r \lambda_u}{2\pi K}} = \sqrt{\frac{\varepsilon_{x,n} \lambda_u}{2\pi K}}. \tag{5.28}$$

The matched beam size (5.28) is typically much larger than the optimal beam size $\sigma_{\text{opt}} \sim \sqrt{\lambda_1 L_G / 4\pi}$, especially at X-ray wavelengths; for example, if $\varepsilon_x = \lambda_1/4\pi$, than $\sigma_y/\sigma_{\text{opt}} \approx \sqrt{K/2\rho} \gg 1$. Thus, high-gain X-ray FELs require additional focusing provided by external magnets.

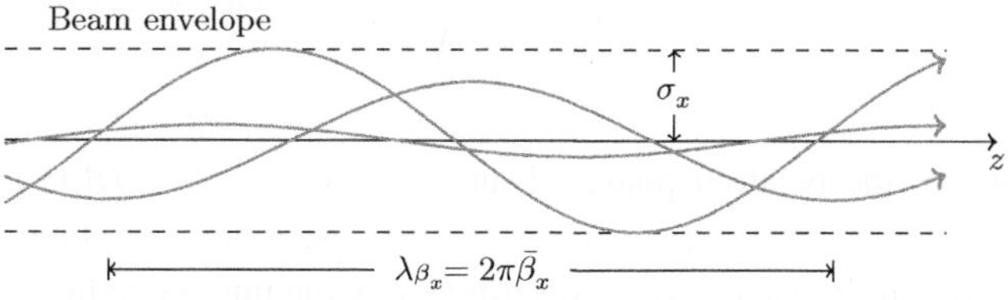

Figure 5.5 Matched beam in a undulator focusing channel.

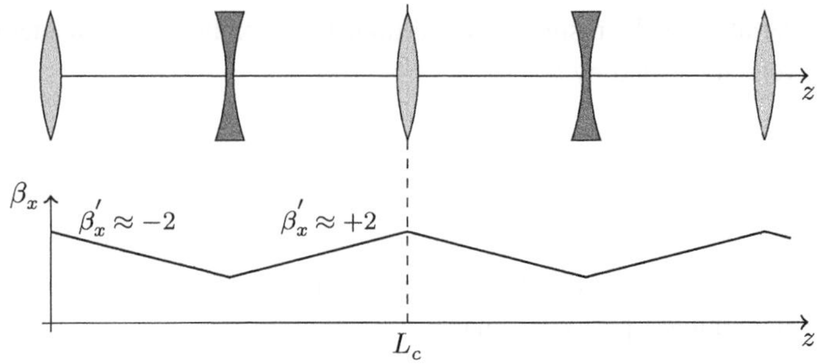

Figure 5.6 Variation of the horizontal beta function along the distance of two FODO cells for a small phase advance per cell. The derivative $\beta_x' \equiv d\beta_x/dz$ is close to the values ± 2, but the deviation of β_x from the average value $\hat{\beta}$ is relatively small. The FODO cell length 2ℓ is assumed to be much smaller than the average beta $\hat{\beta}$.

5.2.2 Betatron Motion in an External Focusing Lattice

As we saw in the previous section, the natural focusing strength $k_{n0} \propto \gamma^{-1}$ is typically too weak to produce a sufficiently small matched beam size, especially for short wavelength FELs that use high-energy electron beams. To further reduce the cross-sectional area, X-ray FELs usually incorporate alternating-gradient focusing such as that provided by a FODO lattice. We discussed stable beam propagation in a FODO lattice in Section 1.1.4, concentrating on a focusing lattice designed to maintain a nearly constant beam size and divergence. While the full FODO dynamics can be easily incorporated into numerical simulations, we will make some additional approximations in order to gain further analytic understanding regarding the effects of finite beam size, divergence, and emittance on the FEL performance.

Our discussion of the single particle transverse dynamics requires a few additional concepts used in accelerator physics. As implied by the fact that the linear dynamics is governed by a second-order differential equation, linear particle optics can be mapped to a generalized simple harmonic oscillator, with natural coordinates given by the particle amplitude and phase in both transverse directions. In accelerator terminology, the transverse degrees of freedom can be written in terms of the Courant–Snyder (Twiss) lattice functions β_x and α_x as (see, e.g., [5])

$$x_\beta(z) = \sqrt{2\mathcal{J}_x\beta_x(z)}\cos\Phi_x(z) \tag{5.29}$$

$$p_x(z) \equiv x_\beta'(z) = -\sqrt{\frac{2\mathcal{J}_x}{\beta_x(z)}}\left[\sin\Phi_x(z) + \alpha_x(z)\cos\Phi_x(z)\right], \tag{5.30}$$

where the betatron phase advance $\Phi_x(z) \equiv \int dz'\ 1/\beta_x(z')$ describes the generalized rotation in phase space, $\beta_x(z)$ sets the oscillation amplitude, and the action $\mathcal{J}_x$ is an invariant for each particle determined by the initial conditions (similar expressions hold in y, too). Equation (5.30) follows from differentiating (5.29) and using the definitions

$d\Phi_x/dz = 1/\beta_x$ and $d\beta_x/dz = 2\langle x'x\rangle = -2\alpha_x$. From (5.30), we can write the transverse velocity squared of an electron as

$$p_x^2 = \frac{2\mathcal{J}_x}{\beta_x(z)}\left\{\alpha_x^2(z) + [1-\alpha_x^2(z)]\sin^2\Phi_x(z) + \alpha_x(z)\sin[2\Phi_x(z)]\right\}$$
$$\approx \frac{2\mathcal{J}_x}{\beta_x(z)}\left\{1 \pm \sin[2\Phi_x(z)]\right\}, \tag{5.31}$$

where the final line results from the fact that the particular FODO lattice used for FELs has $\alpha_x(z) \approx \pm 1$ as we showed in (1.34). The beam has positive α_x (negative correlation $\langle xx'\rangle$) in the drift space following the focusing quadrupole, while after the defocusing quad $\alpha_x \approx -1$. Furthermore, since the average beta-function is much greater than the drift length, $\bar{\beta} \gg \ell$, the phase $\Phi_x(z)$ changes only a small amount over the drift space, so that we can average the final term to zero if the gain length is much larger than drift space [6]. Thus, we can approximate the angle squared in the transverse degrees of freedom by

$$p_x^2(z) \approx \frac{2\mathcal{J}_x}{\bar{\beta}_x} = \text{constant}, \qquad p_y^2(z) \approx \frac{2\mathcal{J}_y}{\bar{\beta}_y} = \text{constant}. \tag{5.32}$$

Although the $\mathcal{J}_{x,y}$ are constants of motion for each electron, different electrons generally have different transverse actions. In fact, the ensemble average of $\mathcal{J}$ over all the electrons is the RMS transverse emittance of the beam, i.e.,

$$\langle \mathcal{J}_x\rangle = \varepsilon_x \qquad \langle \mathcal{J}_y\rangle = \varepsilon_y. \tag{5.33}$$

While (5.29)–(5.30) is a complete representation of the transverse motion with (5.32) a reasonably accurate approximation to $\boldsymbol{p}^2$ for most FEL applications, our subsequent analysis requires a more analytically amenable description of the transverse physics. The model motion should reflect the fact that the RMS beam size is nearly constant in the FODO lattice and, more importantly, have a faithful representation of the coupling between the transverse and longitudinal degrees of freedom. While we have seen that the transverse physics is independent of the longitudinal ones, both $\boldsymbol{p}$ and $\boldsymbol{x}$ affect the evolution of the particle phase by changing the mean velocity of the particle. In terms of the approximation (5.32), we have

$$\frac{\bar{v}_z}{c} \approx 1 - \frac{1+K^2/2}{2\gamma^2} - \frac{\mathcal{J}_x + \mathcal{J}_y}{\bar{\beta}}. \tag{5.34}$$

Thus, our approximate motion should at a minimum respect the following characteristics of the true dynamics in the FODO lattice:

1. Result in a stable beam with nearly constant RMS size equal to $\sqrt{\varepsilon_x\bar{\beta}_x}$;
2. Be periodic in z with period given by $2\pi\bar{\beta}_x = 2\pi/k_\beta$;
3. Possess an invariant whose beam average is proportional to ε_x that can be associated with a decrease in $\bar{v}_z$ similar to (5.34).

All three conditions can be satisfied by approximating the particle trajectories by the simple harmonic motion for smooth focusing that we studied in the previous section. For this reason and for analytic tractability, in what follows we will approximate the effect of the FODO lattice by a smooth focusing field whose focusing strength (oscillator frequency) $k_\beta = 1/\bar{\beta}_x$. In this case the transverse energy and action are related by $H_x = \mathcal{J}_x/\bar{\beta}_x = \bar{k}_\beta \mathcal{J}_x$ and $H_y = \mathcal{J}_x/\bar{\beta}_x = \bar{k}_\beta \mathcal{J}_y$.

To see how these approximations work in practice, we include Figure 5.7 that compares some aspects of the single particle dynamics in a FODO lattice with a smooth focusing channel. In Figure 5.7(a) we plot as a solid line the phase space ellipses at the center of the drift space after both the focusing (top) and defocusing (bottom) quadrupole. We also include dots representing particle positions in each half-section after 0 (0.5), 4 (4.5), 8 (8.5), and 12 (12.5) lattice periods. The FODO parameters are based on the LCLS lattice that has $\bar{\beta}_x \approx 18$ m and $\ell \approx 4$ m. For comparison, we also plot the phase space ellipse for a smooth focusing channel/simple harmonic oscillator (SHO) with $k_\beta = 1/\bar{\beta}_x$ as a dashed line, including particle phase space points at the same z locations. The two show reasonable agreement when averaged over several lattice periods.

In addition, we plot the $\bar{\beta}_x p_x^2/2$ for the FODO lattice, and compare it to the invariant action $\mathcal{J}_x$ of the smooth focusing channel/SHO in Figure 5.7(b). Note that p_x^2 for the FODO lattice is only truly invariant in the drift section, and that it oscillates about the mean value given by $\mathcal{J}_x = \beta_x H_x$. Thus, the single particle transverse coupling to θ provided by the smooth focusing lattice only matches that of a FODO lattice when

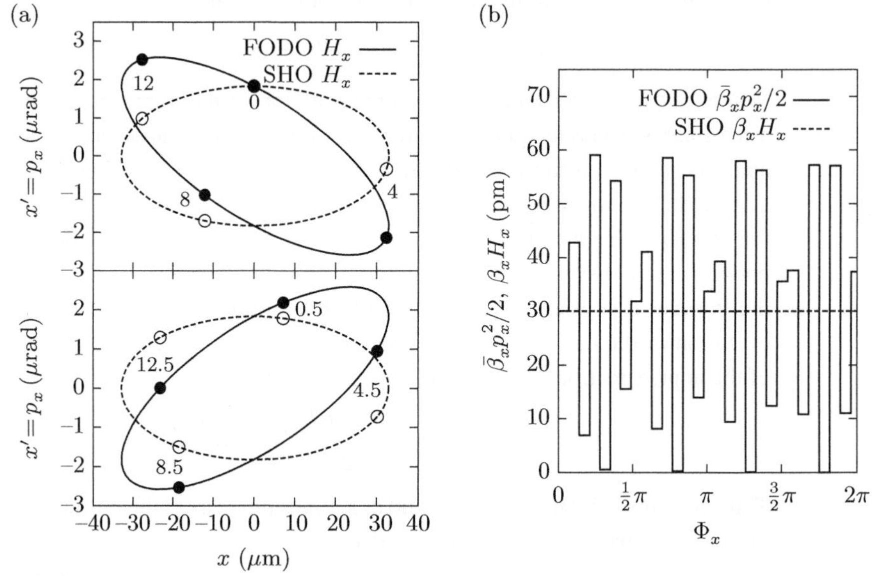

Figure 5.7 Comparison between the particle motion in a FODO lattice and a constant focusing channel (simple harmonic oscillator/SHO). The solid line ellipses in panel (a) are equal-energy contours drawn at the center of the each drift space, with the top being just after the focusing quad, and the bottom just after the defocusing quad. The dotted lines are energy action lines for the constant focusing lattice (simple harmonic oscillator/SHO). Filled and unfilled dots represent one single particle position plotted every four FODO periods. Panel (b) plots the particle $\bar{\beta}_x p_x^2/2$ in the FODO lattice as a solid line, whose average (dashed line) equals the phase space area $\beta_x H_x = (\beta_x p_x^2 + x^2/\beta_x)/2$ for the constant focusing channel/SHO.

averaged over many focusing cells. Furthermore, the difference from the mean changes sign every half period such that $p_x^2/2$ averaged over any full period approximately equals the energy H_x.

As shown in Figure 5.7, the particle trajectories in the smooth focusing channel only roughly mimic those of the FODO lattice. Nevertheless, in the next few sections we will present analysis based upon this approximate motion that results in rather accurate semi-analytic predictions regarding the FEL gain and mode shape. This is because these radiation characteristics do not depend directly on the individual particle orbits, but rather on certain averaged properties of the particle beam distribution. Since the low-order moments of a beam in a FODO lattice are accurately represented by an ensemble of electrons moving in a smooth focusing channel, they yield remarkably accurate predictions for the FEL performance. The matched beam size (as shown in Figure 5.5) and divergence are given in terms of the emittance ε_x as

$$\sigma_x = \sqrt{\frac{\varepsilon_x}{k_{\beta,x}}}, \qquad \sigma_{x'} = \sqrt{\varepsilon_x k_{\beta,x}}. \tag{5.35}$$

The equations for the matched beam and coupling to θ in a FODO lattice are identical in form to those describing the natural focusing that we derived in the previous section. Thus, the smooth focusing approximation for the 3D FEL equations of motion can be used in either setting provided that the focusing strength $k_\beta = 1/\bar{\beta}$ is appropriately interpreted. We now turn to completing the derivation of the FEL equations in 3D.

5.3 3D Equations of the FEL

In this section, we derive 3D governing equations for an FEL. Our goal is to obtain the coupled Maxwell–Klimontovich system for the radiation and electron beam. The radiation field will be governed by the paraxial wave equation driven by an appropriately defined FEL current, while the characteristic curves of the Klimontovich equation are given by the single particle equations in 3D. These particle equations generalize the 1D FEL pendulum equations to include transverse effects.

5.3.1 Maxwell Equation

The derivation of the paraxial wave equation is a straightforward 3D generalization of the 1D arguments laid out in Chapter 3. Rewriting Equation (2.10) for N_e electrons, the paraxial equation for the $\hat{x}$-component of the electric field is

$$\left[\frac{\partial}{\partial z} + \frac{ik}{2}\boldsymbol{\phi}^2\right]\tilde{E}_\omega(\boldsymbol{\phi}; z) = \sum_{j=1}^{N_e} \frac{e(\beta_{x,j} - \phi_x)}{4\pi\epsilon_0 c\lambda^2} e^{ik[ct_j(z) - z - \boldsymbol{\phi}\cdot\boldsymbol{x}_j(z)]}. \tag{5.36}$$

The source in Equation (5.36) has two contributions: one from the time derivative of the current $\sim \beta_x$, and the other from the transverse derivative of the charge $\sim \phi_x$. We estimate the relative size of these two source terms by noting that FELs are characterized by

$$\beta_x \sim \frac{K}{\gamma}, \qquad \phi_x \sim \sqrt{\frac{\lambda_1}{\lambda_u}} \begin{cases} \sqrt{1/N_u} & \text{in the low-gain regime} \\ \sqrt{\rho} & \text{in the high-gain regime.} \end{cases} \tag{5.37}$$

Thus, for $K \gtrsim 1$ the ratio of the charge to current source is small:

$$\frac{|\phi_x|}{|\beta_x|} \sim \frac{2}{K\sqrt{2+K^2}} \begin{cases} \sqrt{1/N_u} & \text{in the low-gain regime} \\ \sqrt{\rho} & \text{in the high-gain regime.} \end{cases} \tag{5.38}$$

Additionally, the charge density term does not resonantly drive the field at lowest order. In fact, it can be shown that the resonant coupling to the charge density is actually $O(\rho)$ [or $O(1/N_u)$] smaller than that to the current density J_x. Thus, we can drop the charge density term from the Maxwell equation.

The procedure of reducing the 3D wave equation (5.36) to describe the FEL is exactly the same as we did in Section 3.4.1: we replace the particle time t_j with the slowly varying phase θ_j by subtracting off the fast figure-eight motion via

$$\theta_j(z) \equiv (k_u + k_1)z - ck_1 \left[t_j(z) - \frac{K^2}{ck_1(4+2K^2)} \sin(2k_u z) \right], \tag{5.39}$$

average over an undulator period, and simplify the resulting equation by defining an analog to the phase-shifted electric field Equation (3.67). For this last step we will also choose to transform from the angular to the spatial representation of the field, so that we define[3]

$$E_\nu(\boldsymbol{x}; z) \equiv ck_1 e^{-i\Delta\nu k_u z} \int d\boldsymbol{\phi}\, \tilde{E}_\omega(\boldsymbol{\phi}; z) e^{ik\boldsymbol{x}\cdot\boldsymbol{\phi}}. \tag{5.40}$$

Using definition (5.40) in the wave equation (5.36), we find that the paraxial wave equation (5.36) for the slowly varying amplitude $E_\nu(\boldsymbol{x}; z)$ is

$$\left[\frac{\partial}{\partial z} + i\Delta\nu k_u - \frac{i}{2hk_1}\nabla_\perp^2 \right] E_\nu(\boldsymbol{x}; z) = -\kappa_h \frac{k_1}{2\pi} \sum_{j=1}^{N_e} e^{-i\nu\theta_j(z)} \delta[\boldsymbol{x} - \boldsymbol{x}_j(z)], \tag{5.41}$$

where we recall that the normalized frequency difference $\Delta\nu \equiv \nu - 1$ and the coupling $\kappa_h \equiv eK[\mathrm{JJ}]_h/4\epsilon_0\gamma_r$. Note that we approximate $1/k$ by $1/hk_1$ in the coefficient of the transverse derivatives, as the difference between these is assumed small. Additionally, it is easy to verify that the 3D Maxwell equation (5.41) reduces to its 1D counterpart (3.68) if we assume that E_ν is independent of transverse coordinates and set $\delta[\boldsymbol{x} - \boldsymbol{x}_j(z)] \to \mathscr{A}_{\mathrm{tr}}^{-1}$.

5.3.2 3D Pendulum Equations for the Electron Motion

Most of the heavy lifting required to derive the single particle equations of motion has already been done, partly in the 1D analysis of Section 3.2 and partly when we discussed

[3] While we do use the angular representation for the low-gain theory, the high-gain analysis employs the spatial representation since it is nearly universal in the high-gain literature.

the transverse degrees of freedom earlier in this chapter. Here, we review some of the salient points and collect the final 3D equations.

The rate of work done on the electrons by the radiation is given by $\boldsymbol{F} \cdot \boldsymbol{v}$, which for the FEL equals the product of the undulator wiggle velocity along $\hat{x}$ and the transverse electric field. Changing the independent variable from $t \to z$, we have

$$\begin{aligned} mc^2 \frac{d\gamma}{dz} &= -e\frac{dx}{dz}E_x \\ &= \frac{eK}{\gamma}\cos(k_u z)\left[\int d\nu\, E_\nu(\boldsymbol{x};z)e^{i\nu(k_1 z-\omega_1 t)}e^{i\Delta\nu k_u z} + c.c.\right]. \end{aligned} \tag{5.42}$$

Replacing the Lorentz factor with the normalized energy deviation from resonance $\eta \equiv (\gamma - \gamma_r)/\gamma_r$, and the particle time with θ using (5.39), we average over the fast oscillations in the undulator to obtain

$$\frac{d\eta}{dz} = \sum_{h\text{ odd}} \chi_h \int d\nu\, E_\nu(\boldsymbol{x};z)e^{i\nu\theta} + c.c. \tag{5.43}$$

where again $\nu \approx h$ and $\chi_h \equiv eK[\text{JJ}]_h/(2\gamma_r^2 mc^2)$.

To determine the rate of change of the ponderomotive phase, we calculate

$$\frac{d}{dz}\theta = \frac{d}{dz}\left[(k_u + k_1)z - ck_j\bar{t}(z)\right] = (k_u + k_1) - k_1\frac{c}{\bar{v}_z}, \tag{5.44}$$

where $\bar{v}_z$ is the mean particle velocity averaged over an undulator period. The particle is slowed from its maximum longitudinal speed $\sqrt{1-1/\gamma^2}$ due to the transverse motion, which now includes both an average of the fast wiggle oscillation in the undulator field and the slow betatron dynamics given by (5.24):

$$\frac{c}{\bar{v}_z} \approx 1 + \frac{1}{2\gamma^2} + \frac{\overline{x'^2 + y'^2}}{2} \approx 1 + \frac{1 + K^2/2}{2\gamma_r^2}(1 - 2\eta) + \frac{1}{2}\left(\boldsymbol{p}^2 + k_\beta^2\boldsymbol{x}^2\right), \tag{5.45}$$

where the focusing strength k_β is determined by either the natural undulator focusing or by the external FODO lattice if the latter is much stronger. The first two constant terms in (5.45) cancel the $(k_u + k_1)$ in the phase equations (5.44) due to the resonance condition, leaving only the deviations from the ideal on-resonance, on-axis trajectory in the phase evolution

$$\frac{d\theta}{dz} = 2k_u\eta - \frac{k_1}{2}\left(\boldsymbol{p}^2 + k_\beta^2\boldsymbol{x}^2\right) \tag{5.46}$$

$$= 2k_u\eta - k_1 H_\perp. \tag{5.47}$$

Here, the focusing k_β is the inverse of the average betatron function $\bar{\beta}_{x,y}$, and we use the smooth focusing approximation to characterize the transverse betatron oscillation, in which case $(\boldsymbol{p}^2 + k_\beta^2\boldsymbol{x}^2)/2 = k_\beta\mathcal{J} = H_\perp$ is a constant of motion. The transverse degrees of freedom therefore obey

$$\frac{d\boldsymbol{x}}{dz} = \boldsymbol{p} \qquad \frac{d\boldsymbol{p}}{dz} = -k_\beta^2\boldsymbol{x}. \tag{5.48}$$

Collecting the 3D particle equations for convenience, we have

$$\frac{d\theta}{dz} = 2k_u\eta - \frac{k_1}{2}(\boldsymbol{p}^2 + k_\beta^2\boldsymbol{x}^2), \tag{5.49}$$

$$\frac{d\eta}{dz} = \chi_h \int d\nu \, e^{i\nu\theta} E_\nu(\boldsymbol{x};z) + c.c., \tag{5.50}$$

$$\frac{d\boldsymbol{x}}{dz} = \boldsymbol{p}, \qquad \frac{d\boldsymbol{p}}{dz} = -k_\beta^2\boldsymbol{x}, \tag{5.51}$$

where $\chi_h = eK[\mathrm{JJ}]_h/(2\gamma_r^2 mc^2)$ and we have assumed that the radiation force is dominated by one harmonic.

5.3.3 Coupled Maxwell–Klimontovich Equations

We describe the electrons with their microscopic distribution in phase space. In a manner similar to the 1D analysis, we define the electron Klimontovich distribution function

$$\begin{aligned} F(\theta,\eta,\boldsymbol{x},\boldsymbol{p};z) = \frac{k_1}{I/ec}\sum_{j=1}^{N_e} \delta[\theta-\theta_j(z)]\delta[\eta-\eta_j(z)] \\ \times\, \delta[\boldsymbol{x}-\boldsymbol{x}_j(z)]\delta[\boldsymbol{p}-\boldsymbol{p}_j(z)], \end{aligned} \tag{5.52}$$

where I/ec is the peak line density. Note that integrating (5.52) over $\boldsymbol{p}$ and setting $\delta(\boldsymbol{x}-\boldsymbol{x}_j) \to \mathscr{A}_{\mathrm{tr}}^{-1}$ yields the 1D F defined in (4.1). The evolution of the Klimontovich distribution function is governed by the continuity equation

$$\frac{\partial F}{\partial z} + \frac{d\theta}{dz}\frac{\partial F}{\partial \theta} + \frac{d\eta}{dz}\frac{\partial F}{\partial \eta} + \frac{d\boldsymbol{x}}{dz}\cdot\frac{\partial F}{\partial \boldsymbol{x}} + \frac{d\boldsymbol{p}}{dz}\cdot\frac{\partial F}{\partial \boldsymbol{p}} = 0, \tag{5.53}$$

with the equations of motion given by Equations (5.49), (5.50), and (5.51).

The rest of this chapter will be devoted to analyzing the 3D FEL equations in the small signal regime, applicable for linear FEL gain before saturation. To do this, we again divide the distribution function F into a smooth background $\bar{F}$ that is independent of θ and a fluctuation δF that describes the θ-scale variations. This δF contains the particle shot noise and the FEL-generated micro-bunching, while assuming $\bar{F}$ to be uniform in θ (the coasting beam approximation) is valid if the e-beam current, energy spread, emittance, etc. is nearly constant over a coherence length. In exact analogy to the 1D case of Section 4.1, we use the decomposition $F = \bar{F} + \delta F$, determine the equation for the smooth perturbation $\bar{F}$ by integrating over θ, and isolate the fast microbunching near harmonic h by Fourier transforming Equation (5.53) with $\delta F \to F_\nu$. We then linearize the system by neglecting the higher-order terms $\sim E_\nu F_\nu$ to obtain the following linear system:

$$\left\{\frac{\partial}{\partial z} + \boldsymbol{p}\cdot\frac{\partial}{\partial \boldsymbol{x}} - k_\beta^2\boldsymbol{x}\cdot\frac{\partial}{\partial \boldsymbol{p}}\right\}\bar{F} = 0, \tag{5.54}$$

$$\left\{\frac{\partial}{\partial z} + \boldsymbol{p}\cdot\frac{\partial}{\partial \boldsymbol{x}} - k_\beta^2\boldsymbol{x}\cdot\frac{\partial}{\partial \boldsymbol{p}} + i\nu\left[2\eta k_u - \frac{k_1}{2}(\boldsymbol{p}^2 + k_\beta^2\boldsymbol{x}^2)\right]\right\}F_\nu = -\chi_h E_\nu \frac{\partial \bar{F}}{\partial \eta}. \tag{5.55}$$

We have assumed the field E_ν and, hence, the microbunching F_ν is localized about the one of the odd harmonic resonant frequencies $\nu \approx h$. The linear system (5.55)–(5.54) applies before saturation when the optical power $P \ll \rho P_{\text{beam}}$.

A significant difference between the 3D equation (5.54) and its 1D counterpart is the fact that the background distribution $\bar{F}$ now has nontrivial z-dependence due to the transverse degrees of freedom. To solve for $\bar{F}(\boldsymbol{x},\boldsymbol{p};z)$, we integrate along the transverse particle trajectories, which are the characteristic curves of (5.54). The physics of this solution was discussed in Section 1.1.5: the value of $\bar{F}$ is transported along the single particle trajectories, so that [7]

$$\bar{F}(\boldsymbol{x},\boldsymbol{p};z) = \bar{F}[\boldsymbol{x}_0(\boldsymbol{x},\boldsymbol{p},z;s),\boldsymbol{p}_0(\boldsymbol{x},\boldsymbol{p},z;s);s] \tag{5.56}$$

for all s. Here, the initial coordinates $(\boldsymbol{x}_0,\boldsymbol{p}_0)$ are functions of the present coordinates $(\boldsymbol{x},\boldsymbol{p})$ and z that satisfy the transverse equations of motion subject to the initial conditions $(\boldsymbol{x}_0,\boldsymbol{p}_0) = (\boldsymbol{x},\boldsymbol{p})$ at $z = s$. The trajectory equations in the smooth focusing lattice are

$$\begin{bmatrix} \boldsymbol{x}_0(\boldsymbol{x},\boldsymbol{p},z;s) \\ \boldsymbol{p}_0(\boldsymbol{x},\boldsymbol{p},z;s) \end{bmatrix} = \begin{bmatrix} \cos[k_\beta(z-s)] & -\sin[k_\beta(z-s)]/k_\beta \\ k_\beta \sin[k_\beta(z-s)] & \cos[k_\beta(z-s)] \end{bmatrix} \begin{bmatrix} \boldsymbol{x} \\ \boldsymbol{p} \end{bmatrix} \tag{5.57}$$

$$\equiv \mathsf{T}_{z\to s} \begin{bmatrix} \boldsymbol{x} \\ \boldsymbol{p} \end{bmatrix}. \tag{5.58}$$

Having solved for $\bar{F}$ at lowest order, we conclude this section by collecting the linearized equations for the field E_ν and density perturbation F_ν. We insert the definition of the electron distribution function into the paraxial wave equation (5.41), include the Liouville equation (5.55), and extend both to harmonic interaction to find that the linear Maxwell–Klimontovich equations for the FEL in 3D are

$$\left[\frac{\partial}{\partial z} + i\Delta\nu k_u - \frac{i}{2hk}\nabla_\perp^2\right] E_\nu = -\kappa_h n_e \int d\boldsymbol{p} d\eta \, F_\nu \tag{5.59}$$

$$\left\{\frac{\partial}{\partial z} + \boldsymbol{p}\cdot\frac{\partial}{\partial \boldsymbol{x}} - k_\beta^2 \boldsymbol{x}\cdot\frac{\partial}{\partial \boldsymbol{p}} + i\left[2\nu\eta k_u - \frac{k}{2}(\boldsymbol{p}^2 + k_\beta^2\boldsymbol{x}^2)\right]\right\} F_\nu = -\chi_h E_\nu \frac{\partial \bar{F}}{\partial \eta}. \tag{5.60}$$

We will solve the linear FEL equations (5.59)–(5.60) with $\bar{F}$ given by (5.56) in both the low-gain and high-gain regimes using two different mathematical techniques. The low-gain solution will be obtained by integrating over the unperturbed trajectories/characteristics of δF and E_ν, namely, those ignoring the coupling between the distribution function and the electromagnetic field. This solution will be valid if the E_ν does not change significantly during its interaction with the undulator, meaning that this solution will generalize the 1D low-gain results of Section 3.3 and Section 4.2.

The high-gain analysis will more closely resemble that of Section 3.4, and will focus on understanding the 3D FEL solution with the largest growth rate. While this method could in principle include the low-gain results as a special case, we will simplify the discussion by restricting our attention to only the growing modes.

5.4 Solution in the Low-Gain Regime

In this section we derive a formula for the linear gain including the effects of beam emittance and energy spread when the external focusing can be approximated by the constant focusing parameter k_β. This derivation follows that of Ref. [7], Here we integrate the low-gain FEL equations along the particle trajectories (characteristic curves), and to consolidate notation we will abbreviate the transverse phase space coordinates as $\boldsymbol{Z} \equiv (\boldsymbol{x}, \boldsymbol{p})$. In this case the derivatives on the left-hand side of the Klimontovich equation (5.60) can be considered as a total derivative along the trajectory

$$\left(\frac{\partial}{\partial z} + \boldsymbol{p} \cdot \frac{\partial}{\partial \boldsymbol{x}} - k_\beta^2 \boldsymbol{x} \cdot \frac{\partial}{\partial \boldsymbol{p}}\right) F_\nu = \left(\frac{\partial}{\partial z} + \frac{d\boldsymbol{Z}}{dz} \cdot \frac{\partial}{\partial \boldsymbol{Z}}\right) F_\nu = \left.\frac{d}{dz}\right|_{\boldsymbol{Z}} F_\nu. \tag{5.61}$$

We now use the conserved transverse energy $H_\perp = k_\beta \mathcal{J} = (\boldsymbol{p}^2 + k_\beta^2 \boldsymbol{x}^2)/2$ and the total derivative $d/dz|_{\boldsymbol{Z}}$ to write the Klimontovich equation for F_ν as

$$e^{-i(2\nu k_u \eta - kH_\perp)z} \left.\frac{d}{dz}\right|_{\boldsymbol{Z}} e^{i(2\nu k_u \eta - kH_\perp)z} F_\nu = -\chi_h E_\nu \frac{\partial}{\partial \eta} \bar{F}. \tag{5.62}$$

We isolate the derivative, integrate over z, and simplify the expression to obtain

$$\begin{aligned} F_\nu = e^{-i(2\nu k_u \eta - kH_\perp)z} F_\nu^0 - \chi_h \int_0^z ds \; e^{-i(2\nu k_u \eta - kH_\perp)(z-s)} \\ \times E_\nu[\boldsymbol{x}_0(\boldsymbol{Z}, z; s); s] \frac{\partial}{\partial \eta} \bar{F}[\eta, \boldsymbol{Z}_0(\boldsymbol{Z}, z; s); s]. \end{aligned} \tag{5.63}$$

Recall that the coordinate $\boldsymbol{Z}_0(\boldsymbol{Z}, z; s)$ denotes the solution to the transverse equations (5.48) subject to the condition that $\boldsymbol{Z}_0(\boldsymbol{Z}, z; z) = \boldsymbol{Z}$. We will find it convenient to write the initial F_ν^0 in terms of its value at the middle of the undulator where in the low-gain regime we anticipate the electron and radiation beams to come to a waist. Thus, we take $F_\nu^0 = F_\nu[\eta, \boldsymbol{Z}_0(\boldsymbol{Z}, z; L_u/2); L_u/2]$, since this is equal to $F_\nu[\eta, \boldsymbol{Z}_0(\boldsymbol{Z}, z; 0); 0]$. The background distribution function is given by (5.56) with the coordinates $(\boldsymbol{x}_0, \boldsymbol{p}_0)$ given by (5.58). To determine how the radiation evolves, we consider the transverse Fourier transform of the paraxial equation (5.59). Using the angular representation of the field

$$\mathcal{E}_\nu(\boldsymbol{\phi}; z) \equiv \frac{1}{\lambda^2} \int d\boldsymbol{x} \; E_\nu(\boldsymbol{x}; z) e^{-ik\boldsymbol{x}\cdot\boldsymbol{\phi}}, \tag{5.64}$$

the paraxial wave equation is

$$\left\{\frac{\partial}{\partial z} + i\Delta\nu k_u + \frac{ik}{2}\boldsymbol{\phi}^2\right\} \mathcal{E}_{\nu,\boldsymbol{\phi}}(z) = -\kappa_h n_e \int d\eta d\boldsymbol{p} \; \frac{1}{\lambda^2} \int d\boldsymbol{x} \; F_\nu(\eta, \boldsymbol{x}, \boldsymbol{p}; z) e^{-ik\boldsymbol{x}\cdot\boldsymbol{\phi}}. \tag{5.65}$$

In reciprocal (frequency-angular) space, the paraxial wave equation has the homogeneous solution

$$\mathcal{G}_{\boldsymbol{\phi}}(z) = e^{-i(\Delta\nu k_u + k\boldsymbol{\phi}^2/2)z}. \tag{5.66}$$

Thus, the solution to (5.65) is

$$\mathcal{E}_{\nu,\boldsymbol{\phi}}(z) = \mathcal{G}_{\boldsymbol{\phi}}\left(z - \tfrac{L_u}{2}\right) \mathcal{E}^0_{\nu,\boldsymbol{\phi}}\left(\tfrac{L_u}{2}\right) - \frac{\kappa_h n_e}{\lambda^2} \int_0^z ds\ \mathcal{G}_{\boldsymbol{\phi}}(z-s) \int d\eta d\boldsymbol{p} d\boldsymbol{x}\ F_\nu(\eta, \boldsymbol{x}, \boldsymbol{p}; s) e^{-ik\boldsymbol{x}\cdot\boldsymbol{\phi}}, \tag{5.67}$$

where we express the initial value of the field $\mathcal{E}^0$ in terms of that in the middle of the undulator. Inserting the solution for the perturbed distribution function F_ν into (5.67) yields an integral equation for $\mathcal{E}_\nu$:

$$\begin{aligned}
\mathcal{E}_{\nu,\boldsymbol{\phi}}(L_u) &= \mathcal{G}_{\boldsymbol{\phi}}\left(\tfrac{L_u}{2}\right) \mathcal{E}^0_{\nu,\boldsymbol{\phi}}\left(\tfrac{L_u}{2}\right) \\
&\quad - \frac{\kappa_h n_e}{\lambda^2} \int_0^{L_u} dz\ \mathcal{G}_{\boldsymbol{\phi}}(L_u - z) \int d\eta d\boldsymbol{Z}\ e^{-ik\boldsymbol{\phi}\cdot\boldsymbol{x}} e^{-i(2\nu k_u \eta - kH_\perp)z} F^0_\nu \\
&\quad + \frac{n_e \kappa_h \chi_h}{\lambda^2} \int_0^{L_u} dz\ \mathcal{G}_{\boldsymbol{\phi}}(L_u - z) \int d\eta d\boldsymbol{Z}\ e^{-ik\boldsymbol{\phi}\cdot\boldsymbol{x}} \\
&\quad \times \int_0^z ds\ e^{-i(2\nu k_u \eta - kH_\perp)(z-s)} \int d\boldsymbol{\phi}_1\ e^{ik\boldsymbol{\phi}_1 \cdot \boldsymbol{x}_0(\boldsymbol{Z},z;s)} \\
&\quad \times \mathcal{E}_{\nu,\boldsymbol{\phi}_1}(s) \frac{\partial}{\partial\eta} \bar{F}[\eta, \boldsymbol{Z}_0(\boldsymbol{Z},z;s); s].
\end{aligned} \tag{5.68}$$

The first term in (5.68) represents the free-space propagation of the initially incident radiation; the second term yields the spontaneous undulator radiation generated by the electrons moving on their unperturbed particle trajectories via F^0_ν; the third term is proportional to the product of the radiation and the smooth distribution function and gives rise to FEL gain. To acquire additional insight into these terms, we will make a number of simplifying manipulations. First, we consider the spontaneous radiation, and change the integration variables from $(\boldsymbol{x}, \boldsymbol{p}) \to (\boldsymbol{x}_0, \boldsymbol{p}_0)$ using the relation

$$\boldsymbol{Z}(\boldsymbol{Z}_0, s; z) = \mathsf{T}^{-1}_{z\to s} \boldsymbol{Z}_0 = \mathsf{T}_{s\to z} \boldsymbol{Z}_0 \tag{5.69}$$

from Equation (5.58). Since the transformation is merely a rotation, the determinant is unity and $d\boldsymbol{p}d\boldsymbol{x} = d\boldsymbol{p}_0 d\boldsymbol{x}_0$, while $F^0_\nu = F_\nu(\eta, \boldsymbol{Z}_0; L_u/2)$.

Second, we consider the smooth distribution $\bar{F}$ from the gain term in (5.68). In the linear gain regime $\bar{F}$ is transported along the unperturbed trajectories as we discussed in Section 1.1.5 and illustrated in Figure 1.5. Hence, at any location σ we have

$$\begin{aligned}
\bar{F}[\eta, \boldsymbol{Z}_0(\boldsymbol{Z}, z; s); s] &= \bar{F}[\eta, \boldsymbol{Z}_\sigma(\boldsymbol{Z}, z; \sigma); \sigma] && (5.70) \\
&= \bar{F}[\eta, \boldsymbol{Z}_1(\boldsymbol{Z}, z; L_u/2); L_u/2]. && (5.71)
\end{aligned}$$

Physically, Equation (5.70) states that the value of $\bar{F}$ at location z can be obtained by transporting its value at location σ along trajectories whose initial and final coordinates are at σ and z, respectively; in particular, in Equation (5.71) we choose the coordinates $\boldsymbol{Z}_1$ whose initial condition is at the middle of the undulator. We then change the integration variable in the gain term from $\mathbf{Z}$ to $\mathbf{Z}_1$ using the coordinate relationships $\mathbf{Z}(\mathbf{Z}_1, L_u/2; z) = \mathsf{T}_{L_u/2\to z}\mathbf{Z}_1$ and $\mathbf{Z}_0(\mathbf{Z}_1, L_u/2; s) = \mathsf{T}_{L_u/2\to s}\mathbf{Z}_1$, in analogy with Equation (5.69).

Third, we introduce the field

$$U_\nu(\eta, \mathbf{Z}_1, \boldsymbol{\phi}; z) = e^{-i(\Delta\nu k_u + k\phi^2/2)(L_u/2-z)} e^{-i(2\nu k_u\eta - kH_\perp)z} \\ \times e^{-ik\boldsymbol{\phi}\cdot\{\boldsymbol{x}_1\cos[k_\beta(L_u/2-z)] - \boldsymbol{p}_1\sin[k_\beta(L_u/2-z)]/k_\beta\}}, \tag{5.72}$$

which is related to the spontaneous undulator radiation formula (2.93) that we computed in Chapter 2.4.2; the integral of U_ν over the undulator length is proportional to the spontaneous undulator field.

Fourth, we apply our assumption of low-gain to make the replacement

$$\begin{aligned} \mathcal{E}_{\nu,\boldsymbol{\phi}_1}(s) &= \mathcal{G}_{\boldsymbol{\phi}_1}(s)\mathcal{E}^0_{\nu,\boldsymbol{\phi}_1}(0) + \ldots \\ &= \mathcal{G}_{\boldsymbol{\phi}_1}(s)\mathcal{G}_{\boldsymbol{\phi}_1}(-L_u/2)\mathcal{E}^0_{\nu,\boldsymbol{\phi}_1}(L_u/2) + \ldots \end{aligned} \tag{5.73}$$

to lowest order. Using (5.71), (5.72) and (5.73), and inserting the definition of F^0_ν, we find that (5.68) simplifies to

$$\begin{aligned} \mathcal{E}_{\nu,\boldsymbol{\phi}}(L_u) &= \mathcal{G}_{\boldsymbol{\phi}}\left(\tfrac{L_u}{2}\right)\mathcal{E}^0_{\nu,\boldsymbol{\phi}}\left(\tfrac{L_u}{2}\right) \\ &\quad - \mathcal{G}_{\boldsymbol{\phi}}\left(\tfrac{L_u}{2}\right)\frac{\kappa_h k_1}{2\pi\lambda^2}\sum_{j=1}^{N_e}\int_0^{L_u} dz\, e^{-i\nu\theta_j} U_\nu(\eta_j, \boldsymbol{x}_j, \boldsymbol{p}_j, \boldsymbol{\phi}; z) \\ &\quad + \mathcal{G}_{\boldsymbol{\phi}}\left(\tfrac{L_u}{2}\right)\frac{n_e\kappa_h\chi_h}{\lambda^2}\int d\boldsymbol{\phi}_1 d\eta\, d\mathbf{Z}_1\, \mathcal{E}^0_{\nu,\boldsymbol{\phi}_1}\left(\tfrac{L_u}{2}\right)\frac{\partial\bar{F}}{\partial\eta} \\ &\quad \times \int_0^{L_u} dz \int_0^z ds\, U_\nu(\eta, \mathbf{Z}_1, \boldsymbol{\phi}; z) U^*_\nu(\eta, \mathbf{Z}_1, \boldsymbol{\phi}_1; s). \end{aligned} \tag{5.74}$$

Again, the three terms above can be understood as the incident radiation, the spontaneous (or undulator) radiation, and the first-order field amplification, respectively. Equation (5.74) is therefore the 3D generalization of the 1D solution (4.15). Since both the derivation and the resulting expression equation (5.74) are rather complicated, it's a good idea at this point to check whether (5.74) reproduces the 1D formula. In this limit the electric field as well as the electron distribution are uniform in the transverse direction, meaning that they are directed entirely along the forward direction. Hence, the 1D electric field $\mathcal{E}_\nu$ and electron distribution function $\bar{F}$ are given by

$$\mathcal{E}_\nu(\boldsymbol{\phi}; z) \to \mathcal{E}_\nu(z)\delta(\boldsymbol{\phi}) \qquad\qquad \bar{F}(\eta, \boldsymbol{x}, \boldsymbol{p}) \to V(\eta)\delta(\boldsymbol{p}). \tag{5.75}$$

Additionally, in 1D $k_\beta \to 0$ so that the undulator field

$$U_\nu(\eta, \boldsymbol{x}, \boldsymbol{p}, \boldsymbol{\phi}; z) \to e^{-i(2\nu\eta - \Delta\nu)z}. \tag{5.76}$$

By making the replacements (5.75) and (5.76) in the 3D gain formula (5.74), it is straightforward to reproduce the 1D formulae derived in Section 4.2.

It is important to recognize that the gain term for $\mathcal{E}$ is complex, meaning that it both amplifies the field and shifts it in phase. The former is important both for gain and in determining the group velocity of an FEL pulse, while the latter phase shift contributes to the phase velocity longitudinally and also to refractive guiding transversely. These effects can be fruitfully interpreted as resulting from a complex refractive index associated with the FEL interaction (see, e.g., [8, 9, 2, 10].

To determine the gain, we consider the output power given by

$$P_{\text{out}} \propto \int d\boldsymbol{\phi} \, \left|\mathcal{E}_{\nu,\boldsymbol{\phi}}(L_u)\right|^2. \tag{5.77}$$

The absolute square of $\mathcal{E}^0_{\nu,\boldsymbol{\phi}}\left(\frac{L_u}{2}\right)$ is proportional to the input power P_{in}, while the absolute square of the second term in (5.74) yields the spontaneous undulator radiation. The spontaneous radiation can be shown to be identical to (2.93) in the limit of zero focusing $k_\beta \to 0$, and we drop it for the remainder of the calculation. The cross terms involving the spontaneous radiation lead to a sum over particle phases that average to zero. Thus, the lowest-order gain arises from the product between the incident field (the first term in (5.74)) and the amplitude gain [the third term in (5.74)]. Using the fact that $\mathcal{G}(L_u/2)$ is a pure phase and that when integrating over $\boldsymbol{\phi}$ and $\boldsymbol{\phi}_1$ we can write (using abbreviated notation)

$$\begin{aligned}
&\mathcal{E}_{\boldsymbol{\phi}}\mathcal{E}^*_{\boldsymbol{\phi}_1}\int_0^{L_u} dz \int_0^{z} ds \; U^*_{\boldsymbol{\phi}}(z)U_{\boldsymbol{\phi}_1}(s) + \mathcal{E}^*_{\boldsymbol{\phi}}\mathcal{E}_{\boldsymbol{\phi}_1}\int_0^{L_u} dz \int_0^{z} ds \; U_{\boldsymbol{\phi}}(z)U^*_{\boldsymbol{\phi}_1}(s) \\
&= \mathcal{E}_{\boldsymbol{\phi}}\mathcal{E}^*_{\boldsymbol{\phi}_1}\int_0^{L_u} dz \int_0^{z} ds \; U^*_{\boldsymbol{\phi}}(z)U_{\boldsymbol{\phi}_1}(s) + \mathcal{E}_{\boldsymbol{\phi}}\mathcal{E}^*_{\boldsymbol{\phi}_1}\int_0^{L_u} ds \int_0^{s} dz \; U_{\boldsymbol{\phi}_1}(s)U^*_{\boldsymbol{\phi}}(z)
\end{aligned} \tag{5.78}$$

$$= \mathcal{E}_{\boldsymbol{\phi}}\mathcal{E}^*_{\boldsymbol{\phi}_1}\int_0^{L_u} dz \int_0^{L_u} ds \; U^*_{\boldsymbol{\phi}}(z)U_{\boldsymbol{\phi}_1}(s), \tag{5.79}$$

where the second term on the first line interchanges $\boldsymbol{\phi} \leftrightarrow \boldsymbol{\phi}_1$ and the dummy integration variables $z \leftrightarrow s$. The last line can be proven using the diagram in Figure 5.8, and we find that the output power can be written as

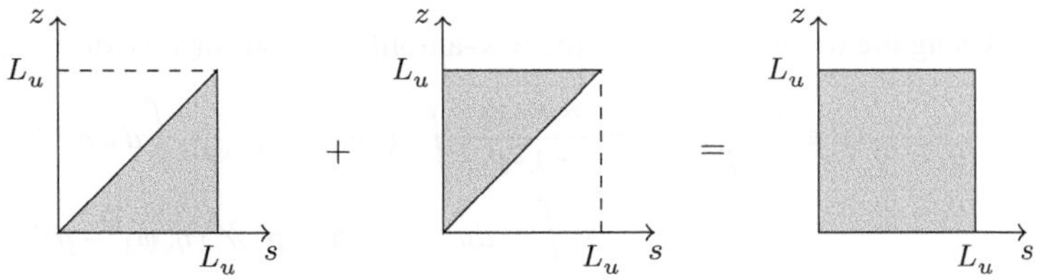

Figure 5.8 Diagram showing how the sum of the integrations in Equation (5.78) equals the integration over the entire region.

$$P_{\text{out}} = P_{\text{in}} + \frac{n_e \kappa_h \chi_h}{\lambda^2} \int d\boldsymbol{\phi} d\boldsymbol{\phi}_1 \, \mathcal{E}^0_{\nu\boldsymbol{\phi}}\left(\tfrac{L_u}{2}\right) \mathcal{E}^{0*}_{\nu\boldsymbol{\phi}_1}\left(\tfrac{L_u}{2}\right) \int d\eta d\boldsymbol{p} d\boldsymbol{x}$$
$$\times \int_0^{L_u} dz \, ds \, U_\nu(\eta, \boldsymbol{Z}, \boldsymbol{\phi}; z) U^*_\nu(\eta, \boldsymbol{Z}, \boldsymbol{\phi}_1; s) \frac{\partial}{\partial \eta} \bar{F}(\eta, \boldsymbol{Z}; L_u/2). \tag{5.80}$$

The gain formula (5.80) is a generalization of Madey's theorem [11] that includes energy spread, the transverse effects of beam emittance and diffraction, and smooth transverse focusing. To make the connection more explicit, integrating by parts with respect to η shows that the gain involves

$$-\frac{\partial}{\partial \eta} U_\nu(\eta, \boldsymbol{Z}, \boldsymbol{\phi}; z) U^*_\nu(\eta, \boldsymbol{Z}, \boldsymbol{\phi}_1; s) = 2\frac{\partial}{\partial \nu} U_\nu(\eta, \boldsymbol{Z}, \boldsymbol{\phi}; z) U^*_\nu(\eta, \boldsymbol{Z}, \boldsymbol{\phi}_1; s),$$

where the equality follows from the definition (5.72). Thus, the FEL gain is proportional to the derivative of the spontaneous radiation spectrum with respect to frequency, which is the essence of Madey's theorem.

5.4.1 Low-Gain Expression for No Transverse Focusing

The expression for the gain further simplifies when we can ignore the transverse focusing so that $k_\beta = 0$. This limit is particularly appropriate for low-gain devices because in this case gain-guiding is not effective, so that maximal energy transfer occurs when the beam mode matches the vacuum radiation mode, which is achieved when the (unfocused) e-beam spreading matches that of the radiation. To ignore the natural undulator focusing requires that $k_n L_u \lesssim 1$, i.e.,

$$\frac{2\pi K N_u}{\sqrt{2}\gamma} \lesssim 1. \tag{5.81}$$

For $k_\beta \to 0$, we have

$$U_\nu(\eta, \boldsymbol{x}, \boldsymbol{p}, \boldsymbol{\phi}; z) = e^{-ik\boldsymbol{\phi}\cdot\boldsymbol{x}} e^{-ik(\boldsymbol{\phi}-\boldsymbol{p})^2/2} e^{i(2\nu\eta k_u - \Delta\nu k_u)z} \tag{5.82}$$
$$= e^{-ik\boldsymbol{\phi}\cdot\boldsymbol{x}} \, U_\nu(\eta, \boldsymbol{0}, \boldsymbol{0}, \boldsymbol{\phi} - \boldsymbol{p}; z), \tag{5.83}$$

so that we define the reduced undulator field

$$\mathscr{U}_\nu(\eta, \boldsymbol{\phi} - \boldsymbol{p}) = \int_0^{L_u} dz \, e^{-ik\boldsymbol{\phi}\cdot\boldsymbol{x}} \, U_\nu(\eta, \boldsymbol{0}, \boldsymbol{0}, \boldsymbol{\phi} - \boldsymbol{p}; z). \tag{5.84}$$

Using the definition (5.84), the low-amplitude gain for no external focusing is given by

$$G \equiv \frac{P_{\text{out}} - P_{\text{in}}}{P_{\text{in}}} = \frac{n_e \kappa_h \chi_h}{2\pi \lambda^2 P_{\text{in}}} \int d\boldsymbol{\phi} d\boldsymbol{\phi}_1 \, \mathcal{E}^0_{\nu\boldsymbol{\phi}} \mathcal{E}^{0*}_{\nu\boldsymbol{\phi}_1} \int d\boldsymbol{x} \, e^{ik(\boldsymbol{\phi}-\boldsymbol{\phi}_1)\cdot\boldsymbol{x}}$$
$$\times \int d\eta d\boldsymbol{p} \, \mathscr{U}_\nu(\eta, \boldsymbol{\phi} - \boldsymbol{p}) \mathscr{U}_\nu(\eta, \boldsymbol{\phi}_1 - \boldsymbol{p})^* \frac{\partial}{\partial \eta} \bar{F}^0. \tag{5.85}$$

The gain formula (5.85) can be written in terms of a convolution over the brightness functions $\mathcal{B}$ of the undulator, radiation, and electron beam. The electron beam "brightness" is given by the distribution function $\bar{F}$, while the brightness of any radiation field R is determined from its Wigner function

$$\mathcal{B}_R(\boldsymbol{x},\boldsymbol{\phi}) = \int d\boldsymbol{\xi}\; R(\boldsymbol{\phi}+\tfrac{1}{2}\boldsymbol{\xi})^* R(\boldsymbol{\phi}-\tfrac{1}{2}\boldsymbol{\xi}) e^{-ik\boldsymbol{x}\cdot\boldsymbol{\xi}}, \tag{5.86}$$

so that

$$R(\boldsymbol{\phi}+\tfrac{1}{2}\boldsymbol{\xi})^* R(\boldsymbol{\phi}-\tfrac{1}{2}\boldsymbol{\xi}) = \frac{1}{\lambda^2}\int d\boldsymbol{x}\; \mathcal{B}_R(\boldsymbol{x},\boldsymbol{\phi}) e^{ik\boldsymbol{x}\cdot\boldsymbol{\xi}}. \tag{5.87}$$

By appropriately changing the integration variables, the gain (5.85) can be written as

$$G = \frac{n_e\kappa_h\chi_h}{\lambda^2}\frac{\int d\eta d\boldsymbol{p} d\boldsymbol{\phi} d\boldsymbol{x} d\boldsymbol{y}\; \mathcal{B}_E(\boldsymbol{y},\boldsymbol{\phi})\mathcal{B}_U(\eta,\boldsymbol{x}-\boldsymbol{y},\boldsymbol{\phi}-\boldsymbol{p})\frac{\partial}{\partial\eta}\bar{F}(\eta,\boldsymbol{x},\boldsymbol{p})}{\int d\boldsymbol{\phi} d\boldsymbol{y}\; \mathcal{B}_E(\boldsymbol{y},\boldsymbol{\phi})}, \tag{5.88}$$

where $\mathcal{B}_E$, $\mathcal{B}_U$, and $\bar{F}$ are the brightness functions of the radiation, undulator, and electron beam, respectively. As such, the gain (5.88) is essentially a convolution over the three distribution functions in phase space.

The formula (5.88) can be put into a useful form by assuming a Gaussian distribution for the input electron and radiation beams. Thus, we take

$$\bar{F}(\eta,\boldsymbol{x},\boldsymbol{p};L_u/2) = \frac{e^{-(\eta-\eta_0)^2/2\sigma_\eta^2}}{\sqrt{2\pi}\,\sigma_\eta}\frac{e^{-p^2/2\sigma_p^2}e^{-x^2/2\sigma_x^2}}{(2\pi)^2\sigma_p^2\sigma_x^2} \tag{5.89}$$

$$\mathcal{B}_E(\boldsymbol{y},\boldsymbol{\phi}) = \frac{P_{\text{in}}}{(2\pi)^2\sigma_y^2\sigma_\phi^2}e^{-\phi^2/2\sigma_\phi^2}e^{-y^2/2\sigma_r^2}. \tag{5.90}$$

Calculating the gain is a long exercise in careful Gaussian integration; we will merely quote the result. Defining the detuning $x = (\eta_0 - \Delta\nu/2)k_uL_u$ (x_0 is identical to that defined in Section 4.2) and the (convolved) spatial and angular widths via

$$\Sigma_x^2 \equiv \sigma_x^2 + \sigma_r^2 \qquad\qquad \Sigma_\phi^2 \equiv \sigma_p^2 + \sigma_\phi^2, \tag{5.91}$$

the gain at the harmonic h is given by

$$G = \frac{j_{C,h}}{2}\frac{\sigma_x^2}{\Sigma_x^2}\int_{-1/2}^{1/2} ds \int_{-1/2}^{1/2} dz\; e^{-2[2\pi hN_u(z-s)\sigma_\eta]^2} \times \frac{(z-s)\{\sin[2x_0(z-s)] - i\cos[2x_0(z-s)]\}}{1 + zs\frac{L_u^2\Sigma_\phi^2}{\Sigma_x^2} - i(z-s)\left[kL_u\Sigma_\phi^2 + \frac{L_u}{4k\Sigma_x^2}\right]}. \tag{5.92}$$

This gain formula generalizes the one obtained in Ref. [12], which was derived in the limit that all electrons move parallel to the optical axis (meaning that the emittance is zero). In the 1D limit ($\Sigma_\phi^2 \to 0$ and $L_u/k\Sigma_x^2 \ll 1$) G in Equation (5.85) reduces to the 1D gain formula including energy spread (4.29), so that the integrand is the 3D generalization of the 1D gain function $\bar{g}(x_0)/2$. If we also have $\sigma_\eta \to 0$, then Equation (5.85) becomes the ideal, linear gain formula (4.28) in Section. 4.2 [or Equation (3.38) in Section 3.3].

We plot some typical gain results using formula (5.4.1) in Figure 5.9, where we recall that the electron beam focusing parameter Z_β is defined to be the beta-function at the

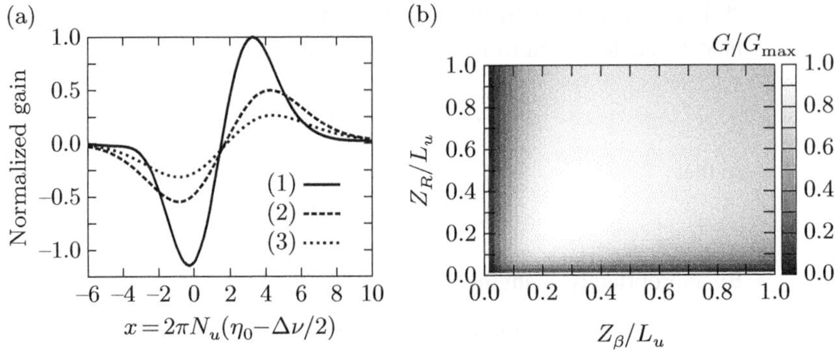

Figure 5.9 (a) Gain curves as a function of the scaled frequency detuning x for (1) $\varepsilon_x = \lambda_1/4\pi$, $\sigma_\eta = 1/6N_u$; (2) $\varepsilon_x = \lambda_1/4\pi$, $\sigma_\eta = 1/3N_u$; and (3) $\varepsilon_x = \lambda_1/2\pi$, $\sigma_\eta = 1/3N_u$. The radiation Rayleigh range Z_R and beam focusing Z_β have been chosen to maximize G, which has been normalized to the maximum gain of (1). (b) Normalized gain, maximized over $\Delta\nu$, plotted as a function of Z_β/L_u and Z_R/L_u for $\varepsilon_x = \lambda_1/4\pi$ and $\sigma_\eta = 1/2N_u$.

waist, $Z_\beta \equiv \sigma_x^2/\varepsilon_x$, while the radiation Rayleigh length $Z_R \equiv 4\pi\sigma_r^2/\lambda_1$. Panel (a) shows three different gain curves as a function of the scaled frequency difference from resonance $x_0 = 2\pi N_u(\eta_0 - \Delta\nu/2)$. The gain decreases as the electron beam brightness decreases, while all are quite similar in shape to the 1D result $\propto \frac{d}{dx}(\sin x/x)^2$. The gain degradation becomes appreciable when the beam emittance becomes larger than that of the radiation, $\varepsilon_x \gtrsim \lambda_1/4\pi$, and/or when the energy spread increases beyond the natural low-gain bandwidth, i.e., when $\sigma_\eta \gtrsim 1/6N_u$.

Figure 5.9(b) plots the maximum normalized gain as a function of the electron and radiation focusing parameters Z_β and Z_R. For a given electron beam focusing parameter, the gain is largest when the radiation mode shape matches that of the electrons, i.e., when $Z_\beta = Z_R$. Furthermore, the global maximum of the gain approximately occurs when the electron beam focusing matches that of the natural undulator radiation, $Z_R \approx L_u/2\pi \approx 0.16L_u$.

5.5 Solution in the High-Gain Regime

A high-quality beam in a sufficiently long undulator can produce FEL radiation that grows exponentially along the length of the undulator. In this high-gain regime the growing modes associated with the fundamental frequency tend to dominate the FEL dynamics. Hence, in what follows we will focus on growing solutions whose wavelength is centered at λ_1.

To analyze a high-gain device, it is convenient for our analysis to scale the field variables according to

$$a_\nu = \frac{\chi_1}{2k_u\rho^2}E_\nu = \frac{eK[\mathrm{JJ}]}{4\gamma_r^2 mc^2 k_u\rho^2}E_\nu, \qquad f_\nu = \frac{2k_u\rho^2}{k_1}F_\nu. \tag{5.93}$$

The electromagnetic scaling is identical to that of Chapter 3, so that we anticipate saturation when

$$\frac{P_{\text{rad}}}{\rho P_{\text{beam}}} = \frac{\lambda_1}{cT}\left\langle \int d\nu d\boldsymbol{x}\ \left|a_\nu(\hat{z})\right|^2 \right\rangle \sim 1, \tag{5.94}$$

where $P_{\text{beam}} = (I/e)\gamma_r mc^2$ is the electron beam power for a beam of peak current I. The distribution function scaling preserves $\int d\hat{p}\ f_\nu$ if we also introduce the dimensionless coordinates

$$\hat{z} = 2\rho k_u z \qquad \hat{\eta} = \frac{\eta}{\rho},$$
$$\hat{\boldsymbol{x}} = \boldsymbol{x}\sqrt{2k_1 k_u \rho} \qquad \hat{\boldsymbol{p}} = \boldsymbol{p}\sqrt{\frac{k_1}{2k_u\rho}}. \tag{5.95}$$

The scaled propagation distance and energy are identical to that introduced in the 1D analysis, while the transverse scaling is set so that the scaled RMS radiation size and divergence are of order unity, with $\sigma_r\sigma_{r'} \sim 1/k_1$. In terms of these dimensionless variables, the governing FEL equations (5.59)–(5.60) are

$$\left(\frac{\partial}{\partial\hat{z}} + i\frac{\Delta\nu}{2\rho} + \frac{\hat{\nabla}_\perp^2}{2i}\right) a_\nu(\hat{\boldsymbol{x}};\hat{z}) = -\int d\hat{\eta}d\hat{\boldsymbol{p}}\, f_\nu(\hat{\eta},\hat{\boldsymbol{x}},\hat{\boldsymbol{p}};\hat{z}) \tag{5.96}$$

$$\left(\frac{\partial}{\partial\hat{z}} + i\nu\dot{\theta} + \hat{\boldsymbol{p}}\cdot\frac{\partial}{\partial\hat{\boldsymbol{x}}} - \hat{k}_\beta^2\hat{\boldsymbol{x}}\cdot\frac{\partial}{\partial\hat{\boldsymbol{p}}}\right) f_\nu = -a_\nu\frac{\partial\bar{f}_0}{\partial\hat{\eta}}, \tag{5.97}$$

where the phase derivative

$$\dot{\theta} = \frac{d\theta}{d\hat{z}} = \hat{\eta} - \frac{\hat{p}^2 + \hat{k}_\beta^2\hat{x}^2}{2} \tag{5.98}$$

describes the inhomogeneous effects of energy spread and emittance, and $\hat{k}_\beta = k_\beta/(2k_u\rho)$ is the scaled focusing strength.

5.5.1 Van Kampen's Normal Mode Expansion

In our previous analysis of the high-gain FEL, we found that there are typically three independent linear solutions in the 1D problem, one of which is the exponentially growing mode that dominates the long-distance dynamics. We would like to find a similar set of linear modes for the FEL in 3D, although in this case the mathematics is a bit more involved. First, there is the complication that in 3D there can be many (possibly infinitely many) growing modes. Second, the linear system (5.96)–(5.97) is not self-adjoint, so that the modes we find are not necessarily orthogonal. One approach to solve this problem is to introduce bi-orthogonal modes as in Ref. [13], while Refs. [14, 15] apply a "multilayer" Laplace transform technique to solve the initial value problem for a beam with vanishing emittance. Here we will use Van Kampen's normal mode expansion [16] to deal with these mathematical issues in a systematic way, developing enough of the mathematics to determine the mode profiles and growth rates of a 3D FEL in the linear regime.

To begin, we introduce the state vector

$$\Psi = \begin{bmatrix} a_\nu(\hat{\boldsymbol{x}};\hat{z}) \\ f_\nu(\hat{\eta},\hat{\boldsymbol{x}},\hat{\boldsymbol{p}},\hat{z}) \end{bmatrix}, \tag{5.99}$$

and define the scalar product of two state vectors by

$$(\Psi_1,\Psi_2) \equiv \int d\hat{\boldsymbol{x}}\; a_{1\nu}a_{2\nu} + \int d\hat{\boldsymbol{x}}d\hat{\boldsymbol{p}}d\eta\, f_{1\nu}f_{2\nu}. \tag{5.100}$$

Equations (5.96) and (5.97) can then be written as

$$\left(\frac{\partial}{\partial\hat{z}} - i\mathbf{M}\right)\Psi = 0, \tag{5.101}$$

where action of the matrix operator **M** on Ψ is

$$\mathbf{M}\Psi(\hat{z}) = \begin{bmatrix} \left(-\frac{\Delta\nu}{2\rho} + \frac{1}{2}\hat{\nabla}_\perp^2\right)a_\nu + i\int d\hat{\boldsymbol{p}}d\hat{\eta}\, f_\nu \\ ia_\nu\frac{\partial\bar{f}_0}{\partial\hat{\eta}} + \left\{-\nu\dot{\theta} + i\left(\hat{\boldsymbol{p}}\cdot\frac{\partial}{\partial\hat{\boldsymbol{x}}} - \hat{k}_\beta^2\hat{\boldsymbol{x}}\cdot\frac{\partial}{\partial\hat{\boldsymbol{p}}}\right)\right\}f_\nu \end{bmatrix}. \tag{5.102}$$

The matrix equation (5.101) is a linear system that generalizes the 1D FEL equations (4.11) and (4.12). The solution can be written as a linear superposition of modes with z dependence $e^{-i\mu_\ell z}$, where the μ_ℓs are the eigenvalues of the operator **M**. Since **M** is not Hermitian, μ_ℓ can be complex and the associated eigenvectors are not orthogonal. We will address the latter issue later, but for now we extract the z-dependence and assume a solution of the form

$$\Psi_\ell(\hat{z}) = e^{-i\mu_\ell\hat{z}}\begin{bmatrix} \mathcal{A}_\ell(\hat{\boldsymbol{x}}) \\ \mathcal{F}_\ell(\hat{\boldsymbol{x}},\hat{\boldsymbol{p}},\hat{\eta}) \end{bmatrix}. \tag{5.103}$$

We now consider one eigenvalue and eigenvector indexed by ℓ and defined by the eigenvalue equation $(\mu_\ell + \mathbf{M})\Psi_\ell = 0$ given by Equation (5.101), which we rewrite as,

$$\begin{bmatrix} \mu_\ell\mathcal{A}_\ell + \left(-\frac{\Delta\nu}{2\rho} + \frac{1}{2}\hat{\nabla}_\perp^2\right)\mathcal{A}_\ell + i\int d\hat{\boldsymbol{p}}d\hat{\eta}\;\mathcal{F}_\ell \\ \mu_\ell\mathcal{F}_\ell + i\mathcal{A}_\ell\frac{\partial\bar{f}_0}{\partial\hat{\eta}} + \left\{-\nu\dot{\theta} + i\left(\hat{\boldsymbol{p}}\cdot\frac{\partial}{\partial\hat{\boldsymbol{x}}} - \hat{k}_\beta^2\hat{\boldsymbol{x}}\cdot\frac{\partial}{\partial\hat{\boldsymbol{p}}}\right)\right\}\mathcal{F}_\ell \end{bmatrix} = 0. \tag{5.104}$$

To find an equation for the radiation mode $\mathcal{A}_\ell$ alone, we must first eliminate the electron distribution $\mathcal{F}_\ell$. We do this by solving for $\mathcal{F}_\ell$ using the second row of Equation (5.104) as follows. First, we introduce the parameter s such that

$$\frac{d\mathcal{F}_\ell}{ds} = \frac{d\hat{\boldsymbol{x}}}{ds}\cdot\frac{\partial\mathcal{F}_\ell}{\partial\hat{\boldsymbol{x}}} + \frac{d\hat{\boldsymbol{p}}}{ds}\cdot\frac{\partial\mathcal{F}_\ell}{\partial\hat{\boldsymbol{p}}} = \hat{\boldsymbol{p}}\cdot\frac{\partial\mathcal{F}_\ell}{\partial\hat{\boldsymbol{x}}} - \hat{k}_\beta^2\hat{\boldsymbol{x}}\cdot\frac{\partial\mathcal{F}_\ell}{\partial\hat{\boldsymbol{p}}}. \tag{5.105}$$

Thus, s can be thought of as the parameter along the transverse motion, and the following procedure is closely related to our low-gain analysis in which we integrated along the transverse characteristic curves associated with motion in the focusing channel. Next, we rewrite the second row of (5.104) as

$$e^{-i(\nu\dot{\theta}-\mu_\ell)s}\frac{d}{ds}\left[e^{i(\nu\dot{\theta}-\mu_\ell)s}\mathcal{F}_\ell\right] = -\mathcal{A}_\ell\frac{\partial\bar{f}_0}{\partial\hat{\eta}}, \tag{5.106}$$

which can be easily integrated to yield

$$\mathcal{F}_\ell = -e^{-i(\nu\dot{\theta}-\mu_\ell)s}\int^{s} ds'\, e^{i(\nu\dot{\theta}-\mu_\ell)s'}\mathcal{A}_\ell[\hat{\boldsymbol{x}}_0(s-s')]\frac{\partial}{\partial\hat{\eta}}\bar{f}_0[\hat{\boldsymbol{p}}_0^2+\hat{k}_\beta^2\hat{\boldsymbol{x}}_0^2,\eta]. \tag{5.107}$$

Here, the transverse motion is given by the solutions (5.58); for example, the position $\hat{\boldsymbol{x}}_0(s-s') = \hat{\boldsymbol{x}}\cos[k_\beta(s-s')] - (\hat{\boldsymbol{p}}/\hat{k}_\beta)\sin[k_\beta(s-s')]$. To simplify (5.107), we now recall that $\hat{\boldsymbol{p}}^2 + \hat{k}_\beta^2\hat{\boldsymbol{x}}^2$ is conserved, which implies that the matched $\bar{f}_0$ is independent of s' and that $\dot{\theta} = \hat{\eta} - (\hat{\boldsymbol{p}}^2 + \hat{k}_\beta^2\hat{\boldsymbol{x}}^2)/2$ is a constant of the unperturbed motion. We find it convenient to change the integration variable to $\tau \equiv s' - s$ and define the coordinate $\hat{\boldsymbol{x}}_+(\tau) \equiv \hat{\boldsymbol{x}}\cos(\hat{k}_\beta\tau) + (\hat{\boldsymbol{p}}/\hat{k}_\beta)\sin(\hat{k}_\beta\tau)$. For the growing solution we can ignore the initial conditions and extend the integration over τ to $-\infty$, in which case Equation (5.107) simplifies to

$$\mathcal{F}_\ell = -\frac{\partial\bar{f}_0}{\partial\hat{\eta}}\int_{-\infty}^{0} d\tau\, \mathcal{A}_\ell(\hat{\boldsymbol{x}}_+)e^{i(\nu\dot{\theta}-\mu_\ell)\tau}. \tag{5.108}$$

Substituting this into the first row of Equation (5.104) yields the mode equation for the radiation profile and growth rate

$$\begin{aligned}&\left(\mu_\ell - \frac{\Delta\nu}{2\rho} + \frac{1}{2}\hat{\nabla}_\perp^2\right)\mathcal{A}_\ell(\hat{\boldsymbol{x}})\\ &\quad - i\int d\hat{\boldsymbol{p}}d\hat{\eta}\int_{-\infty}^{0} d\tau\, e^{i(\nu\dot{\theta}-\mu_\ell)\tau}\frac{d\bar{f}_0}{d\hat{\eta}}\mathcal{A}_\ell(\hat{\boldsymbol{x}}_+) = 0.\end{aligned} \tag{5.109}$$

The expression (5.109) is often called the dispersion relation [17, 3]. Solutions to (5.109) determine the transverse eigenmodes and corresponding eigenvalues for the radiation field as a function of the frequency difference from resonance $\Delta\nu$. Thus, the dispersion relation describes the essential properties of the radiation in terms of mode profiles and growth rates. In the following section we discuss in more detail both the physics of and some approximate solutions to Equation (5.109). Prior to this, we show how to solve the 3D FEL initial value problem once the eigenvectors and eigenvalues are known.

Assuming that the set of eigenvectors is complete, the solution to the initial value problem can be found by expanding the initial state $\Psi(0)$ as a sum of the Ψ_ℓ:

$$\Psi(0) = \sum_\ell C_\ell\Psi_\ell. \tag{5.110}$$

If one can find another basis set $\{\Phi_k\}$ such that the scalar product $(\Phi_k, \Psi_\ell) = \delta_{k,\ell}$, than each expansion coefficient C_ℓ can be extracted by projecting the state $\Psi(0)$ onto Φ_ℓ. Van Kampen's method constructs such an orthogonal basis set $\{\Phi_k\} = \{\Psi_k^\dagger\}$ using the adjoint eigenvalue equation [16, 18]

$$\left(\mu_\ell^\dagger + \mathbf{M}^\dagger\right)\Psi_\ell^\dagger = 0. \tag{5.111}$$

Here, $\mu_\ell^\dagger$ and $\Psi_\ell^\dagger = (A_\ell^\dagger, \mathcal{F}_\ell^\dagger)$ are the eigenvalues and eigenvectors of the adjoint operator $\mathbf{M}^\dagger$, defined by the scalar product relationship

$$\left(\mathbf{M}^\dagger \Psi_\ell^\dagger, \Psi\right) = \left(\Psi_\ell^\dagger, \mathbf{M}\Psi\right) \tag{5.112}$$

for all Ψ_ℓ. This implies that $\mathbf{M}^\dagger$ acting on $\Psi_\ell^\dagger$ is

$$\mathbf{M}^\dagger \Psi_\ell^\dagger = \begin{bmatrix} \left(-\frac{\Delta\nu}{2\rho} + \frac{1}{2}\hat{\nabla}_\perp^2\right) \mathcal{A}_\ell^\dagger + i\int d\hat{\boldsymbol{p}} d\hat{\boldsymbol{x}} d\hat{\eta}\, \frac{\partial \bar{f}_0}{\partial \hat{\eta}} \mathcal{F}_\ell^\dagger \\ i\mathcal{A}_\ell^\dagger + \left\{-\nu\dot{\theta} - i\left(\hat{\boldsymbol{p}} \cdot \frac{\partial}{\partial \hat{\boldsymbol{x}}} - \hat{k}_\beta^2 \hat{\boldsymbol{x}} \cdot \frac{\partial}{\partial \hat{\boldsymbol{p}}}\right)\right\} \mathcal{F}_\ell^\dagger \end{bmatrix}. \tag{5.113}$$

Putting this into Equation (5.111) and solving for $\mathcal{F}_\ell^\dagger$ in a manner similar to what we did for $\mathcal{F}_\ell$, we find that

$$\mathcal{F}_\ell^\dagger = -\frac{\partial \bar{f}_0}{\partial \eta} \int_{-\infty}^{0} d\tau\; \mathcal{A}_\ell^\dagger(\hat{\boldsymbol{x}}_-) e^{i(\nu\dot{\theta} - \mu_\ell^\dagger)\tau}, \tag{5.114}$$

where $\hat{\boldsymbol{x}}_- = \hat{\boldsymbol{x}}\cos(\hat{k}_\beta \tau) - (\hat{\boldsymbol{p}}/\hat{k}_\beta)\sin(\hat{k}_\beta \tau)$. We use (5.114) to solve for the adjoint radiation mode and find that $\mathcal{A}_\ell^\dagger$ satisfies the same dispersion relation (5.109) but with $\hat{\boldsymbol{x}}_+ \to \hat{\boldsymbol{x}}_-$. Since the matched beam distribution $\bar{f}_0$ is symmetric in $\hat{\boldsymbol{p}}$, by changing the integration variable $\hat{\boldsymbol{p}} \to -\hat{\boldsymbol{p}}$ we find that these two dispersion relations are actually identical. Hence we set $\mathcal{A}_\ell^\dagger = \mathcal{A}_\ell$ and $\mu_\ell^\dagger = \mu_\ell$.

By virtue of the adjoint definition (5.112), one immediately derives

$$(\mu_\ell - \mu_k)\left(\Psi_k^\dagger, \Psi_\ell\right) = \left(\Psi_k^\dagger, \mathbf{M}\Psi_\ell\right) - \left(\mathbf{M}^\dagger \Psi_k^\dagger, \Psi_\ell\right) = 0. \tag{5.115}$$

If the normal modes are not degenerate, i.e., if $\mu_k \neq \mu_\ell$ for any $k \neq \ell$, then the set of adjoint eigenvectors are orthogonal to the eigenvectors Ψ_ℓ in precisely the manner required to extract the expansion coefficients from (5.110). Specifically, for a discrete set of eigenvectors, the Van Kampen orthogonality condition is

$$\left(\Psi_k^\dagger, \Psi_l\right) = \delta_{k,\ell}\left(\Psi_l^\dagger, \Psi_\ell\right). \tag{5.116}$$

In a similar manner, one can write down an orthogonality condition for a continuous set of eigenvectors using the Dirac δ-function instead of the Kronecker $\delta_{k,\ell}$. Assuming that the set Ψ_ℓ is complete, we can use (5.116) to express any state vector Ψ as a sum of eigenvectors by projecting the components of Ψ onto the adjoint basis. The expansion of the initial state $\Psi(0)$ given by (5.110) can therefore be written as

$$\Psi(\hat{z}) = \sum_\ell C_\ell \Psi_\ell e^{-i\mu_\ell \hat{z}} = \sum_\ell \frac{\left(\Psi_\ell^\dagger, \Psi(0)\right)}{\left(\Psi_\ell^\dagger, \Psi_\ell\right)} \Psi_\ell e^{-i\mu_\ell \hat{z}}. \tag{5.117}$$

For the FEL, the initial state vector $\Psi(0)$ consists of the external signal $a_\nu(0)$ and the νth Fourier component of the bunching $f_\nu(0)$, which will describe the shot noise along with any externally imposed signal. Thus, we have

$$\left(\Psi_\ell^\dagger, \Psi(0)\right) = \int d\hat{\boldsymbol{x}}\ a_\nu(\hat{\boldsymbol{x}};0)\mathcal{A}_\ell(\hat{\boldsymbol{x}}) - \int d\hat{\boldsymbol{x}} d\hat{\boldsymbol{p}} d\hat{\eta}\, f_\nu(\hat{\eta},\hat{\boldsymbol{x}},\hat{\boldsymbol{p}};0) \int_{-\infty}^{0} d\tau\ \mathcal{A}_\ell(\hat{\boldsymbol{x}}_-)e^{i(\nu\dot{\theta}-\mu_\ell)\tau}, \tag{5.118}$$

and

$$\left(\Psi_\ell^\dagger, \Psi_\ell\right) = \int d\hat{\boldsymbol{x}}\ \mathcal{A}_\ell(\hat{\boldsymbol{x}})\mathcal{A}_\ell(\hat{\boldsymbol{x}}) + \int d\hat{\boldsymbol{x}} d\hat{\boldsymbol{p}} d\hat{\eta}\ \frac{\partial \bar{f}_0}{\partial \hat{\eta}} \int_{-\infty}^{0} d\tau d\tau'\ \mathcal{A}_\ell(\hat{\boldsymbol{x}}_+)\mathcal{A}_\ell(\hat{\boldsymbol{x}}'_-)e^{i(\nu\dot{\theta}-\mu_\ell)(\tau+\tau')}. \tag{5.119}$$

5.5.2 Dispersion Relation with Four Scaled Parameters

In what follows, we assume that the smooth background distribution is uniform in θ (coasting beam), and is Gaussian distributed in energy, the transverse positions, and the transverse angles:

$$\bar{f}_0(\hat{\boldsymbol{p}}^2 + \hat{k}_\beta^2\hat{\boldsymbol{x}}^2) = \frac{1}{(2\pi)^2\hat{k}_\beta\hat{\sigma}_x^2}\exp\left(-\frac{\hat{\boldsymbol{p}}^2 + \hat{k}_\beta^2\hat{\boldsymbol{x}}^2}{2\hat{k}_\beta^2\hat{\sigma}_x^2}\right)\frac{1}{\sqrt{2\pi}\hat{\sigma}_\eta}\exp\left(-\frac{\hat{\eta}^2}{2\hat{\sigma}_\eta^2}\right), \tag{5.120}$$

where

$$\hat{\sigma}_x = \sigma_x\sqrt{2k_1k_u\rho} \quad \text{and} \quad \hat{\sigma}_\eta = \sigma_\eta/\rho \tag{5.121}$$

are the scaled RMS transverse beam size and the scaled RMS energy spread, respectively. The electron beam emittance is given by

$$\varepsilon_x = \hat{\sigma}_x^2\hat{k}_\beta/k_1. \tag{5.122}$$

We rewrite the dispersion relation (5.109) by performing the $\hat{\eta}$ integral with the help of integration by parts. Dropping the subscripts ℓ for clarity, the dispersion relation for a beam that is Gaussian in energy is

$$\left(\mu - \frac{\Delta\nu}{2\rho} + \frac{1}{2}\hat{\nabla}_\perp^2\right)\mathcal{A}(\hat{\boldsymbol{x}}) - \frac{1}{2\pi\hat{k}_\beta^2\hat{\sigma}_x^2}\int_{-\infty}^{0} d\tau\ \tau e^{-\hat{\sigma}_\eta^2\tau^2/2 - i\mu\tau} \times \int d\hat{\boldsymbol{p}}\ \mathcal{A}[\hat{\boldsymbol{x}}_+(\hat{\boldsymbol{x}},\hat{\boldsymbol{p}},\tau)]\exp\left[-\frac{1+i\tau\hat{k}_\beta^2\hat{\sigma}_x^2}{2\hat{k}_\beta^2\hat{\sigma}_x^2}\left(\hat{\boldsymbol{p}}^2 + \hat{k}_\beta^2\hat{\boldsymbol{x}}^2\right)\right] = 0. \tag{5.123}$$

The complex growth rate μ and the mode profile $\mathcal{A}(\hat{\boldsymbol{x}})$ are completely determined by four scaled parameters [3], which here we have defined as $\hat{\sigma}_x$, $\hat{k}_\beta$, $\hat{\sigma}_\eta$, and $\Delta\nu/2\rho$. Let us first discuss the physical interpretation of these four scaled parameters.

1. $\hat{\sigma}_x$ represents the *diffraction* effect. To see this, consider

$$\hat{\sigma}_x^2 = \sigma_x^2 2k_1k_u\rho = \frac{2\pi\sigma_x^2}{\lambda_1}\frac{4\pi\rho}{\lambda_u} = \frac{2}{\sqrt{3}}\frac{Z_R}{L_{G0}}, \tag{5.124}$$

where $Z_R = \pi\sigma_x^2/\lambda_1$ is the Rayleigh length of radiation whose transverse size is defined by the electron beam size. To mitigate the diffraction effect, we need

$$Z_R \gtrsim L_{G0} \quad \text{or} \quad \hat{\sigma}_x \gtrsim 1. \tag{5.125}$$

2. $\hat{\sigma}_x^2\hat{k}_\beta^2$ represents the *emittance* or *angular spread* effect. From the resonance condition equation (5.6), the beam angular spread $\sigma_{x'} = k_\beta\sigma_x$ inevitably introduces a spread in the resonant wavelength given by

$$(\hat{\sigma}_x\hat{k}_\beta)^2 = \left(\sigma_{x'}^2\frac{k_1}{2k_u}\right)\frac{1}{\rho} \to 4\pi\sqrt{3}\frac{\Delta\lambda_1}{\lambda_1}\frac{L_{G0}}{\lambda_u} = \frac{1}{\rho}\frac{\Delta\lambda}{\lambda_1}. \tag{5.126}$$

The resonance condition can be maintained if the induced wavelength spread $\Delta\lambda/\lambda_1 < \rho$, i.e., if

$$(\hat{\sigma}_x\hat{k}_\beta)^2 \approx \frac{4\pi L_G}{\lambda_1\bar{\beta}}\varepsilon_x < 1. \tag{5.127}$$

This is the same requirement that we derived in (5.8). Furthermore, in order for the radiation to be transversely coherent, the electron beam emittance should not be significantly larger than the radiation emittance. Rewriting Equation (5.122) as $\hat{\sigma}_x^2\hat{k}_\beta = \varepsilon_x/2\varepsilon_r$ with $\varepsilon_x/2\varepsilon_r \lesssim 1$, we see that both of these requirements can be satisfied if the scaled focusing $\hat{k}_\beta \lesssim 1$.

3. $\hat{\sigma}_\eta$ represents the effect of *energy spread*. The variation of the resonant wavelength due to the beam energy spread must be less than the natural FEL bandwidth, i.e.,

$$\hat{\sigma}_\eta = \frac{\Delta\gamma}{\gamma\rho} \to \frac{\Delta\lambda_1}{2\lambda_1}\frac{1}{\rho} = 2\pi\sqrt{3}\frac{\Delta\lambda_1}{\lambda_1}\frac{L_{G0}}{\lambda_u} \lesssim 1. \tag{5.128}$$

If this condition is satisfied, than the change in the ponderomotive phase over one gain length due to off-resonant electron energies is much less than unity.

4. $\Delta\nu/2\rho$ represents the *detuning* effect, which quantifies the deviation of the wavelength from the 1D resonant condition. The frequency bandwidth with significant gain is

$$\sigma_\nu \sim \frac{\Delta\lambda_1}{\lambda_1} \sim \rho. \tag{5.129}$$

The first parameter $\hat{\sigma}_x$ quantifies the effectiveness of gain guiding, while the last three represent deviations from the resonance condition due to emittance, energy spread, and frequency detuning, respectively. The gain is adversely affected when the total normalized frequency difference is $\gtrsim \rho$.

5.5.3 Gain Guiding and Transverse Coherence

Solving the dispersion relation (5.123) is in general not analytically possible. We will discuss numerical approaches to determining the growing FEL eigenvalues and eigenmodes of realistic electron distributions in the next section. Here, however, we present an idealized electron beam model that can be solved exactly, as this will provide useful insight into the general structure of the solutions, along with qualitatively accurate 3D eigenmode profiles and growth rates.

Our simple model will focus on the transverse effects of gain guiding, ignoring the gain degrading effects associated with nonzero emittance and energy spread. Thus, we assume that the electron beam has zero divergence ($\sigma_p = k_\beta = 0$), vanishing energy spread, and a finite transverse envelope $U(\hat{\boldsymbol{x}})$. In this model the smooth background distribution function is given by

$$\bar{f}_0(\hat{\eta}, \hat{\boldsymbol{x}}, \hat{\boldsymbol{p}}) = \delta(\hat{\eta})\delta(\hat{\boldsymbol{p}})U(\hat{\boldsymbol{x}}). \tag{5.130}$$

Inserting (5.130) into the dispersion relation (5.123) and changing to the polar coordinates $\hat{r} \equiv |\hat{\boldsymbol{x}}|$ and $\phi \equiv \tan^{-1}(\hat{y}/\hat{x})$ yields

$$\frac{1}{2}\left[\frac{1}{\hat{r}}\frac{\partial}{\partial \hat{r}}\left(\hat{r}\frac{\partial}{\partial \hat{r}}\right) + \frac{1}{\hat{r}^2}\frac{\partial^2}{\partial \phi^2}\right]\mathcal{A}_\ell(\hat{\boldsymbol{x}}) + \left[\mu_\ell - \frac{\Delta\nu}{2\rho} - \frac{U(\hat{\boldsymbol{x}})}{\mu_\ell^2}\right]\mathcal{A}_\ell(\hat{\boldsymbol{x}}) = 0. \tag{5.131}$$

The Equation (5.131) is essentially the 2D Schrödinger equation with complex coefficients, and a few exact solutions exist for certain transverse beam envelopes $U(\hat{\boldsymbol{x}})$. One example given in the literature [1, 13] is for a finite but uniform (square) distribution. A particularly elegant solution also exists for a parabolic approximation to a Gaussian [19], namely

$$U(\hat{\boldsymbol{x}}) = 1 - \frac{\hat{r}^2}{2\hat{\sigma}_x^2} = 1 - \frac{|\boldsymbol{x}|^2}{2\sigma_x^2}. \tag{5.132}$$

The parabolic beam profile equation (5.132) should be limited to $|\boldsymbol{x}|^2 < 2\sigma_x^2$ with $U = 0$ otherwise to avoid negative currents, as was done in Ref. [14]. However, this significantly complicates the eigenmode equation, and we will show that the eigenmodes obtained using (5.132) equal those of the more physical current distribution when $\hat{\sigma}_x \gg 1$. Furthermore, since the solutions are quite similar and much simpler for finite $\hat{\sigma}_x$, we assume that the quadratic profile is valid everywhere.

For the current profile (5.132), separation of variables in cylindrical coordinates implies that $\mathcal{A}_\ell(\hat{\boldsymbol{x}}) = \mathcal{A}_{\ell,m}(\hat{r})e^{im\phi}$. We introduce the coordinate $y = i\hat{r}^2/\mu_\ell\hat{\sigma}_x$, noting that $\Re(y) \geq 0$ where (5.131) is valid, namely, for the growing modes with $\Im(\mu) > 0$. Expressing the dispersion relation for the parabolic beam (5.132) in terms of y, we have

$$\left[\frac{\partial}{\partial y}\left(y\frac{\partial}{\partial y}\right) - \frac{m^2}{4y} + \frac{\mu_\ell\hat{\sigma}_x}{2i}\left(\mu_\ell - \frac{\Delta\nu}{2\rho} - \frac{1}{\mu_\ell^2}\right) - \frac{y}{4}\right]\mathcal{A}_{\ell,m}(y) = 0. \tag{5.133}$$

Since the potential is quadratic, this equation is a complex version of the 2D simple harmonic oscillator Schrödinger equation, albeit in different variables, and we only briefly present its solution here. The behavior of the mode at large y can be easily determined from (5.133), being given by $\mathcal{A}'' = \mathcal{A}/4$, so that $\mathcal{A} \sim e^{-y/2}$ in the limit $|y| \to \infty$. To put the eigenmode equation (5.133) in standard form, we extract the exponential dependence and eliminate the term $\sim m^2/4y$ by defining $\alpha_{\ell,m}(y)$ via

$$\mathcal{A}_{\ell,m}(y) = y^{m/2}e^{-y/2}\alpha_{\ell,m}(y). \tag{5.134}$$

Inserting this definition into Equation (5.133) results in Laguerre's differential equation for the field $\alpha_{\ell,m}(y)$:

$$y\alpha''_{\ell,m}(y) + (1 + m - y)\alpha'_{\ell,m}(y) \\ -\frac{1}{2}\left[m + 1 + i\mu_\ell\hat{\sigma}_x\left(\mu_\ell - \frac{\Delta\nu}{2\rho} - \frac{1}{\mu_\ell^2}\right)\right]\alpha_{\ell,m}(y) = 0. \tag{5.135}$$

The only non-singular solution for integer m is the generalized Laguerre function, $\alpha_{\ell,m}(y) = \mathrm{L}_\ell^m(y)$, where -2ℓ is equal to the term in square brackets from (5.135). It can be shown that $\mathrm{L}_\ell^m(y) \sim e^y/y^{\ell+m+1}$ if ℓ is not a positive integer, in which case the field $\mathcal{A}$ would diverge exponentially ($\sim e^{y/2}$) at infinity. Thus, requiring bounded solutions implies that ℓ is a nonnegative integer, meaning that the eigenvalue μ_ℓ must satisfy

$$\mu_\ell^2\left(\mu_\ell - \frac{\Delta\nu}{2\rho}\right) - 1 = \frac{i\mu_\ell}{\hat{\sigma}_x}(2\ell + m + 1), \text{ for } \ell \geq 0,\ \ell \in \mathbb{Z} \text{ and } m \in \mathbb{Z}. \tag{5.136}$$

We plot the growth rates for several mode numbers as a function of $1/\hat{\sigma}_x$ in Figure 5.10(a). Note that as $\hat{\sigma}_x \to \infty$, the growth rates for all modes become degenerate, with each satisfying the cold, 1D dispersion relation. The eigenmodes themselves are given by the Gauss–Laguerre functions

$$\mathcal{A}_{\ell,m}(\hat{r}) = \left(\frac{i\hat{r}^2}{\mu_\ell\hat{\sigma}_x}\right)^{m/2} \mathrm{L}_\ell^m\left(\frac{i\hat{r}^2}{\mu_\ell\hat{\sigma}_x}\right)\exp\left(-\frac{i\hat{r}^2}{2\mu_\ell\hat{\sigma}_x}\right). \tag{5.137}$$

Figure 5.10(b) shows a few of the lowest-order modes. The RMS width of the fundamental mode $\mathcal{A}_{0,0}$ can be easily determined using $|\mathcal{A}_{0,0}|^2 \propto e^{-r^2/2\sigma_r^2}$; we have

$$\sigma_r^2 = \frac{|\mu_\ell|^2\,\sigma_x}{2\Im(\mu_\ell)\sqrt{2\rho k_1 k_u}}, \tag{5.138}$$

and the radiation mode size is proportional to the geometric mean of the FEL mode size $\sim (\rho k_1 k_u)^{-1/2}$ and the electron beam size σ_x. Thus, for sufficiently large electron beams

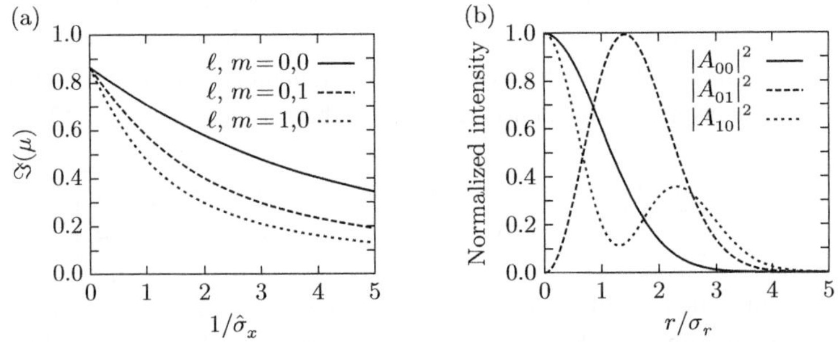

Figure 5.10 Transverse mode structure for the parabolic beam. (a) Growth rate as a function of the inverse of the scaled electron beam size, maximized with respect to $\Delta\nu$. All modes approach the 1D growth rate $\sqrt{3}/2 \approx 0.866$ as $\hat{\sigma}_x \to \infty$, and decrease as the beam size decreases. The gain of the lowest-order Gaussian mode G_{00} decreases the slowest, so that at finite beam size it tends to dominate the other transverse modes. In panel (b) we plot the intensity profile of the three lowest-order modes, each being normalized to its peak value with $\hat{\sigma}_x \gg 1$. The mode shapes look quite similar to those shown in Figure 5.11, which were numerically computed for the LCLS.

that satisfy $\hat{\sigma}_x \gg 1$, the radiation modes have $\sigma_r \ll \sigma_x$. In this limit our parabolic approximation equation (5.132) is valid even though the current is negative when $|\hat{\boldsymbol{x}}|^2 > 2\sigma_x^2$, and we find that the low-order Gauss–Laguerre eigenmodes (5.137) and growth rates (5.136) are nearly identical to those of a completely nonnegative current density.

5.5.4 Numerically Solving the Dispersion Relation

Computing the growing FEL modes including nonzero emittance and energy spread requires a more general approach. In the next section we write down semi-analytic, approximate solutions using the variational technique. Here, on the other hand, we present a numerical method for solving the dispersion relation (5.123). To facilitate these analyses, we open this section by simplifying (5.123), after which we discuss how one can solve the resulting equation numerically.

To solve the dispersion relation equation (5.123), we first make a change of variable

$$\hat{\boldsymbol{x}}' = \hat{\boldsymbol{x}} \cos(\hat{k}_\beta \tau) + \frac{\hat{\boldsymbol{p}}}{\hat{k}_\beta} \sin(\hat{k}_\beta \tau), \tag{5.139}$$

which is accompanied by the following transformations:

$$d\hat{\boldsymbol{p}} = d\hat{\boldsymbol{x}}' \frac{\hat{k}_\beta^2}{\sin^2(\hat{k}_\beta \tau)},$$

$$\hat{\boldsymbol{p}}^2 + \hat{k}_\beta^2 \hat{\boldsymbol{x}}^2 = \frac{\hat{k}_\beta^2}{\sin^2(\hat{k}_\beta \tau)} \left[\hat{\boldsymbol{x}}^2 + \hat{\boldsymbol{x}}'^2 - 2\hat{\boldsymbol{x}} \cdot \hat{\boldsymbol{x}}' \cos(\hat{k}_\beta \tau) \right].$$

Equation (5.123) then becomes

$$\left(\mu - \frac{\Delta\nu}{2\rho} + \frac{1}{2}\hat{\nabla}_\perp^2 \right) \mathcal{A}(\hat{\boldsymbol{x}}) - \int_{-\infty}^{0} d\tau \frac{\tau e^{-\hat{\sigma}_\eta^2 \tau^2/2 - i\mu\tau}}{2\pi \hat{\sigma}_x^2 \sin^2(\hat{k}_\beta \tau)} \int d\hat{\boldsymbol{x}}' \, \mathcal{A}(\hat{\boldsymbol{x}}')$$
$$\times \exp\left[-\frac{1 + i\hat{k}_\beta^2 \hat{\sigma}_x^2 \tau}{2\hat{\sigma}_x^2} \frac{\hat{\boldsymbol{x}}^2 + \hat{\boldsymbol{x}}'^2 - 2\hat{\boldsymbol{x}} \cdot \hat{\boldsymbol{x}}' \cos(\hat{k}_\beta \tau)}{\sin^2(\hat{k}_\beta \tau)} \right] = 0.$$

The typical electron beam is approximately axially symmetric, so that we will find it it convenient to employ the polar coordinates (R, φ) with

$$R = \frac{|\boldsymbol{x}|}{\sigma_x} = \frac{|\hat{\boldsymbol{x}}|}{\hat{\sigma}_x}, \tag{5.140}$$

while φ is the azimuthal angle. Denoting the azimuthal mode number by m, we write

$$\mathcal{A}(\hat{\boldsymbol{x}}) = \mathcal{A}_m(R) e^{im\varphi}, \tag{5.141}$$

in which case the transverse differentials and derivatives are given by

$$\int d\hat{\boldsymbol{x}}' = \hat{\sigma}_x^2 \int_0^\infty R' dR' \int_0^{2\pi} d\varphi, \qquad \hat{\nabla}_\perp^2 = \frac{1}{\hat{\sigma}_x^2} \left[\frac{1}{R} \frac{d}{dR} \left(R \frac{d}{dR} \right) - \frac{m^2}{R^2} \right].$$

Finally, using the definition of the modified Bessel function of order m

$$I_m(\xi) = \frac{1}{2\pi}\int_0^{2\pi} d\varphi\, e^{im\varphi+\xi\cos\varphi}, \tag{5.142}$$

the dispersion relation can be written as

$$\left\{\mu - \frac{\Delta\nu}{2\rho} + \frac{1}{2\hat{\sigma}_x^2}\left[\frac{1}{R}\frac{d}{dR}\left(R\frac{d}{dR}\right) - \frac{m^2}{R^2}\right]\right\}\mathcal{A}_m(R) = \int_0^\infty R'dR'\, G_m(R,R')\mathcal{A}_m(R'), \tag{5.143}$$

where

$$\begin{aligned} G_m(R,R') = &\int_{-\infty}^0 d\tau\, \frac{\tau}{\sin^2(\hat{k}_\beta\tau)} I_m\left[\frac{RR'(1+i\hat{k}_\beta^2\hat{\sigma}_x^2\tau)\cos(\hat{k}_\beta\tau)}{\sin^2(\hat{k}_\beta\tau)}\right] \\ &\times \exp\left[-\frac{\hat{\sigma}_\eta^2\tau^2}{2} - i\mu\tau - \frac{(R^2+R'^2)(1+i\hat{k}_\beta^2\hat{\sigma}_x^2\tau)}{2\sin^2(\hat{k}_\beta\tau)}\right]. \end{aligned} \tag{5.144}$$

Variational [3, 20] and matrix [20] methods can be used to solve the eigenvalue equation (5.143) for the mode profiles and associated growth rates. In the following, we briefly illustrate the matrix approach to (5.143), reserving the variational solution to the subsequent section. To obtain a more convenient formula for the matrix solutions, we follow Ref. [20] and introduce the Hankel transform pair for the radial mode $\mathcal{A}_m(R)$:

$$A_m(Q) = \int_0^\infty RdR\, J_0(QR)\mathcal{A}_m(R), \qquad \mathcal{A}_m(R) = \int_0^\infty QdQ\, J_0(QR)A_m(Q).$$

Using the formulae

$$\int_0^\infty xdx\, e^{-\alpha x^2} I_\nu(\beta x)J_\nu(\gamma x) = \frac{1}{2\alpha}e^{\frac{\beta^2-\gamma^2}{4\alpha}} J_\nu(\beta\gamma/2\alpha),$$

$$\int_0^\infty xdx\, e^{-\alpha x^2} J_\nu(\beta x)J_\nu(\gamma x) = \frac{1}{2\alpha}e^{-\frac{\beta^2-\gamma^2}{4\alpha}} I_\nu(\beta\gamma/2\alpha),$$

Equation (5.143) can be converted to an integral equation in Q space [20]

$$A_m(Q) = \int_0^\infty Q'dQ'\, T_m(Q,Q')A_m(Q') \tag{5.145}$$

with the kernel

$$T_m(Q,Q') = \frac{1}{\mu - \Delta\nu/2\rho - Q^2/2\hat{\sigma}_x^2} \int_{-\infty}^{0} \frac{\tau d\tau}{(1+i\hat{k}_\beta^2\hat{\sigma}_x^2\tau)^2} I_m\left[\frac{QQ'\cos(\hat{k}_\beta\tau)}{1+i\hat{k}_\beta^2\hat{\sigma}_x^2\tau}\right]$$
$$\times \exp\left[-\frac{\hat{\sigma}_\eta^2\tau^2}{2} - i\mu\tau - \frac{Q^2+Q'^2}{2(1+i\hat{k}_\beta^2\hat{\sigma}_x^2\tau)}\right].$$

A numerical solution to (5.145) can be found by approximating the integral operator with a matrix in the following way. We evaluate the kernel $T_m(Q,Q')$ on a finite number of suitably chosen points $Q_1, Q_2, \ldots, Q_N$ for both Q and Q', in which case Equation (5.145) can be approximated by the matrix equation

$$[\mathbf{I} - \mathsf{T}_m(\mu)]\boldsymbol{A}_m = 0. \tag{5.146}$$

Here $\mathbf{I}$ is the identity matrix, while the matrix element approximation to the integrand $QdQ\, T_m(Q,Q')$ is

$$\mathsf{T}_m^{nn'} = Q_{n'}(Q_{n'} - Q_{n'-1})T_m(Q_n, Q_{n'}), \text{ for } n, n' = 1,2,\ldots,N \text{ and } Q_0 = 0.$$

The vector $\boldsymbol{A}_m$ represents the eigenmode $A_m(Q)$ at the discrete points $Q = Q_1, Q_2, \ldots, Q_N$. For a given detuning $\Delta\nu$, one can use numerical iteration to determine the eigenvalue $\mu_{\ell m}$ of the matrix $(\mathbf{I} - \mathsf{T}_m)$, where ℓ represents the radial mode number that denotes the number of nodes in the radial direction. Which eigenvalue is found will depend on the initial guess for both μ and the transverse profile. For example, the largest growing eigenvalue can usually be found by using the μ determined from 1D theory as the initial value. Each eigenvalue $\mu_{\ell m}$ will have an associated eigenmode $A_{\ell m}(Q)$, from which the transverse profile $\mathcal{A}_{\ell m}(R)$ can be found via Hankel transformation.

As a concrete example, we plot in Figure 5.11 the three lowest-order modes for the LCLS's original design parameters from the Conceptual Design Report [3] (these differ from the achieved parameters in Table 8.1). In this case, we have $\hat{\sigma}_x = 2.8$, $\hat{\sigma}_\eta = 0.45$, and $\hat{k}_\beta = 0.29$. The fundamental guided mode (the A_{00} mode) has a complex growth rate $\mu_{00} = -1.2 + 0.42i$ at the optimal detuning $\Delta\nu = -2.0\rho$. The next two lowest-order modes A_{01} and A_{10} have complex growth rates [20] $\mu_{01} = -1.3 + 0.26i$ and $\mu_{10} = -1.3 + 0.10i$, respectively. The lowest-order mode is nearly Gaussian, while the higher-order profiles are qualitatively similar to those of the analytic model of Section 5.5.3 plotted in Figure 5.10.

Among all transverse modes, the one with the largest imaginary part of the growth rate is called the fundamental mode. We denote its eigenvalue by μ_{00}. The fundamental mode typically possesses a near-Gaussian profile whose width is comparable to that of the electron beam. This shape typically has the highest gain because it reduces the effects of diffraction while maximizing the spatial overlap with the electron beam. Higher-order modes suffer from greater diffraction and/or reduced FEL coupling, implying that the

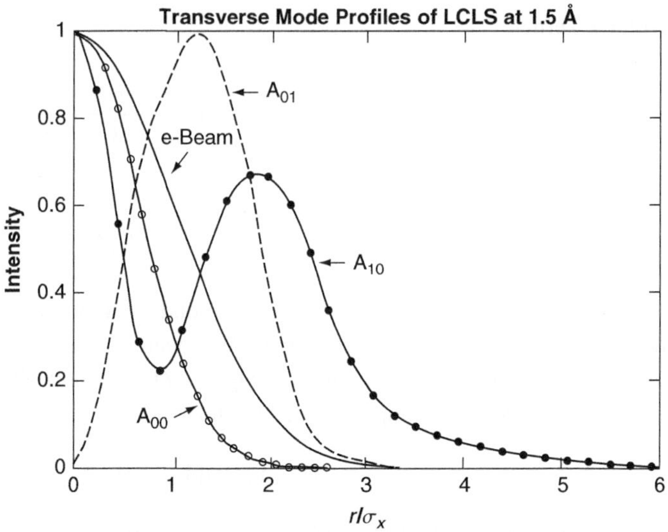

Figure 5.11 Three transverse mode profiles with largest gain for LCLS parameters. Courtesy of M. Xie.

gain lengths associated with these modes are much longer. In the high-gain limit, we may keep only the fundamental mode in Equation (5.117) and arrive at [22, 23]

$$a_\nu(R,\hat{z}) = \frac{A_{00}(R)e^{-i\mu_{00}\hat{z}}}{\left(\Psi_{00}^\dagger,\Psi_{00}\right)}\left[\int d\hat{\boldsymbol{x}}\;A_{00}(\hat{\boldsymbol{x}})a_\nu(\hat{\boldsymbol{x}};0)\right.$$
$$\left. + \int d\hat{\boldsymbol{x}}d\hat{\boldsymbol{p}}d\hat{\eta}\;f_\nu(\hat{\eta},\hat{\boldsymbol{x}},\hat{\boldsymbol{p}};0)\int_{-\infty}^{0} d\tau\;A_{00}(\hat{\boldsymbol{x}}_-)e^{i(\dot{\theta}-\mu_{00})\tau}\right], \tag{5.147}$$

where the normalization is derived from Equation (5.119) using the fundamental mode. The first term in the square brackets describes the process of coherent amplification, which amplifies the radiation input signal $a_\nu(\hat{\boldsymbol{x}};0)$. The second term describes the process of self-amplified spontaneous emission (SASE), which starts from the white noise associated with the discrete particle nature of the electron beam. Equation (5.147) for the parallel e-beam (with vanishing emittance) reduces to those of Refs. [17, 13].

The fact that SASE is dominated by a single transverse mode means that it is transversely coherent (diffraction limited). In contrast to the spontaneous undulator radiation that comprises a sum of all discrete independent radiators, the FEL process naturally develops coherent bunching after several gain lengths. Both the bunching and the generated radiation become dominated by the fundamental mode when its growth outstrips the other modes, the result of which is radiation that is transversely coherent. In particular, the angular divergence of SASE in the high-gain regime is characterized by the diffraction angle determined by its transverse size, while the spontaneous undulator radiation has an angular divergence that is determined by the convolution of the radiation and the e-beam angular spreads.

The precise 3D properties of SASE are determined by the second term in Equation (5.147). Although the ensemble average of $f_\nu(\hat{\eta},\hat{\boldsymbol{x}},\hat{\boldsymbol{p}};0)$ vanishes, the average radiation intensity can be computed by using the relation

$$\begin{aligned}&\left\langle f_\nu(\hat{\eta},\hat{\boldsymbol{x}},\hat{\boldsymbol{p}};0) f_\nu^*(\hat{\eta}',\hat{\boldsymbol{x}}',\hat{\boldsymbol{p}}';0)\right\rangle \\ &\quad = \frac{2\rho^3 k_1^3 k_u cT}{\pi^2 n_0}\delta(\hat{\eta}-\hat{\eta}')\delta(\hat{\boldsymbol{x}}-\hat{\boldsymbol{x}}')\delta(\hat{\boldsymbol{p}}-\hat{\boldsymbol{p}}')\bar{f}_0(\hat{\eta},\hat{\boldsymbol{x}},\hat{\boldsymbol{p}},\hat{\eta};0).\end{aligned} \tag{5.148}$$

The power spectrum of the fundamental mode is obtained by integrating over the transverse dimensions. Much like in the 1D case, we approximate the spectral shape by a Gaussian, so that in the high-gain regime we have

$$\frac{dP}{d\omega} = g_A\left(\left.\frac{dP}{d\omega}\right|_0 + g_s\frac{\rho\gamma mc^2}{2\pi}\right)\exp\left[\frac{z}{L_G} - \frac{(\omega-\omega_m)^2}{2\sigma_\omega^2}\right], \tag{5.149}$$

where $dP/d\omega|_0$ is the input power spectrum and $\rho\gamma mc^2/2\pi$ is the 1D SASE noise power spectrum associated with the electron beam shot noise. The coupling factor g_A represents the overlap of the fundamental mode with the input coherent radiation, the product $g_A g_s$ gives the effective start-up noise of SASE in units of $\rho\gamma mc^2/2\pi$, while the FEL gain length and RMS bandwidth are given by L_G and σ_ω, respectively. These are extensions of the 1D quantities introduced in Chapter 4.3.2; in the cold, 1D limit we found that $g_A = 1/9$, $g_s = 1$, while the FEL gain length and RMS bandwidth are

$$\text{Cold, 1D}: \; L_G = L_{G0} = \frac{\lambda_u}{4\pi\sqrt{3}} \qquad \sigma_\omega = \sqrt{\frac{3\sqrt{3}\,\rho}{k_u z}}\,\omega_1. \tag{5.150}$$

For a more general beam distribution, L_G and σ_ω are determined by the 3D FEL dispersion relation, while the coupling factors g_A and g_s can be computed with the Van Kampen mode solution of the initial value problem. It is noted that the effective start-up noise g_s increases with larger energy spread and emittance mainly because of the corresponding increase in FEL gain length.

To obtain the FEL power, we integrate (5.149) over frequency; assuming that the bandwidth of the external seed associated with $dP/d\omega|_0$ is much less than that of SASE, we have

$$P = g_A\left(P_0 + g_s\sigma_\omega\frac{\rho\gamma mc^2}{\sqrt{2\pi}}\right)e^{z/L_G}. \tag{5.151}$$

Here, the second term in parentheses is the effective noise power of SASE.

As a numerical example, Figure 5.12 shows the predicted total radiated energy in the LEUTL FEL [27] at $\lambda_1 = 130$ nm using 1D and 3D FEL theory, and compares it to that obtained from two commonly employed time-dependent FEL codes, GINGER and GENESIS. The agreement in the high-gain behavior between the simulations and Equation (5.151) is very good when the proper input coupling and effective noise power (i.e., g_A and g_s) are calculated, so that the exponentially growing phase is dominated by a single growing mode (the fundamental) whose ensemble averaged initial energy is given by solving the 3D initial value problem. On the other hand, the input

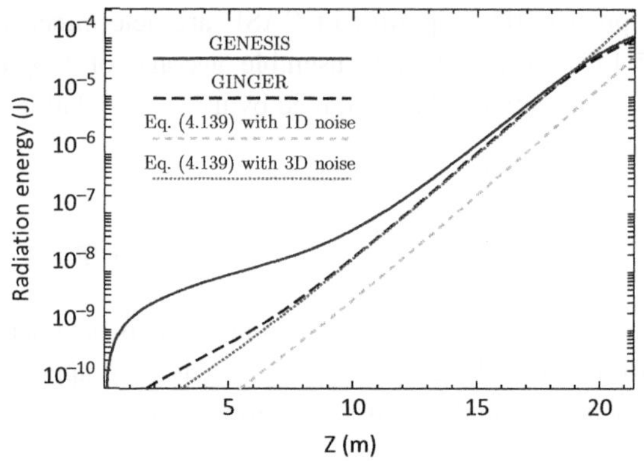

Figure 5.12 GINGER (long-dashed) and GENESIS (solid) simulations of the LEUTL FEL energy at 130 nm versus the undulator distance z, as compared to predictions using Equation (5.151) with 1D noise (short-dashed) and 3D noise (dotted). Adapted from Ref. [25].

signal power computed by the 1D theory is approximately five times smaller than that observed in simulations. The large difference between GENESIS and GINGER for $z \lesssim 11$ m can be attributed to the fact that the radiation field in GENESIS is fully 3D while that of GINGER is assumed to be axially symmetric. Hence, GENESIS resolves many more higher-order transverse modes than does GINGER in the start-up regime, which leads to significantly more radiated power in the early part of the undulator length. This discrepancy decreases as the fraction of power in the fundamental mode increases, and the two codes predict very similar power levels for $z \gtrsim 15$ m.

5.5.5 Variational Solution and Fitting Formulas

The matrix method of the previous section can be used to exactly (up to discretization errors) solve for an arbitrary eigenmode of the dispersion relation (5.143). However, this procedure does require a fair bit of numerical effort to solve the resulting matrix equations. A considerably simpler, approximate scheme for determining the low-order growing FEL modes uses variational tools. Here, we illustrate the utility of the variational method to approximately solve for the fundamental mode.

Our procedure is a specific example of the Rayleigh–Ritz–Galerkin method, and can be thought of as resulting from multiplying the dispersion relation

$$\left\{\mu - \frac{\Delta\nu}{2\rho} + \frac{1}{2\hat{\sigma}_x^2}\left[\frac{1}{R}\frac{d}{dR}\left(R\frac{d}{dR}\right) - \frac{m^2}{R^2}\right]\right\} \mathcal{A}_m(R) = \int_0^\infty R'dR'\; G_m(R,R')\mathcal{A}_m(R') \tag{5.152}$$

by $R\mathcal{A}_m(R)$ and integrating over R from 0 to infinity. We then find that

$$0 = \mathcal{I}[\mathcal{A}] = \int_0^\infty RdR \left\{ \frac{1}{2\hat{\sigma}_x^2} \left[\frac{d\mathcal{A}_m(R)}{dR} \right]^2 - \left[\mu - \frac{\Delta\nu}{2\rho} - \frac{m^2}{R^2} \right] \mathcal{A}_m^2(R) \right\} + \int_0^\infty RdR \int_0^\infty R'dR'\; G_m(R,R')\mathcal{A}_m(R)\mathcal{A}_m(R'), \tag{5.153}$$

and the dispersion relation can be derived by requiring the functional $\mathcal{I}[\mathcal{A}]$ to be stationary under arbitrary variations of the mode profile $\mathcal{A}(R)$ [3, 20]. In other words, if we take the functional (5.153) and set its variation $\delta\mathcal{I} = 0$ (or, equivalently, the functional derivative $\delta\mathcal{I}/\delta\mathcal{A} = 0$), then we find that $\mathcal{A}_m$ and μ must also satisfy the dispersion relation (5.152). Now, we can parametrize the mode profile $\mathcal{A}_m(R)$ in any suitable manner and the variational principle will still apply. Thus, if we approximate the eigenmode by an analytic function that depends on a finite number of parameters, we will reduce the infinite dimensional problem to a more tractable (albeit approximate) one. An important advantage of the variational method is that a first-order approximation of the mode profile $\mathcal{A}_m(R)$ yields a stationary solution for the eigenvalue μ_m that is accurate to second order.

We illustrate the basic program with the simplest and perhaps most useful example: approximating the fundamental mode $\mathcal{A}_0(R)$ by a Gaussian. To approximate the fundamental mode, we insert the Gaussian trial function $\mathcal{A}_0(R) = \exp(-wR^2)$ into Equation (5.153). We denote the fundamental growth rate by μ_{00}, so that after integrating with respect to R and R' we obtain

$$\mathcal{I}(w) = \frac{\mu_{00} - \Delta\nu/2\rho}{4w} - \frac{1}{4\hat{\sigma}_x^2} - \int_{-\infty}^0 d\tau\, \frac{\tau e^{-\hat{\sigma}_\eta^2\tau^2/2 - i\mu_{00}\tau}}{\left[(1 + i\hat{k}_\beta^2\hat{\sigma}_x^2\tau) + 2w\right]^2 - 4w^2\cos^2(\hat{k}_\beta\tau)}. \tag{5.154}$$

Equation (5.153) implies that $\mathcal{I}$ itself vanishes, so that one complex equation for μ_{00} and w comes from setting Equation (5.154) to zero. A second relationship comes from varying $\mathcal{I}$ with respect to w, which implies that $\partial\mathcal{I}/\partial w = 0$ or

$$\frac{\mu_{00} - \Delta\nu/2\rho}{4w^2} = \int_{-\infty}^0 d\tau\, \frac{\left[4(1 + i\hat{k}_\beta^2\hat{\sigma}_x^2\tau) + 8w\sin^2(\hat{k}_\beta\tau)\right]\tau e^{-\hat{\sigma}_\eta^2\tau^2/2 - i\mu_{00}\tau}}{\left\{\left[(1 + i\hat{k}_\beta^2\hat{\sigma}_x^2\tau) + 2w\right]^2 - 4w^2\cos^2(\hat{k}_\beta\tau)\right\}^2}. \tag{5.155}$$

Solving these two equations enables one to determine μ_{00} and, hence, the power gain length. Applying this solution to the previous LCLS example, we plot the imaginary (growing) part of the eigenvalue μ_{00} as a function of the scaled detuning $\Delta\nu/2\rho$ in Figure 5.13. We see that the growth rate maximizes at $\Delta\nu \approx -2\rho$ where its value is approximately one-half the ideal growth rate predicted by the cold, 1D theory, and almost exactly what the matrix methods of Section 5.5.4 predict, $\Im(\mu_{00}) = 0.42$.

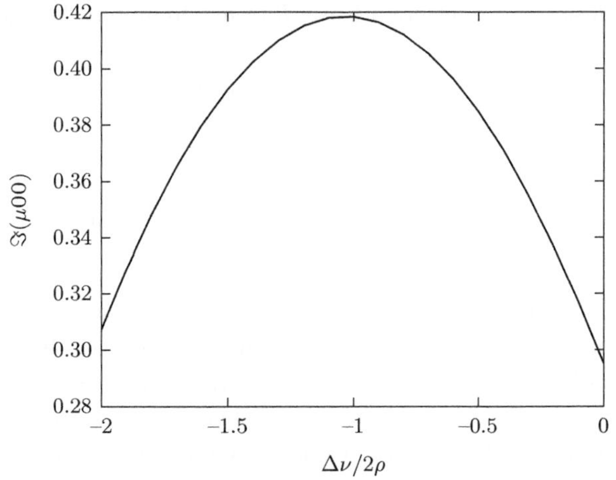

Figure 5.13 The LCLS fundamental mode growth rate Im[μ_{00}] versus detuning $\Delta\nu/2\rho$.

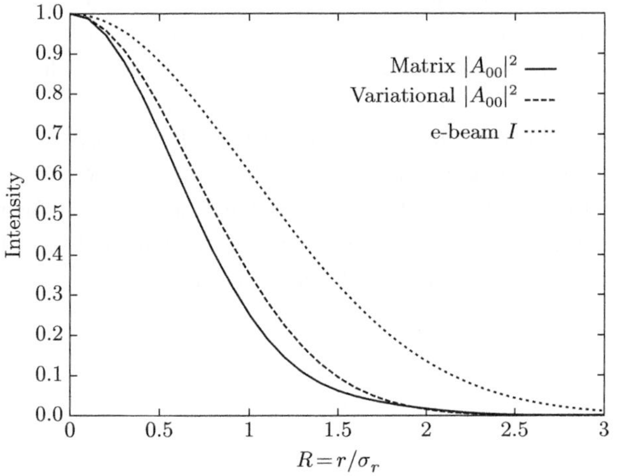

Figure 5.14 LCLS fundamental mode profile as found from the matrix approach (solid line) and from the variational method (dashed line). For comparison, the electron density profile is shown as the dotted line.

Additionally, at optimal detuning, the variational Gaussian approximation to the fundamental guided mode is quite similar to that obtained using the more accurate matrix approach as shown in Figure 5.14.

When designing a high-gain FEL, the power gain length is one of the most important figures of merit, as it determines the undulator length required to reach saturation. Deviations of the gain length from the ideal, 1D limit also indicate to what extent effects such as the finite beam emittance and energy spread play a role in the dynamics. For this reason, it is convenient to parametrize the 3D FEL gain length L_G by

$$L_G = L_{G0}\frac{\sqrt{3}/2}{\Im(\mu_{00})} = L_{G0}(1+\Lambda), \tag{5.156}$$

where $L_{G0} = \lambda_u/(4\sqrt{3}\pi\rho)$ is the 1D power gain length and Λ takes into account effects of energy spread, emittance and diffraction. Based on numerical analysis of the variational solution (5.154)–(5.155), Ming Xie [26] developed a very useful fitting formula for Λ in terms of four scaled parameters:

$$\Lambda = a_1\eta_d^{a_2} + a_3\eta_\varepsilon^{a_4} + a_5\eta_\gamma^{a_6} + a_7\eta_\varepsilon^{a_8}\eta_\gamma^{a_9} + a_{10}\eta_d^{a_{11}}\eta_\gamma^{a_{12}} + a_{13}\eta_d^{a_{14}}\eta_\varepsilon^{a_{15}} + a_{16}\eta_d^{a_{17}}\eta_\varepsilon^{a_{18}}\eta_\gamma^{a_{19}}. \tag{5.157}$$

The scaled parameters are defined in a slightly different manner than those introduced in Section 5.5.2, being given by

$$\eta_d = \frac{L_{G0}}{2k_1\sigma_x^2} = \frac{1}{2\sqrt{3}\hat{\sigma}_x^2} \qquad \text{diffraction parameter,} \tag{5.158}$$

$$\eta_\varepsilon = 2\frac{L_{G0}}{\bar{\beta}}k_1\varepsilon_x = \frac{2}{\sqrt{3}}\hat{k}_\beta^2\hat{\sigma}_x^2 \qquad \text{angular spread parameter,} \tag{5.159}$$

$$\eta_\gamma = 4\pi\frac{L_{G0}}{\lambda_u}\sigma_\eta = \frac{\hat{\sigma}_\eta}{\sqrt{3}} \qquad \text{energy spread parameter.} \tag{5.160}$$

The fourth parameter, the frequency detuning, has been optimized to yield the shortest gain length. The fitting coefficients are

$a_1 = 0.45,$	$a_2 = 0.57,$	$a_3 = 0.55,$	$a_4 = 1.6,$	$a_5 = 3,$
$a_6 = 2,$	$a_7 = 0.35,$	$a_8 = 2.9,$	$a_9 = 2.4,$	$a_{10} = 51,$
$a_{11} = 0.95,$	$a_{12} = 3,$	$a_{13} = 5.4,$	$a_{14} = 0.7,$	$a_{15} = 1.9,$
$a_{16} = 1140,$	$a_{17} = 2.2$	$a_{18} = 2.9,$	$a_{19} = 3.2.$	

The discrepancy between Xie's fitting formula (5.157) and numerical solutions of the FEL eigenmode equation is typically less than 10 percent. The positive fitting coefficients quantitatively show that all three scaled beam parameters in Equation (5.157) should be kept small to avoid a large gain reduction; this agrees with the qualitative arguments we made in Section 5.5.2 regarding e-beam requirements. Furthermore, in the limit that the transverse parameters η_d, $\eta_\varepsilon \to 0$, Equation (5.157) matches the quadratic dependence of the growth length on σ_η/ρ that we gave in (4.62).

Another useful practical formula obtained through numerical simulations [26] predicts the saturation power to be given by

$$P_{\text{sat}} \approx 1.6\left(\frac{L_{G0}}{L_G}\right)^2 \rho P_{\text{beam}} = \frac{1.6}{(1+\Lambda)^2}\rho P_{\text{beam}}, \tag{5.161}$$

where $P_{\text{beam}} = (\gamma_r mc^2/e)I_{\text{beam}}$ is the total electron beam power for a beam of energy $\gamma_r mc^2$ and peak current I_{beam}. In practical units,

$$P_{\text{beam}}[\text{TW}] = \gamma_r mc^2[\text{GeV}]I[\text{kA}]. \tag{5.162}$$

An alternate fitting formula for the gain length was developed by Saldin and collaborators [27]; they found that for X-ray wavelengths and when the energy spread is small, $\sigma_\eta \ll \rho$, the power gain length at the optimized beta function may be estimated as

$$L_G \approx 1.2\left(\frac{I_A}{I}\right)^{1/2}\left(\frac{\varepsilon_n^5\lambda_u^5}{\lambda_1^4}\right)^{1/6}\frac{(1+K^2/2)^{1/3}}{K[\mathrm{JJ}]} \propto \frac{1}{\mathcal{B}_\perp^{1/2}}\frac{1}{\varepsilon_n^{1/6}}. \tag{5.163}$$

This equation is not only useful in its own right, but it shows how the gain length scales with the invariant 5D electron beam brightness $\mathcal{B}_\perp$; we define the invariant 5D brightness as the number of electrons per unit time per unit invariant phase space area given by the normalized emittance ε_n,

$$\mathcal{B}_\perp \equiv \frac{I/ec}{4\pi^2\varepsilon_n^2}. \tag{5.164}$$

References

[1] G. T. Moore, "The high-gain regime of the free electron laser," *Nucl. Instrum. Methods Phys. Res., Sect. A*, vol. 239, p. 19, 1985.

[2] E. T. Scharlemann, A. M. Sessler, and J. S. Wurtele, "Optical guiding in a free-electron laser," *Phys. Rev. Lett.*, vol. 54, p. 1925, 1985.

[3] L.-H. Yu, S. Krinsky, and R. L. Gluckstern, "Calculation of a universal scaling function for free-electron laser gain," *Phys. Rev. Lett.*, vol. 64, p. 3011, 1990.

[4] E. T. Scharlemann, "Wiggle plane focusing in linear wigglers," *J. Appl. Phys.*, vol. 58, p. 2154, 1985.

[5] H. Wiedemann, *Particle Accelerator Physics I and II*, 2nd ed. Berlin: Springer-Verlag, 1999.

[6] S. Reiche, "Compensation of FEL gain reduction by emittance effects in a strong focusing lattice," *Nucl. Instrum. Methods Phys. Res., Sect. A*, vol. 445, p. 90, 2000.

[7] K.-J. Kim, "FEL gain taking into account diffraction and electron beam emittance; generalized Madey's theorem," *Nucl. Instrum. Methods Phys. Res., Sect. A*, vol. 318, p. 489, 1992.

[8] N. M. Kroll, P. L. Morton, and M. N. Rosenbluth, "Free-electron lasers with variable parameter wigglers," *IEEE J. Quantum Electron.*, vol. 17, p. 1436, 1981.

[9] P. Sprangle and C. M. Tang, "Three-dimensional nonlinear theory of the free electron laser," *Appl. Phys. Lett.*, vol. 39, p. 677, 1981.

[10] P. Sprangle, A. Ting, and C. M. Tang, "Radiation focusing and guiding with application to the free-electron laser," *Phys. Rev. Lett.*, vol. 59, p. 202, 1987.

[11] J. M. J. Madey, "Relationship between mean radiated energy, mean squared radiated energy, and spontaneous power spectrum in a power series expansion of the equations of motion in a free-electron laser," *Nuovo Cim. B*, vol. 50, p. 64, 1979.

[12] G. T. Moore, "Modes for gain maximization of the free-electron laser in the low-gain, small-signal regime," *Nucl. Instrum. Methods Phys. Res., Sect. A*, vol. 250, p. 418, 1986.

[13] S. Krinsky and L.-H. Yu, "Output power in guided modes for amplified spontaneous emission in a single-pass free-electron laser," *Phys. Rev. A*, vol. 35, p. 3406, 1987.

[14] E. L. Saldin, E. A. Schneidmiller, and M. V. Yurkov, "On a linear theory of an FEL amplifier with an axisymmetric electron beam," *Optics Comm.*, vol. 97, p. 272, 1993.

[15] —, *Physics of Free-Electron Lasers*. Berlin: Springer, 2000.

[16] N. Van Kampen, "On the theory of stationary waves in plasmas," *Physica*, vol. 21, p. 949, 1955.

[17] K.-J. Kim, "Three-dimensional analysis of coherent amplification and self-amplified spontaneous emission in free electron lasers," *Phys. Rev. Lett.*, vol. 57, p. 1871, 1986.

[18] K. M. Case, "Plasma oscillations," *Annals of Physics*, vol. 7, p. 349, 1959.

[19] M. Xie, "Theory of optical guiding in free-electron lasers," Ph.D. dissertation, Stanford University, 1988.

[20] —, "Exact and variational solutions of 3D eigenmodes in high gain FELs," *Nucl. Instrum. Methods Phys. Res., Sect. A*, vol. 445, p. 59, 2000.

[21] J. Galayda *et al.*, "Linac Coherent Light Source (LCLS) Conceptual Design Report," SLAC, Report SLAC-R-593, 2002.

[22] M. Xie, "Grand initial value problem of high gain free electron lasers," *Nucl. Instrum. Methods Phys. Res., Sect. A*, vol. 475, p. 51, 2001.

[23] Z. Huang and K.-J. Kim, "Solution to the initial value problem for a high-gain FEL via Van Kampen's method," *Nucl. Instrum. Methods Phys. Res., Sect. A*, vol. 475, p. 59, 2001.

[24] S. Milton *et al.*, "Exponential gain and saturation of a self-amplified spontaneous emission free-electron laser," *Science*, vol. 292, p. 2037, 2001.

[25] Z. Huang and K.-J. Kim, "Review of X-ray free-electron laser theory," *Phys. Rev. ST Accel. Beams*, vol. 10, p. 034801, 2007.

[26] M. Xie, "Design optimization for an X-ray free-electron laser driven by SLAC linac," in *Proceedings of the 1995 Particle Accelerator Conference*. Piscataway, NJ: IEEE, p. 183, 1995.

[27] E. L. Saldin, E. A. Schneidmiller, and M. V. Yurkov, "Design formulas for short-wavelength FELs," *Optics Comm.*, vol. 235, p. 415, 2004.

6 Harmonic Generation in High-Gain FELs

The ability to generate radiation at higher harmonics of the fundamental frequency is an important aspect of high-gain FELs, particularly for generating the highest possible photon energies and for producing coherent light at harmonics of an external seed laser. In this chapter we consider two types of harmonic generation. First, we analyze the abundant nonlinear harmonic radiation that is naturally produced in a planar undulator near FEL saturation, showing how high-gain FELs can serve as a bright source of radiation whose wavelength is much shorter than that of the fundamental. Since SASE FELs are typically used to generate the harmonic emission, the resulting output field is temporally incoherent. Second, we discuss methods that use powerful external lasers to manipulate the electron beam in a nonlinear way so as to create density modulations at a higher harmonic of the laser frequency. These higher harmonic density modulations can then be used to coherently "seed" a short-wavelength FEL, thereby increasing the temporal coherence over that of SASE. Hence, these methods essentially use an electron beam to frequency up-convert laser light to soft X-rays, where practical limits appear to restrict the final photon energy to $\lesssim 1$ keV. We specifically describe two such methods of coherent harmonic generation: high-gain harmonic generation (HGHG) and echo-enabled harmonic generation (EEHG).

6.1 Nonlinear Harmonic Generation

To some extent we have already shown how the coupling to odd harmonics arises in Chapter 3. Here, we give a more detailed derivation, discuss the FEL dynamics, and indicate to what extent harmonic emission may provide a useful source of short-wavelength radiation. The emission of odd harmonics in a planar undulator is due to the figure-eight motion in the electron's comoving frame, which we have shown comes about because the axial velocity v_z oscillates at twice the undulator period. This figure-eight motion means that a Fourier decomposition of the electron trajectory has additional harmonic content, which can in turn lead to harmonic emission.

Our mathematical analysis in Section 3.4.1 has already showed the consequences of the figure-eight motion, finding that the source current can resonantly drive odd radiation harmonics of the fundamental. Obviously, the source current resulting from this undulator-averaging process is significant only when ν is close to an odd integer, i.e., when $\nu \approx h = 2n - 1 = 1, 3, 5, \ldots$. Thus, the field amplitude $E_h(\Delta\nu_h, \boldsymbol{x}, z)$ near the hth harmonic obeys the equation

$$\left(\frac{\partial}{\partial z} + \frac{1}{2ihk_1}\nabla_\perp^2\right) E_h(\Delta\nu_h, \boldsymbol{x}, z)$$
$$= \frac{ek_1 K[\text{JJ}]_h}{4\pi\epsilon_0\gamma_r} e^{i\Delta\nu_h k_u z} \int d\theta\, e^{-i\nu\theta} \sum_{j=1}^{N_e} \delta(\boldsymbol{x} - \boldsymbol{x}_j)\delta(\theta - \theta_j), \tag{6.1}$$

where we recall that the harmonic Bessel function factor was defined as

$$[\text{JJ}]_h \equiv (-1)^{(h-1)/2} \left[J_{(h-1)/2}(h\xi) - J_{(h+1)/2}(h\xi)\right]. \tag{6.2}$$

Thus, in the forward z direction, the electric field consists of a series of nearly monochromatic waves around the odd harmonic frequencies hck_1 [1], with the frequency detuning $\Delta\nu_h \equiv \nu - h \ll 1$.

In a planar undulator with a strong magnetic field such that $K > 1$, the spontaneous emission at the fundamental resonant frequency and its higher harmonics induce bunching at their respective wavelength scales, which in turn leads to amplified emission at these frequencies [1]. However, the linear amplification of the higher harmonics is always much smaller than that of the fundamental, because both the coupling strength $\propto K[\text{JJ}]_h$ is weaker and the gain-degrading effects of emittance and energy spread become more significant at shorter wavelengths. Thus, emission at the fundamental frequency is heavily favored in a high-gain FEL because its gain length is significantly shorter than that of the harmonics.

Nevertheless, coherent harmonic emission is produced when the radiation at the fundamental bunches the electron beam strongly, since this produces a density modulation with significant Fourier content at the higher harmonics. We illustrate this strong FEL-induced bunching in Figure 6.1. To study the generation of the nonlinear harmonics, we first use a simple 1D model [2] that extends the collective FEL equations used in Section 3.4.4, and subsequently discuss certain aspects of the 3D physics. A comprehensive 3D analysis based on the coupled Maxwell–Vlasov equations is given in Ref. [3], which is extended to include even harmonics in Ref. [4].

Our analysis of coherent harmonic emission begins by considering the FEL system including two radiation wavelengths, so that the electron interacts with radiation at both the fundamental and the third harmonic, while each harmonic is driven by the appropriate particle bunching. Using the scaled variables of Section 3.4, we write down the following systems of equations:

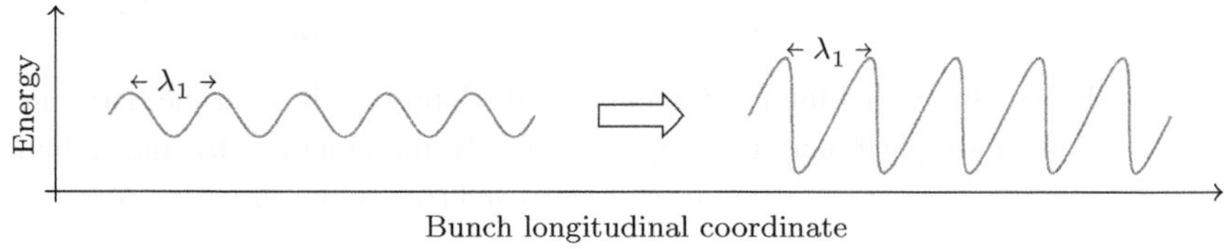

Figure 6.1 Transformation of energy modulation to spatial bunching which includes rich harmonic content.

$$\frac{d\theta_j}{d\hat{z}} = \hat{\eta}_j \qquad \frac{d\hat{\eta}_j}{d\hat{z}} = a_1 e^{i\theta} + a_3 e^{3i\theta} + c.c. \tag{6.3}$$

$$\frac{da_1}{d\hat{z}} = -\langle e^{-i\theta_j}\rangle \qquad \frac{da_3}{d\hat{z}} = -\frac{[\mathrm{JJ}]_3^2}{[\mathrm{JJ}]_1^2}\langle e^{-3i\theta_j}\rangle, \tag{6.4}$$

where $a_h(\hat{z}) = -eK_hE_h(\hat{z})/(4\gamma^2mc^2k_u\rho^2)$. For simplicity, we have neglected the effects of slippage (assumed a vanishing detuning $\Delta\nu_h$) and of transverse $\boldsymbol{x}$-dependence. We now introducing the harmonic bunching factors

$$b_h = \langle e^{-ih\theta_j}\rangle\,, \tag{6.5}$$

which we will treat as small such that $|b_h| \sim O(\epsilon^h)$ with $\epsilon \ll 1$. To be consistent we must also assume that a_h is an $O(\epsilon^h)$ quantity, and the harmonic collective equations become

$$\frac{d^3a_1}{d\hat{z}^3} = -\frac{d^2b_1}{d\hat{z}^2} = \frac{d}{d\hat{z}}i\langle e^{-i\theta_j}\hat{\eta}_j\rangle \approx i\left\langle e^{-i\theta_j}\frac{d\hat{\eta}_j}{d\hat{z}}\right\rangle \approx ia_1, \tag{6.6}$$

$$\frac{d^2b_2}{d\hat{z}^2} \approx -2i\left\langle e^{-2i\theta_j}\frac{d\hat{\eta}_j}{d\hat{z}}\right\rangle \approx -2ia_1b_1 \approx i\frac{d}{d\hat{z}}a_1^2, \tag{6.7}$$

$$\begin{aligned}\frac{d^3a_3}{d\hat{z}^3} &= -\frac{[\mathrm{JJ}]_3^2}{[\mathrm{JJ}]_1^2}\frac{d^2b_3}{d\hat{z}^2} = \frac{3i[\mathrm{JJ}]_3^2}{[\mathrm{JJ}]_1^2}\frac{d}{d\hat{z}}\langle e^{-3i\theta_j}\hat{\eta}_j\rangle \\ &\approx \frac{3i[\mathrm{JJ}]_3^2}{[\mathrm{JJ}]_1^2}\left\langle e^{-3i\theta_j}\frac{d\hat{\eta}_j}{d\hat{z}}\right\rangle \approx \frac{3i[\mathrm{JJ}]_3^2}{[\mathrm{JJ}]_1^2}(a_3 + a_1b_2).\end{aligned} \tag{6.8}$$

In Equations (6.6), (6.7), and (6.8) we have dropped the higher-order terms $\sim \epsilon^2$, $\sim \epsilon^3$, and $\sim \epsilon^4$, respectively, and in all three equations we have assumed that the electron beam energy spread is insignificant, $\sigma_\eta \ll \rho$. Equation (6.6) yields the usual cubic equation for the fundamental field, which has a dominant, exponentially-growing solution given by

$$a_1(\hat{z}) = Ae^{-i\mu_{00}\hat{z}}, \quad \mu_{00} = \frac{-1+\sqrt{3}i}{2}, \tag{6.9}$$

where we write the growth rate as μ_{00} to avoid any possible confusion with exponential growth at the third harmonic (the growing root was labeled μ_3 in Chapter 4, concerning 1D FEL theory). The second harmonic bunching evolves as

$$\frac{db_2}{d\hat{z}} \approx ia_1^2 \quad \Rightarrow \quad b_2 = -\frac{1}{2\mu_{00}}A^2e^{-2i\mu_{00}\hat{z}}. \tag{6.10}$$

By assuming that the fundamental fields dominate those at the harmonics, we find that b_2 is a simple function of a_1. To simplify the equation for the radiation at the third harmonic, we insert the solutions (6.9) and (6.10) into Equation (6.8):

$$\frac{d^3a_3}{d\hat{z}^3} - \frac{3i[\mathrm{JJ}]_3^2}{[\mathrm{JJ}]_1^2}a_3 = -\frac{3i[\mathrm{JJ}]_3^2}{2\mu_{00}[\mathrm{JJ}]_1^2}A^3e^{-3i\mu_{00}\hat{z}}. \tag{6.11}$$

Early in the linear growth phase, $\hat{z} \sim 1$ and the right hand-side is approximately zero since $|a_1(0)| \ll 1$. Solving the remaining homogeneous equation yields a cubic equation for the growth rate of the third harmonic. The growing root has a smaller growth rate (imaginary part) than that of the fundamental since $3[\mathrm{JJ}]_3^2/[\mathrm{JJ}]_1^2 < 1$. Additionally, including the effects of emittance and energy spread makes the linear harmonic growth rate smaller yet, meaning that the harmonic field has experienced relatively few exponentiations over the distance required for saturation of the fundamental. On the other hand, near FEL saturation, where $\hat{z} \gg 1$ and the fundamental field $\left|a_1(0)e^{-i\mu_{00}\hat{z}}\right| \sim 1$, the source term on the right-hand side can dominate both the weak linear gain and the spontaneous emission of a_3. In this case, we see that the growth rate of the nonlinear third harmonic is three times that of the fundamental, being given by

$$a_3 \approx -\frac{1}{18\mu_{00}}\frac{[\mathrm{JJ}]_3^2}{[\mathrm{JJ}]_1^2}A^3 e^{-3i\mu_{00}\hat{z}}. \tag{6.12}$$

Using the relationships for the power in the fundamental $P_1 = |a_1|^2\rho P_{\text{beam}}$ and the harmonic power $P_3 = [\mathrm{JJ}]_1^2/[\mathrm{JJ}]_3^2|a_3|^2\rho P_{\text{beam}}$, we have

$$\frac{P_3}{\rho P_{\text{beam}}} \approx 0.003\frac{[\mathrm{JJ}]_3^2}{[\mathrm{JJ}]_1^2}\left(\frac{P_1}{\rho P_{\text{beam}}}\right)^3 \propto \exp(3z/L_G), \tag{6.13}$$

where L_G is the power gain length of the fundamental radiation.

In general, one can extend our previous arguments to show that for sufficiently large z the power associated with the nonlinear harmonic h scales as

$$P_h \propto P_1^h \propto \exp(hz/L_G) \propto e^{hz/L_G}. \tag{6.14}$$

Thus, P_h grows with a gain length that is inversely proportional to h near saturation. The power scaling (6.14) has been confirmed with FEL simulations that include harmonic emission [5]. Note, however, that the maximal P_h is a decreasing function of harmonic number h; Equation (6.12) implies that $P_3/P_1 \sim 0.1\%$–1% at saturation.

A 3D analysis [3] shows that the electromagnetic field of the nonlinear third harmonic is transversely coherent since its transverse profile is governed by the gain-guided fundamental mode of the $h = 1$ radiation field. We show the transverse profiles of the third harmonic field, the fundamental radiation, and the electron beam for parameters of the LCLS in Figure 6.2. The transverse phase space area of the electromagnetic field scales with the wavelength $\lambda/4\pi$. Near saturation, roughly 1 percent of the power in the fundamental is typically emitted in the third harmonic. While transversely coherent, SASE produces harmonics that are less coherent longitudinally and have a more spiky intensity profile than the fundamental. Thus, the nonlinear harmonics have significantly larger statistical power fluctuations from shot to shot.

Finally, if the radiation power at the fundamental frequency can be suppressed, the harmonic radiation can grow exponentially to its saturation power level. This is referred to as harmonic lasing [6, 7]. Contrary to nonlinear harmonic generation, harmonic lasing can provide more intense, stable, and narrow-band FEL pulses to extend the wavelength reach of an X-ray FEL facility. To illustrate how one might suppress the fundamental

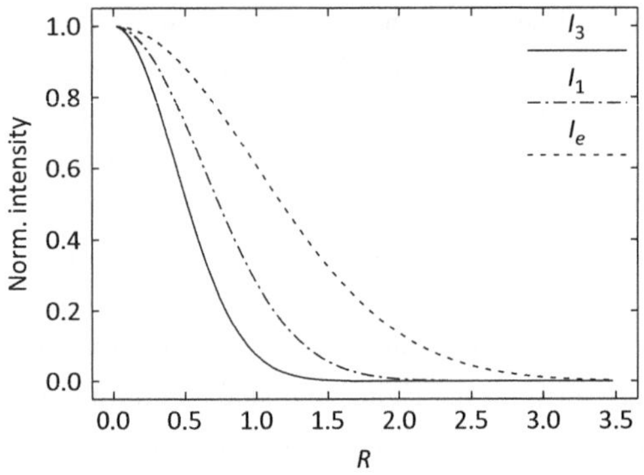

Figure 6.2 Transverse intensity profiles of the third harmonic (I_3), the fundamental radiation (I_1), and the electron beam (I_e) as functions of the radius in units of electron beam size, using LCLS parameters with λ_1 = 1.5 Å.

in favor of, for example, the third harmonic, we imagine an FEL that is composed of many undulator sections separated by short drift spaces. If we set the length of the drift space such that the field slips ahead of the electrons by a distance $(n\lambda_1 + \lambda_1/3)$ with n an integer, the field at the fundamental will suffer significant destructive interference, while the third harmonic field will be relatively unaffected. To do this over some range of wavelengths requires a variable delay or "phase shifter," but we will defer further extensions and applications of these ideas to the Refs. [6, 7].

6.2 High-Gain Harmonic Generation

Although SASE FELs are characterized by approximately one transverse mode, they typically have poor temporal coherence because SASE relies on amplifying the spontaneous emission that is seeded by random shot noise in the electron beam. When the electron bunch is much longer than the coherence length $c\sigma_\tau = c/2\sigma_\omega$ (as is usually the case), the radiation comprises many longitudinal modes. For very short electron bunches, on the other hand, the e-beam may amplify only one FEL mode, but the shot-to-shot variations in EM energy approaches unity. In principle one can improve longitudinal coherence and stability by seeding the FEL with a coherent signal at the wavelength of interest, but coherent lasers do not exist at the shortest (X-ray) wavelengths. One strategy to overcome this limitation and improve the longitudinal coherence at short wavelengths is to use a laser to imprint a coherent signal on the electron beam, and then manipulate the e-beam's phase space to produce a coherent density modulation that contains frequency harmonics of the laser. This density modulation can then be used to seed the FEL interaction, thereby generating intense radiation at a higher harmonic of the original seed laser.

The high-gain harmonic-generation (HGHG) FEL [8] accomplishes this conversion of laser energy at one frequency ω_1 to density modulations at a higher frequency $h\omega_1$ by adding an additional undulator and dispersive section designed to manipulate the electron beam phase space as we diagram in Figure 6.3. Since the purpose of this preliminary

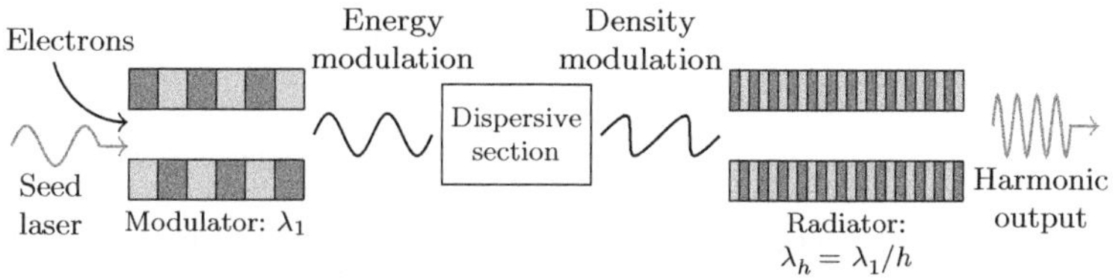

Figure 6.3 Schematic of a HGHG FEL. The laser produces an energy modulation on the electron beam at wavelength λ_1 in the modulator undulator as shown, which is converted into a nonlinear density modulation with harmonic content in the dispersive section. The bunched electron beam is then used to produce FEL radiation at wavelength λ_1/h in the radiator undulator.

undulator is to impose an energy modulation on the beam, it is typically referred to as the modulator. In the modulator, the seed laser co-propagates with the electron beam, thereby imprinting an energy modulation on the electron beam at the laser wavelength. The energy modulation is then converted into a coherent temporal density modulation by passing the electron beam through a dispersive section. As shown in Figure 6.3, the dispersive section acts to bunch the beam in a manner similar to the FEL bunching that we studied in the previous section. Since the generated density modulation will have Fourier content at harmonics of the laser frequency, the resulting electron beam can be used to coherently seed an FEL whose resonance condition is tuned to a higher harmonic of the laser. Emission in this radiator FEL is initialized by coherent microbunching at the wavelength λ_1/h. If the harmonic radiation generated by the imposed microbunching dominates that produced by spontaneous emission, than the output radiation is longitudinally coherent: it is characterized by a single phase determined by the seed laser and its spectral bandwidth can approach the Fourier transform limit.

A complete treatment of the HGHG FEL from modulator through radiator has been provided in Refs. [8, 9]. Here we will only briefly sketch the effect of the modulator, and focus the production of the density modulation in the dispersive section (magnetic chicane), since this involves the new physics associated with the harmonic production. We neglect entirely the subsequent amplification of the density modulation, as this can be analyzed using the methods we presented in the previous two chapters on high-gain FELs.

Let us denote the initial longitudinal phase-space coordinates of the electrons as (θ_0, η_0). We suppose that the beam is initially uniform in θ_0 with initial distribution function $f_0(\eta_0)$. As we showed using the low-gain perturbation theory in Section 3.3, to lowest-order a short undulator keeps the phase nearly constant while mapping the energy as $\eta = \eta_0 + \Delta\eta \sin\theta_0$, where $\Delta\eta$ is the energy modulation amplitude. Using Equation (3.33) for a Gaussian laser beam gives

$$\Delta\eta = \frac{eE_0K[\mathrm{JJ}]}{2\gamma_r^2 mc^2}L_u = \frac{K[\mathrm{JJ}]}{\gamma_r^2}\sqrt{\frac{L_u^2}{\sigma_r^2}\frac{e^2}{4\pi\epsilon_0 mc^2}\frac{P}{mc^3}} = \frac{K[\mathrm{JJ}]}{\gamma_r^2}\frac{L_u}{\sigma_r}\sqrt{\frac{P}{P_{\mathrm{rel}}}} \tag{6.15}$$

$$\to \frac{2\pi K[\mathrm{JJ}]}{\gamma_r^2}\sqrt{\frac{L_u}{\lambda_1}\frac{P}{P_{\mathrm{rel}}}}, \tag{6.16}$$

where $P_{\text{rel}} = mc^3/r_e \approx 8.7$ GW and Equation (6.16) uses $\sigma_r^2 = \lambda_1 Z_R/4\pi$ with the Rayleigh range $Z_R \to L_u/\pi$. The distribution function after the modulator is

$$f(\eta, \theta_0) = f_0(\eta_0(\eta, \theta_0)) = f_0(\eta - \Delta\eta \sin\theta_0). \tag{6.17}$$

After the modulator, the beam is sent through a dispersive section usually chosen to be a magnetic chicane. A chicane is typically made up of four dipoles designed to correlate the final longitudinal position of a particle to its initial energy. Since higher-energy particles are bent less in the magnetic field than lower-energy particles, electrons of greater energy traverse a shorter distance. Thus, the relative particle phase increases by an amount

$$\theta - \theta_0 = k_1 R_{56} \eta, \tag{6.18}$$

where θ is the final phase, θ_0 is the initial phase, η is the relative energy deviation, and R_{56} is a constant that depends on the chicane configuration. At the end of the chicane, the final bunching at harmonic h is given by

$$b_h = \left\langle e^{-ih\theta} \right\rangle = \int d\eta d\theta \; e^{-ih\theta} f(\eta, \theta). \tag{6.19}$$

To compute the harmonic bunching, it is simplest to change integration coordinates from $(\theta, \eta) \to (\theta_0, \eta_0)$. Recalling that the distribution function is conserved along particle trajectories, we have $f(\theta, \eta) = f_0(\eta_0)$. Furthermore, the transformation is symplectic with unit Jacobian. Inserting the coordinate transformations given by (6.17) and (6.18) into the formula for the harmonic bunching b_h leads to

$$\begin{aligned} b_h &= \int d\eta_0 d\theta_0 \; e^{-ih[\theta_0 + k_1 R_{56}(\eta_0 + \Delta\eta \sin\theta_0)]} f_0(\eta_0) \\ &= \int d\theta_0 \; e^{-ih\theta_0} \sum_\ell e^{-i\ell\theta_0} J_\ell(hk_1 R_{56} \Delta\eta) \int d\eta_0 \, f_0(\eta_0) e^{-ihk_1 R_{56}\eta_0} \\ &= (-1)^h J_h(hk_1 R_{56}\Delta\eta) \int d\eta_0 \, f_0(\eta_0) e^{-ihk_1 R_{56}\eta_0}. \end{aligned} \tag{6.20}$$

If we further assume that $f_0(\eta_0) = \exp[-\eta^2/(2\sigma_\eta^2)]/\sqrt{2\pi}\sigma_\eta$, then we obtain

$$b_h = J_h(hk_1 R_{56}\Delta\eta) \exp\left(-\tfrac{1}{2} h^2 k_1^2 R_{56}^2 \sigma_\eta^2\right), \tag{6.21}$$

with the magnitude of the current modulation amplitude given by $2|b_h|$. The Bessel function $J_h(x)$ maximizes when $x \approx 1.2h$, which yields the optimal chicane strength to be [9]

$$k_1 R_{56}\Big|_{\text{optimal}} \approx \frac{1.2}{\Delta\eta}. \tag{6.22}$$

When the harmonic bunching is maximized, the second factor in Equation (6.21) implies that

$$b_h \propto \exp\left[-0.72 \left(\frac{h\sigma_\eta}{\Delta\eta}\right)^2\right]. \tag{6.23}$$

Thus, $|b_h|$ becomes significantly reduced when $\Delta\eta < h\sigma_\eta$. While the bunching favors a large energy modulation, one cannot impose an arbitrarily large energy modulation since this will degrade the FEL gain in the radiator. Specifically, deviations in energy result in an effective energy spread for the radiator which can be approximated as

$$(\sigma_\eta)_{\text{rad}} \approx \sqrt{\sigma_\eta^2 + \Delta\eta^2/2}. \tag{6.24}$$

To preserve FEL gain of the amplifier, we require that the effective energy spread should be less than the FEL Pierce parameter ρ of the radiator. Therefore, the optimal energy modulation for a single-stage HGHG is

$$\Delta\eta \sim h\sigma_\eta \sim \rho. \tag{6.25}$$

In other words, the useful bunching from an HGHG device has a maximum harmonic number $\sim \rho/\sigma_\eta$. Thus, it is critical to have a small initial uncorrelated energy spread to generate short-wavelength radiation with HGHG. In typical systems, the harmonic conversion for a single-stage HGHG appears to be limited to $h < 10$. To get to shorter wavelengths, one can use the harmonic FEL light from the radiator as a seed in another HGHG section further downstream. Such a multi-staged "HGHG cascade" could potentially reach significantly lower wavelengths, but suffers from a significant complication in the overall design [10, 11, 12].

Alternatively, one can consider different beam manipulation schemes to produce bunching at higher harmonics. In the next section we examine so-called echo-enabled harmonic generation, which uses two modulator chicane pairs to directly generate high harmonic bunching.

6.3 Echo-Enabled Harmonic Generation

In an echo-enabled harmonic generation (EEHG) FEL [13, 14], a second modulator followed by a second chicane is inserted before the radiator. The electron beam interacts twice with two laser pulses in the two modulators, as we illustrate in Figure 6.4. With the proper choice of two dispersive sections, the longitudinal phase space of the electron beam becomes highly nonlinear, leading to density modulations at a very high harmonic number for a modest level of energy modulation.

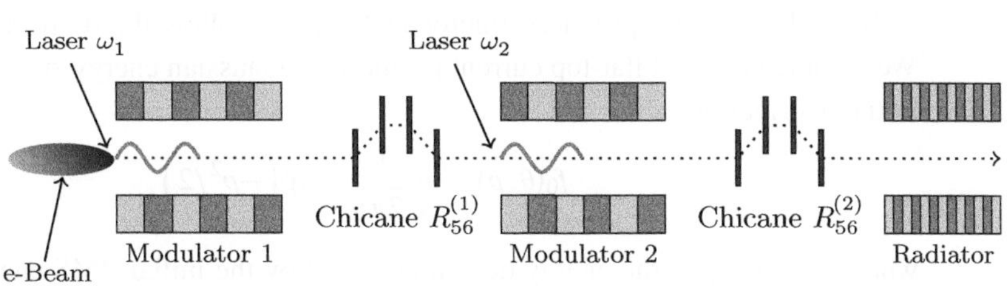

Figure 6.4 Schematic of a EEHG FEL. There are two modulator–chicane pairs: the first prepares a highly nonlinear phase space of many "beamlets," while the second transforms this into harmonic bunching as shown in Figure 6.5.

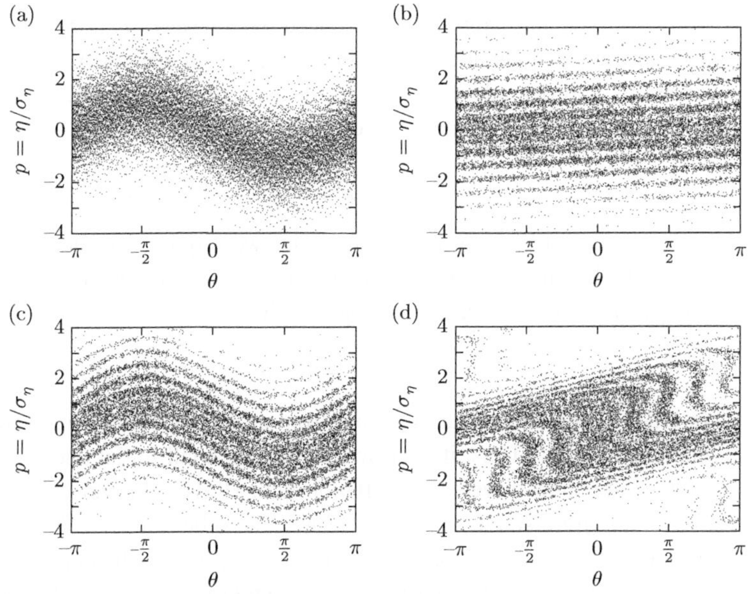

Figure 6.5 Longitudinal phase space evolution illustrates the mechanism for echo microbunching: (a) after first modulator, (b) after the first chicane, (c) after the second modulator, (d) after the second chicane. Adapted from Ref. [13].

Figure 6.5 shows the evolution of the electron beam phase space as it travels through the system whose parameters were discussed in Ref. [13]. These pictures demonstrate the simple physical mechanism behind the echo effect. A large value of dispersion $R_{56}^{(1)}$ after the first modulator leads to a "shredding" of the beam phase space and the generation of multiple "beamlets" along the longitudinal direction. These beamlets appear as a number of approximately horizontal stripes in the top-right picture in Figure 6.5. Each beamlet, which results from the extreme shearing of the initially modulated beam, has an almost uniform density distribution along z and an energy spread that is much smaller than that of the original beam. The role of the second modulator–chicane pair is to orient the beamlets vertically to produce the desired harmonic bunching. This is accomplished in a manner similar to that discussed for HGHG: the second modulator imposes an energy modulation on each beamlet, which is then transformed to a density modulation using a chicane with a relatively modest value of $R_{56}^{(2)}$.

To study the echo-produced microbunching, we follow the derivation of Ref. [14]. We assume an initial flat-top current profile with Gaussian energy spread, and write the initial distribution as

$$f_0(\theta, p) = \frac{1}{\sqrt{2\pi}\theta_b} \exp\Big(-p^2/2\Big), \tag{6.26}$$

where $p = \eta/\sigma_\eta$ is the energy deviation scaled by the initial RMS energy spread, θ is the longitudinal coordinate in units of the reduced laser wavelength $\lambda_1/2\pi$, the scaled bunch length is $\theta_b = ck_1T$, and the normalization of the distribution function is chosen such that

$$\int_{-\infty}^{\infty} dp \int_{-\theta_b/2}^{\theta_b/2} d\theta\, f_0(\theta, p) = 1\,. \tag{6.27}$$

After the first modulator, the energy variable becomes

$$p' = p + A_1 \sin\theta\,, \tag{6.28}$$

where $A_1 = (\Delta\eta)_1/\sigma_\eta$ is the modulation amplitude scaled by the initial RMS energy spread. The modulated beam is then sent through the first chicane that strongly shears the beam phase space. Introducing the normalized chicane strength $B_1 = k_1 R_{56}^{(1)} \sigma_\eta$, the first chicane transforms the longitudinal coordinate as

$$\theta' = \theta + B_1 p'\,. \tag{6.29}$$

Similarly, passage through the second modulator and chicane pair gives rise to the following coordinate mappings

$$p'' = p' + A_2 \sin(\kappa\theta' + \psi) \qquad\qquad \theta'' = \theta' + B_2 p''\,, \tag{6.30}$$

where θ'' and p'' are the phase space variables at the exit from the second chicane, $\kappa = k_2/k_1$ is the ratio of laser wavenumbers in the second modulator to that of the first, and ψ is the relative phase of two modulating lasers.

We determine the EEHG-induced bunching factor at the h harmonic to be given by

$$b_h = \int dp''d\theta''\, f(p'',\theta'') e^{-ih\theta''} = \int_{-\infty}^{\infty} dp \int_{-\theta_b/2}^{\theta_b/2} d\theta\, f_0(p) \exp[-ih\theta''(\theta,p)], \tag{6.31}$$

where we express the integral in terms of the initial coordinates (θ, p) and the initial distribution function. The expression for the final phase θ'' in terms of θ and p can be found from Equations (6.28), (6.29), and (6.30):

$$\begin{aligned} \theta'' &= \theta + (B_1 + B_2)p + A_1(B_1 + B_2)\sin\theta \\ &\quad + A_2 B_2 \sin(\kappa\theta + \kappa B_1 p + \kappa A_1 B_1 \sin\theta + \psi)\,. \end{aligned} \tag{6.32}$$

Inserting this into Equation (6.31), we can expand the exponent using

$$e^{-ihA_1(B_1+B_2)\sin\theta} = \sum_{q=-\infty}^{\infty} e^{iq\theta} J_q[-hA_1(B_1+B_2)]\,, \tag{6.33}$$

and

$$\begin{aligned} &\exp\big[-ihA_2B_2 \sin(\kappa\theta + \kappa B_1 p + \kappa A_1 B_1 \sin\theta + \psi)\big] \\ &= \sum_{m=-\infty}^{\infty} e^{im(\kappa\theta + \kappa B_1 p + \kappa A_1 B_1 \sin\theta + \psi)} J_m(-hA_2B_2) \\ &= \sum_{m=-\infty}^{\infty} J_m(-hA_2B_2) \sum_{\ell=-\infty}^{\infty} e^{im(\kappa\theta + \kappa B_1 p + \psi) + i\ell\theta} J_l(m\kappa A_1 B_1)\,. \end{aligned} \tag{6.34}$$

Collecting all terms that depend on θ in the exponent of Equation (6.31), and performing the integral over the particle phase $\theta_b^{-1}\int d\theta$ yields nonzero harmonics when

$$h = m\kappa + n \;\; \text{with}\; n = q + \ell. \tag{6.35}$$

Thus, Equation (6.31) for the bunching becomes

$$\begin{aligned} b_h = &\int_{-\infty}^{\infty} dp''\, f_0(p) e^{-ihp(B_1+B_2)+im\kappa p B_1} \sum_{m=-\infty}^{\infty} e^{im\psi} J_m(-hA_2B_2) \\ &\times \sum_{l=-\infty}^{\infty} J_l(m\kappa A_1 B_1) \sum_{q=-\infty}^{\infty} J_q[-hA_1(B_1+B_2)]\,. \end{aligned} \tag{6.36}$$

To simplify this expression, we change the summation variable from ℓ to $n = q + \ell$ and use the identity

$$\sum_{q=-\infty}^{\infty} J_q(x) J_{n-q}(y) = J_n(x+y) \tag{6.37}$$

to obtain

$$\begin{aligned} b_h = &\int_{-\infty}^{\infty} dp''\, f_0(p) e^{-ihp(B_1+B_2)+im\kappa p B_1} \sum_{m=-\infty}^{\infty} e^{im\psi} J_m(-hA_2B_2) \\ &\times \sum_{n=-\infty}^{\infty} J_n[m\kappa A_1 B_1 - hA_1(B_1+B_2)] \\ = &\exp\left\{-\tfrac{1}{2}[h(B_1+B_2) - m\kappa B_1]^2\right\} \sum_{m=-\infty}^{\infty} e^{im\psi} J_m(-hA_2B_2) \\ &\times \sum_{n=-\infty}^{\infty} J_n[m\kappa A_1 B_1 - hA_1(B_1+B_2)]\,. \end{aligned} \tag{6.38}$$

We can substitute Equation (6.35) into the last expression to obtain the bunching factor for $h = m\kappa + n$ as

$$\begin{aligned} b_{n,m} = &e^{im\psi} e^{-[(m\kappa+n)B_2+nB_1]^2/2} \\ &\times J_m\left[-(m\kappa+n)A_2B_2\right] J_n\left\{-A_1\left[nB_1 + (m\kappa+n)B_2\right]\right\}\,. \end{aligned} \tag{6.39}$$

The Gaussian prefactor will significantly suppress the bunching unless its argument is small; if we use the same type of device for both dispersive section (e.g., two chicanes), than B_1 and B_2 will have the same sign and the bunching will be insignificant unless n and m have opposites signs. Further analysis shows that the magnitude of the bunching factor attains its maximum when $n = \pm 1$ and rapidly decreases as the absolute value of n increases; we therefore take $n = -1$ and $m > 0$ in Equation (6.35) for a particular harmonic number $h = m\kappa - 1$. Note that because only a single phase $e^{im\psi}$ contributes to the harmonic number h, any phase difference between the two lasers does not affect the magnitude of the bunching; in other words, the harmonic bunching of an optimized

EEHG device is not sensitive to the relative phase between the two lasers. The relevant expression for the maximal bunching at harmonic $h = m\kappa - 1$ is

$$|b_{-1,m}| = |J_m[(m\kappa - 1)A_2B_2]\,J_1(A_1\varpi)|\,e^{-\varpi^2/2}, \tag{6.40}$$

where $\varpi = B_1 - (m\kappa - 1)B_2$.

For $m > 4$, the maximal value of the Bessel function J_m is approximately $0.67/m^{1/3}$, which is achieved when the argument of J_m is

$$(m\kappa - 1)A_2B_2 = m + 0.81m^{1/3}. \tag{6.41}$$

To maximize $J_1(A_1\varpi)e^{-\varpi^2/2}$, we differentiate with respect to ϖ and obtain

$$A_1[J_0(A_1\varpi) - J_2(A_1\varpi)] = 2\varpi J_1(A_1\varpi) = 0. \tag{6.42}$$

Among the infinite roots of Equation (6.42), the two with the smallest value of $|\varpi|$ maximize $J_1(A_1\varpi)e^{-\varpi^2/2}$. These maxima increase linearly with A_1 when the modulation amplitude $A_1 \lesssim 3$, and grow less rapidly for larger values of A_1. When $A_1 = 3$, we have

$$A_1 = 3 : \max_{\varpi}\left[J_1(A_1\varpi)e^{-\varpi^2/2}\right] \approx 0.5, \tag{6.43}$$

and the bunching

$$|b_{\kappa m-1}| \approx \frac{0.3}{m^{1/3}}. \tag{6.44}$$

Thus, the echo bunching factor decreases slowly with the harmonic number $h \approx m$ if $\kappa = 1$ ($k_1 = k_2$), in contrast to HGHG where the bunching factor decreases exponentially with h. To illustrate the EEHG process further, we choose the numerical example suggested in Ref. [14], which also gives the corresponding physical parameters and further discussion. Suppose we design an EEHG system such that the modulating laser frequencies are identical ($\kappa = 1$) with normalized amplitudes $A_1 = 3$ and $A_2 = 1$; in this case, Equations (6.41) and (6.42) determine the ideal chicane strengths B_1 and B_2 as a function of the harmonic number. If we decide to impose bunching at the 24th harmonic, we have $m = 25$, $B_1 = 26.83$, and $B_2 = 1.14$, which leads to a bunching

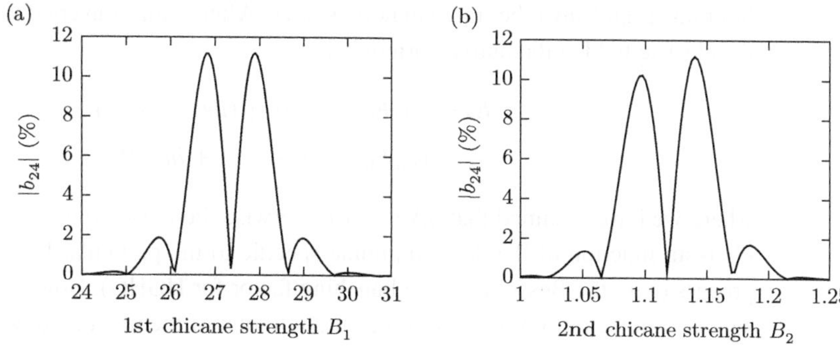

Figure 6.6 (a) Bunching at the 24th harmonic vs. B_1 for $B_2 = 1.14$. (b) Bunching at 24th harmonic vs. B_2 for $B_1 = 26.83$.

magnitude $|b_{24}| = 0.11$. We plot this harmonic bunching as a function of the first chicane strength B_1 in Figure 6.6(a), and the second chicane strength B_2 in Figure 6.6(b). There are two maxima with bunching greater than 10 percent, which reflects the fact that the two roots of (6.42) with smallest $|\varpi|$ give rise to the largest maxima of $|b_h|$.

6.4 Recent Developments in Harmonic Generation

Harmonic generation is a very active research topic, due both to the general appeal of reaching shorter radiation wavelengths and, in particular, for its potential use as an external coherent seed for a high-gain FEL. Chapter 8 has a more detailed discussion of the FEL facility FERMI, which produces coherent VUV and soft X-ray light using the HGHG principle. More recently, a new physical mechanism for strongly enhancing the frequency up-conversion efficiency of HGHG by using an undulator whose K parameter varies transversely was pointed out in Ref. [15].[1] There, theoretical analysis and numerical simulations demonstrated that the local energy spread at a certain phase relative to the seed laser wave can be significantly suppressed, thereby delivering unprecedented frequency up-conversion efficiency at harmonic orders several times larger than 10. Such a harmonic generation technique requires a small transverse emittance beam and a beamline with a relatively large dispersion.

The EEHG scheme has also been experimentally demonstrated in proof-of-principle experiments at optical and UV wavelengths using coherent undulator radiation [16] and in a high-gain FEL [17]. Another set of experiments showed that the EEHG scheme can be used to produce coherent radiation out to the 75th harmonic of the modulating laser [18].

Finally, noise degradation that occurs during the harmonic generation process can be quite significant [19], and must be taken into design considerations. We can understand the role of noise by considering the seed signal

$$E_1 = (E_0 + \Delta E)e^{i\theta + i\Delta\theta} \approx (E_0 + \Delta E)e^{i\theta}(1 + i\Delta\theta), \tag{6.45}$$

where ΔE and $\Delta\theta$ represents the small amplitude and phase noise due to shot noise, laser noise, and any other random noise source. After being converted to the harmonic h, the electric field at the output harmonic is

$$\begin{aligned} E_h &= G_h(E_0 + \Delta E)\exp(ih\theta + ih\Delta\theta) \\ &\approx G_h(E_0 + \Delta E)e^{ih\theta}(1 + ih\Delta\theta), \end{aligned} \tag{6.46}$$

where we have assumed that $h\Delta\theta < 1$ (otherwise the noise dominates the signal), and G_h is an function of the field amplitude specific to the particular harmonic generation process (e.g., the Bessel function bunching factor for HGHG). Now, if we compare the signal-to-noise ratio of the harmonic output to that of the input seed, we find that it has

[1] This kind of undulator is called a transverse gradient undulator (TGU); we discuss some FEL physics in a TGU in Appendix D.

been degraded by a factor of the harmonic number squared,

$$\left(\frac{P_{\text{signal}}}{P_{\text{noise}}}\right)_h = \frac{1}{h^2}\left(\frac{P_{\text{signal}}}{P_{\text{noise}}}\right)_1 . \tag{6.47}$$

Since h can be very large, obtaining coherent harmonic output with large signal-to-noise ratio can put stringent limits on the purity of the initial seed. In addition to the fundamental shot noise of the electron beam [20], the phase and amplitude errors of the seed laser itself may pose some practical limits [21, 22]. The effect of noise on coherence is always more detrimental to higher harmonics, with the signal-to-noise ratio typically scaling as h^{-2}. However, there can be specific situations for which $(P_{\text{signal}}/P_{\text{noise}})_h$ decreases more slowly with h [23]. The detailed treatment of noise and errors in harmonic generation is a developing subject, and we encourage interested readers to consult the references for more information.

References

[1] W. Colson, "The nonlinear wave equation for higher harmonics in free-electron lasers," *IEEE J. Quantum Electron.*, vol. 17, p. 1417, 1981.

[2] R. Bonifacio, L. D. Salvo, and P. Pierini, "Large harmonic bunching in a high-gain free-electron laser," *Nucl. Instrum. Methods Phys. Res., Sect. A*, vol. 293, p. 627, 1990.

[3] Z. Huang and K.-J. Kim, "Three-dimensional analysis of harmonic generation in high-gain free-electron lasers," *Phys. Rev. E*, vol. 62, p. 7295, 2000.

[4] —, "Nonlinear harmonic generation of coherent amplification and self-amplified spontaneous emission," *Nucl. Instrum. Methods Phys. Res., Sect. A*, vol. 475, p. 112, 2001.

[5] H. P. Freund, S. G. Biedron, and S. V. Milton, "Nonlinear harmonic generation in free-electron lasers," *IEEE J. Quantum Electron.*, vol. 36, p. 275, 2000.

[6] B. W. J. McNeil, G. R. M. Robb, M. W. Poole, and N. R. Thompson, "Harmonic lasing in a free-electron-laser amplifier," *Phys. Rev. Lett.*, vol. 96, p. 084801, Mar 2006.

[7] E. A. S. M. V. Yurkov, "Harmonic lasing in X-ray free electron lasers," *Phys. Rev. ST Accel. Beams*, vol. 15, p. 080702, Aug 2012.

[8] L.-H. Yu, "Generation of intense UV radiation by subharmonically seeded single-pass free-electron lasers," *Phys. Rev. A*, vol. 44, p. 5178, 1991.

[9] L.-H. Yu and J. H. Wu, "Theory of high gain harmonic generation: An analytical estimate," *Nucl. Instrum. Methods Phys. Res., Sect. A*, vol. 483, p. 493, 2002.

[10] I. Ben-Zvi, K. M. Yang, and L. H. Yu, "The 'fresh-bunch' technique in FELs," *Nucl. Instrum. Methods Phys. Res., Sect. A*, vol. 318, p. 726, 1992.

[11] M. Farkhondeh, W. S. Graves, F. X. Kaertner, R. Milner, D. E. Moncton, C. Tschalaer, J. B. van der Laan, F. Wang, A. Zolfaghari, and T. Zwart, "The MIT bates X-ray laser project," *Nucl. Instrum. Methods Phys. Res., Sect. A*, vol. 528, p. 553, 2004.

[12] E. Allaria *et al.*, "Two-stage seeded soft-X-ray free-electron laser," *Nature Photonics*, vol. 7, p. 913, 2013.

[13] G. Stupakov, "Using the beam-echo effect for generation of short-wavelength radiation," *Phys. Rev. Lett.*, vol. 102, no. 7, p. 074801, 2009.

[14] D. Xiang and G. Stupakov, "Echo-enabled harmonic generation free electron laser," *Phys. Rev. ST Accel. Beams*, vol. 12, no. 3, p. 030702, Mar 2009.

[15] H. Deng and C. Feng, "Using off-resonance laser modulation for beam-energy-spread cooling in generation of short-wavelength radiation," *Phys. Rev. Lett.*, vol. 111, p. 084801, 2013.

[16] D. Xiang, E. Colby, M. Dunning, S. Gilevich, C. Hast, K. Jobe, D. McCormick, J. Nelson, T. O. Raubenheimer, K. Soong, G. Stupakov, Z. Szalata, D. Walz, S. Weathersby, M. Woodley, and P.-L. Pernet, "Demonstration of the echo-enabled harmonic generation technique for short-wavelength seeded free electron lasers," *Phys. Rev. Lett.*, vol. 105, p. 114801, 2010.

[17] Z. T. Zhao, D. Wang, J. H. Chen, Z. H. C. H. X. Deng, J. G. Ding, C. Feng, Q. Gu, M. M. Huang, T. H. Lan, Y. B. Leng, D. G. Li, G. Q. Lin, B. Liu, E. Prat, X. T. Wang, Z. S. Wang, K. R. Ye, L. Y. Yu, H. O. Zhang, J. Q. Zhang, M. Zhang, M. Zhang, T. Zhang, S. P. Zhong, and Q. G. Zhou, "First lasing of an echo-enabled harmonic generation free-electron laser," *Nature Photonics*, vol. 6, p. 360, 2012.

[18] E. Hemsing, M. Dunning, B. Garcia, C. Hast, T. Raubenheimer, G. Stupakov, and D. Xiang, "Echo-enabled harmonics up to the 75th order from precisely tailored electron beams," *Nature Photonics*, 2016.

[19] E. L. Saldin, E. A. Schneidmiller, and M. V. Yurkov, "Study of a noise degradation of amplification process in a multistage HGHG FEL," *Opt. Commun.*, vol. 202, p. 169, 2002.

[20] Z. Huang, "An analysis of shot noise propagation and amplification in harmonic cascade FELs," in *Proceedings of the 2006 FEL Conference*, p. 130, 2006.

[21] G. Stupakov, Z. Huang, and D. Ratner, "Noise amplification in echo-enabled harmonic generation (EEHG)," in *Proceedings of the 2010 FEL Conference*, p. 278, 2010.

[22] G. Geloni, V. Kocharyan, and E. Saldin, "Analytical studies of constraints on the performance for EEHG FEL seed lasers," 2011, arXiv:1111.1615v1.

[23] D. Ratner, A. Fry, G. Stupakov, and W. White, "Laser phase errors in seeded free electron lasers," *Phys. Rev. ST Accel. Beams*, vol. 15, p. 030702, 2012.

7 FEL Oscillators and Coherent Hard X-Rays

An oscillator FEL is a low-gain device that employs an optical cavity to build up and store the field power produced from successive passes through the undulator. Hence, FEL oscillators operate much like traditional lasers based on atomic transitions: radiation is amplified over many passes through the gain medium, which in our case is supplied by the electron beam in the undulator. The first FEL oscillator was demonstrated at Stanford [1] soon after the invention of the FEL concept [2]. Since then, oscillator devices have been built and operated around the world, generating intense radiation in the IR, optical, and UV wavelengths where low-loss, normal-incidence reflectors and accelerators that can produce the required electron beam quality are readily available (see, e.g., [3, 4]).

A hard X-ray free-electron laser in the oscillator configuration – an X-ray FEL oscillator (XFELO) – will produce highly stable X-ray beams of ultra-high spectral purity and high average brightness, offering unique scientific opportunities complementary to those provided by high-gain X-ray amplifiers. The concept for an XFELO using crystals as low-loss reflectors was presented in 1983 [5] at the same workshop where X-ray SASE was first proposed outside the former Soviet Union [6]. While work in SASE blossomed over the next few decades, the XFELO concept did not receive its due attention until a recent, detailed study showed that an oscillator could be feasible with the low-intensity, ultra-low-emittance electron bunches contemplated for energy recovery linacs [7].

In this chapter we begin by briefly introducing some of the basic operating principles and phenomena that are applicable to a wide range of oscillator FELs. We then discuss many of the physics issues, requirements, and challenges that are unique to an oscillator for X-rays.

7.1 FEL Oscillator Principles

The basic schematic of an FEL oscillator is illustrated in Figure 7.1. Electron bunches from a (usually rf) accelerator pass through an undulator that is located inside a low-loss optical cavity. Starting from an empty cavity, in the first pass the electron beam emits spontaneous undulator radiation that is reflected back into the undulator by the cavity mirrors. In the second pass, the pulse of spontaneous emission meets and overlaps with a second electron bunch at the entrance of the undulator. The radiation and the e-beam

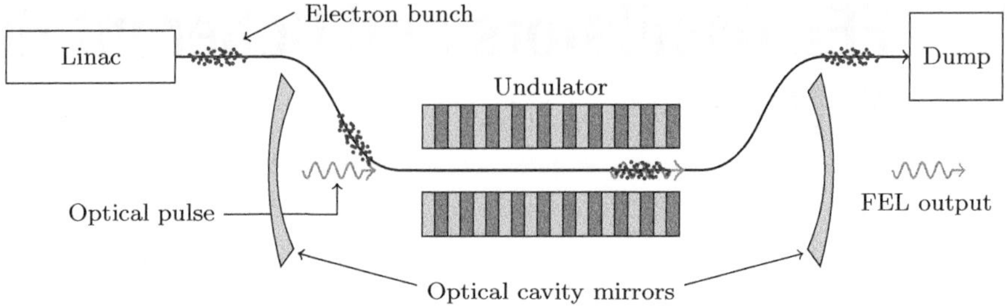

Figure 7.1 Schematic of an FEL oscillator showing its basic operating principle.

interact in the undulator, after which the output field is composed of the spontaneous emission from both the first and second pass, along with an amplified signal due to FEL gain. This process repeats, so that the amplified radiation signal will eventually dominate the output if the gain is larger than the round-trip loss in the cavity.

7.1.1 Power Evolution and Saturation

For a simple mathematical description of the power evolution in an oscillator, let P_n be the power of the optical pulse at the undulator exit after its nth pass, and P_s be the power of spontaneous emission. Then

$$\begin{aligned} P_1 &= P_s \\ P_n &= R(1+G)P_{n-1} + P_s \quad \text{for } n \geq 2, \end{aligned} \tag{7.1}$$

where G is the FEL gain and R is the reflectivity of the optical transport line. The net single pass power amplification is $R(1 + G)$, and evidently the power increases if the single pass gain overcomes the losses such that

$$R(1+G) > 1. \tag{7.2}$$

This is the "lasing" condition for an FEL oscillator. The power after the nth pass is governed by Equations (7.1), whose solution is

$$P_n = \frac{[R(1+G)]^n - 1}{R(1+G) - 1} P_s\,. \tag{7.3}$$

Assuming that $R(1+G) > 1$, we see that the power increases exponentially with n after sufficiently many passes of amplification.

The exponential growth of the intracavity radiation power does not continue indefinitely. Rather, the optical power eventually becomes large enough to trap electrons in the ponderomotive potential and then rotate them to an absorptive phase where they extract energy from the field as we discussed in Section 3.3.2. This in turn reduces the gain from its small signal value, and the system reaches a steady state or "saturates" when the gain decreases to the value G_{sat} given by

$$R(1+G_{\text{sat}}) = 1\,. \tag{7.4}$$

Furthermore, at saturation the power generated during one pass ΔP equals the total losses, so that if the power inside the cavity is P_{sat} we have $\Delta P = (1 - R)P_{\text{sat}}$. In Section 3.3.2 we showed that $\Delta P \approx P_{\text{beam}}/2N_u$, which in turn implies that the intracavity optical power at saturation is

$$P_{\text{sat}} \approx \frac{1}{2N_u(1-R)} P_{\text{beam}} \,. \tag{7.5}$$

The optical elements in the cavity, and in particular the mirrors, must be able to withstand the power P_{sat} for the oscillator to operate stably.

At saturation the power decreases by an amount $(1 - R)\Delta P$ during any complete round-trip cycle; this energy loss can be due to many different mechanisms, including radiation absorption in the mirror material, diffraction at the edges of the optical elements, and transmission out of the cavity for useful purposes. If one had an ideal optical line with no losses, the cavity transmission would equal $(1 - R)$ so that the maximum power that can be coupled out of the oscillator is $(1 - R)P_{\text{sat}} \approx P_{\text{beam}}/2N_u$.

Useful output radiation from an FEL oscillator requires it to operate for some time at saturation. Hence, an oscillator can be driven by a pulsed accelerator only if the number of bunches within each macro-pulse is more than that required to reach saturation. With a CW accelerator, on the other hand, the oscillator can be maintained at a steady state indefinitely. This is a desirable mode of operation, since the FEL then provides a stable source with a higher average photon flux.

7.1.2 Qualitative Description of Longitudinal Mode Development

There is much more physics at work in addition to the power evolution just described. One subtle but important phenomenon is lethargy [8] – the fact that the trailing part of the optical pulse (the tail) is more strongly amplified than the front (the head). This is because the initially unmodulated electron beam must propagate some distance through the undulator to develop the density modulation that provides FEL gain, during which time the electron beam and its gain slips behind the field envelope. As a consequence, the FEL gain is maximized when the cavity length is slightly shorter than that given by the exact synchronism condition (the synchronism condition is when the cavity length equals the distance between successive bunches).

The lethargy effect causes the round-trip time of the pulse envelope to be in general different from the round-trip time of the phase, since the latter is determined essentially by the cavity length. In other words, the phase fronts return to the undulator after a time approximately equal to the round trip time in the cavity, while the peak of the pulse envelope arrives a time of order the slippage time $N_u\lambda_1/c$ after. To be more precise, any delay of the phase fronts is given by the imaginary part of the complex gain, which is small at peak gain as shown in Figure 4.1. This fact will be important later when we discuss a nuclear resonance-stabilized XFELO in Section 7.4.

Temporal coherence in an FEL oscillator is achieved by gain narrowing due to the FEL itself and also through spectral filtering provided by the cavity mirrors if their reflectivity is wavelength-dependent. The FEL-induced spectral gain narrowing occurs because the FEL gain is frequency dependent; alternatively, it can be understood as the slow increase in the coherence length from $N_u\lambda_1$ due to many passes through the undulator. Hence, when the mirror reflectivity is independent of wavelength, we expect that the FEL spectral bandwidth σ_ω decreases with pass number n as

$$\left(\frac{\sigma_\omega}{\omega_1}\right)_n \sim \frac{1}{N_u}\frac{1}{\sqrt{n}}. \tag{7.6}$$

For short electron bunches, gain narrowing stops when $(\sigma_\omega/\omega_1)_n$ becomes the transform limited bandwidth $\lambda_1/(4\pi\sigma_z)$ associated with the RMS length of the electron bunch σ_z. For longer electron bunches that have a current maximum in the center, the nonuniform gain causes the optical pulse profile to also narrow in length/duration, with $(\Delta z)_n^{\text{rms}} \sim \sigma_z/\sqrt{n}$. The spectral and temporal narrowing will stop when the pulse is Fourier transform-limited, i.e., at the pass number $n \sim N_{FT}$ determined by

$$\left(\frac{\sigma_\omega}{\omega_1}\right)_{N_{FT}} (\Delta z)_{N_{FT}}^{\text{rms}} \sim \frac{\lambda_1}{4\pi}, \tag{7.7}$$

from which we determine that the steady state is reached after approximately $N_{FT} \sim 4\pi\sigma_z/\lambda_1 N_u$ passes, and that the limiting bandwidth is [9]

$$\frac{\sigma_\omega}{\omega_1} \sim \sqrt{\frac{\lambda_1}{4\pi N_u \sigma_z}}. \tag{7.8}$$

This limiting mode is known as the dominant supermode [10].

In what follows we will show how the longitudinal supermodes arise from the dynamic interplay between amplification, gain narrowing, FEL lethargy, and spectral filtering from the mirrors. Hence, we will further extend the physics above to include the possibility that the spectral narrowing comes about not only through slippage, but also because the mirrors have a limited bandpass.

7.1.3 Longitudinal Supermodes of the FEL Oscillator

In this section we use the simple low-gain model developed by Elleaume [11] to more fully investigate the supermode longitudinal dynamics. This model divides the evolution during a single round trip into its various components: gain that depends on the current and the propagation/slippage in the undulator, reflection by the mirrors, and propagation in the cavity. Assuming that all of these effects result in small perturbations to the radiation (as is true in the low-gain regime), then we can approximate each as acting individually and in succession on the electric field $E(t)$. In what follows we discuss these longitudinal effects in turn, and then combine them into a single equation describing the linear dynamics of a low-gain oscillator.

We assume that the FEL gain transforms the field via the amplification operator $E \to E + \mathscr{G}[E]$. To develop a simple model for $\mathscr{G}$, we recall that FEL gain depends linearly on the current and that the field interacts with the electron beam within one slippage length $N_u\lambda_1$. In terms of the light-cone coordinate $\tau \equiv z - ct$, this means that the amplification of $E(\tau)$ depends on the interaction between the current and field amplitude for points τ'

satisfying $\tau \leq \tau' \leq \tau + N_u\lambda_1$ (see, e.g., [12]). We will use a very simple description of this process in which we model the gain operation $\mathscr{G}[E]$ as increasing the field by an amount depending on the e-beam current and E-field amplitude at the point one-half the slippage distance $N_u\lambda_1/2$ ahead. Hence, we approximate the amplitude gain from an electron beam with RMS length σ_z as acting via

$$\begin{aligned} E(\tau) \to E(\tau) + \mathscr{G}[E] &\approx E(\tau) + \frac{G}{2} e^{-(\tau + N_u\lambda_1/2)^2/2\sigma_z^2} E\left(\tau + \tfrac{1}{2}N_u\lambda_1\right) \\ &\approx \left[1 + \frac{G}{2}\left(1 - \frac{\tau^2}{2\sigma_z^2}\right)\right] E(\tau) \\ &\quad + \frac{G}{4} N_u\lambda_1 \frac{\partial E}{\partial \tau} + \frac{G}{16}(N_u\lambda_1)^2 \frac{\partial^2 E}{\partial \tau^2}, \end{aligned} \tag{7.9}$$

where for simplicity we assume that $\sigma_z \gg N_u\lambda_1$ and that the amplitude gain $G/2$ is real.[1]

After the FEL interaction, the mirror reduces the field amplitude by the multiplicative factor $\sqrt{R} \equiv \sqrt{1-\alpha} \approx 1 - \alpha/2$, where α is the (assumed real) power loss. In addition, we include the possibility that the reflectivity depends on frequency by modeling it as a Gaussian filter in ω with RMS power bandwidth σ_{refl}. Since we model the mirror filtering as acting on the slowly-varying field envelope, it is centered near $\omega = 0$ and results in the transformation

$$\begin{aligned} E(\tau) \to \int d\omega\, e^{-i\omega\tau/c} R(\omega) E(\omega) &\approx (1 - \alpha/2) \int d\omega\, e^{-i\omega\tau/c} e^{-\omega^2/4\sigma_{\text{refl}}^2} E(\omega) \\ &\approx \left(1 - \frac{\alpha}{2}\right) \int d\omega\, e^{-i\omega\tau/c} \left[1 - \frac{\omega^2}{4\sigma_{\text{refl}}^2}\right] E(\omega) \\ &= \left(1 - \frac{\alpha}{2}\right) E(\tau) + \frac{c^2}{4\sigma_{\text{refl}}^2} \frac{\partial^2}{\partial \tau^2} E(\tau). \end{aligned} \tag{7.10}$$

Finally, we include the possibility that after one round trip through the cavity the arrival time of the radiation pulse and the next electron bunch may differ by an amount ℓ/c; this timing difference, called *detuning* in the FEL community, could be due to adjustments to the cavity length or timing jitter of the electrons; we model it by

$$E(\tau) \to E(\tau + \ell) \approx E(\tau) + \ell \frac{\partial}{\partial \tau} E(\tau). \tag{7.11}$$

A full pass through the oscillator is composed of the transformations (7.9)–(7.11) due to the gain including slippage, the mirror, and the cavity length detuning. Every transformation is written as a sum of the initial field $E(\tau)$ and a perturbation. If each of these perturbing effects is small, then the field at pass $(n+1)$ can be written a sum of the various perturbations acting on the field E_n as follows:

[1] The generalization to complex G and $\sqrt{R}$ is straightforward but messy. For example, the change in power is $|1 + G/2|^2 \approx 1 + (G + G^*)/2$ if G is complex.

$$\begin{aligned} E_{n+1}(\tau) &\approx E_n(\tau) + \frac{G-\alpha}{2}E_n(\tau) - \frac{G\tau^2}{4\sigma_z^2}E_n(\tau) \\ &+ \left(\ell + \frac{GN_u\lambda_1}{4}\right)\frac{\partial E_n}{\partial \tau} + \left[\frac{c^2}{4\sigma_{\text{refl}}^2} + \frac{G(N_u\lambda_1)^2}{16}\right]\frac{\partial^2 E_n}{\partial \tau^2}. \end{aligned} \tag{7.12}$$

Moving E_n to the left-hand side and setting $E_{n+1} - E_n \approx \partial E_n/\partial n$ leads to a linear partial differential equation for the field $E_n(\tau)$. This PDE can be solved by the separation of variables technique, which leads to exponential dependence on n, while the temporal variation is described by Hermite–Gauss functions. We index these linear modes by p and find that the general solution can be written as a sum over the "supermodes"

$$\begin{aligned} E_n^p(\tau) &= \exp\left[\left(\frac{G-\alpha}{2}\right)n - \left(\frac{2D^2\sigma_{\text{filter}}^2}{c^2} + \frac{c(1+2p)\sqrt{G}}{2\sigma_z\sigma_{\text{filter}}}\right)n\right] \\ &\times e^{-2\sigma_{\text{filter}}^2 D\tau/c^2}\exp\left[-\frac{\sqrt{G}c\sigma_{\text{filter}}}{2\sigma_z}\tau^2\right]H_p\left(G^{1/4}\sqrt{\frac{c\sigma_{\text{filter}}}{\sigma_z}}\tau\right), \end{aligned} \tag{7.13}$$

where we have defined the net detuning length $D \equiv \ell + GN_u\lambda_1/4$ and the effective filtering bandwidth σ_{filter} via

$$\frac{1}{\sigma_{\text{filter}}^2} \equiv \frac{1}{\sigma_{\text{refl}}^2} + \frac{GN_u^2\lambda_1^2}{16}. \tag{7.14}$$

The first line in (7.13) indicates that the exponential power growth is reduced from its nominal value $G - \alpha$ (gain minus loss) if the total detuning length $D \neq 0$; this condition shows one effect of lethargy since maximum gain is achieved when the cavity length is reduced slightly from its nominal synchronous length (i.e., $D = 0$ implies that $\ell < 0$). Significant FEL gain requires the total detuning to be within the effective oscillator bandwidth such that $D\sigma_{\text{filter}} \ll 1$. Additionally, setting $D = 0$ shows that the gain approaches the infinite beam limit only if the electron beam is also significantly longer than the inverse bandwidth $1/\sigma_{\text{filter}}$. For shorter electron bunches, only the fraction of current whose spectral content lies within the effective bandpass set by either the mirror σ_{refl} or the slippage $4/(N_u\lambda_1\sqrt{G})$ contributes to the gain.

The RMS width of the pth mode is proportional to the geometric mean of the e-beam size and $1/\sigma_{\text{filter}}$, with temporal width $\sim \sqrt{(1+2p)\sigma_z/c\sigma_{\text{filter}}}$. When the electron bunch is long there are many longitudinal modes with comparable growth rates, and the oscillator output is comprised of a superposition of supermodes whose total bandwidth $\sim 1/N_u$ and temporal duration $\sim \sigma_z$. As the evolution proceeds through many passes, however, the lowest-order ($p = 0$) Gaussian mode with largest gain will eventually become dominant. If the mirror is essentially wavelength independent, $\sigma_{\text{refl}} \gg 1/N_u\lambda_1$, our discussion in the beginning of this chapter applies and the output bandwidth will approach the limiting value (7.8), albeit slowly.

On the other hand, we will see that the crystal mirrors that enable FEL oscillators in the X-ray spectral region have $\sigma_{\text{refl}} \ll 1/N_u\lambda_1$ (typically $N_u \lesssim 3 \times 10^3$ while $\sigma_{\text{refl}}/\omega_1 \sim 10^{-5}$ to 10^{-7}). In this case, $\sigma_{\text{filter}} \to \sigma_{\text{refl}}$ which simplifies some of the preceding discussion. For example, the lowest-order (Gaussian) supermode simplifies to

$$E_n^0(\tau) = \exp\left[\frac{1}{2}\left(G - R - \frac{4\ell^2\sigma_{\text{refl}}^2}{c^2} - \frac{c\sqrt{G}}{\sigma_z\sigma_{\text{refl}}}\right)n\right] e^{-2\sigma_{\text{refl}}^2\ell\tau/c^2} e^{-\tau^2/2\sigma_0^2}, \tag{7.15}$$

where the mean square temporal width $\sigma_0^2 \equiv \sigma_z/(\sqrt{G}c\sigma_{\text{refl}})$. Hence, in this case the gain is reduced from its nominal value if either the cavity length shift ℓ or the electron beam width σ_z is smaller than the inverse bandwidth of the mirror c/σ_{refl}. The steady state temporal width is given by $\sigma_0 \propto \sqrt{\sigma_z/c\sigma_{\text{refl}}}$ with final bandwidth $\propto \sqrt{c\sigma_{\text{refl}}/\sigma_z}$, and it now requires $N_{FT} \sim 2\sigma_{\text{refl}}\sigma_z/c \ll 4\pi\sigma_z/(\lambda_1 N_u)$ passes to reach this steady state.

In addition to modifying the supermode behavior, the additional spectral filtering provided by the mirrors also completely suppresses the sideband/synchrotron instability, eliminating the unstable and chaotic "spiking mode" of operation observed at lower wavelengths [13, 14]. This is because the sideband instability amplifies frequency content near that associated with the synchrotron period, i.e., with frequencies at $\omega = \omega_1 \pm \omega_s$, where

$$\omega_s \sim \frac{\lambda_u}{\lambda_1}ck_s = \frac{\lambda_u}{\lambda_1}\frac{c}{L_u}\sqrt{\epsilon} = \frac{c}{N_u\lambda_1}\sqrt{\epsilon}, \tag{7.16}$$

and we recell that ϵ is the normalized field strength defined in (3.26). At saturation $\epsilon \sim 1$, so that the characteristic frequency of the sideband/"spiking" mode is

$$\omega_s \sim \frac{c}{N_u\lambda_1} \gg \sigma_{\text{refl}}, \tag{7.17}$$

and the narrow bandwidth of the XFELO crystal mirrors effectively filters out the sideband instability.

7.1.4 Transverse Physics of the Optical Cavity

When the gain is small, the transverse mode is typically well described by the vacuum resonator modes of the cavity. We will briefly describe some of the transverse cavity physics in the limit of Gaussian optics, which assumes that angles from the optical axis are small (paraxial) and that optical elements can be treated as producing linear transformations to the field. In Section 1.2.3 we showed that such linear transformations propagate the radiation brightness/Wigner function along rays, which implies that we can analyze the cavity modes using the same matrix formulation that we described for particle beams in Sections 1.1.2–1.1.4. Under these limiting circumstances, the transformations act on the (pseudo)-distribution of rays in the position–angle phase space $(\boldsymbol{x}, \boldsymbol{\phi})$, and the wave behavior can be described by referencing only the propagation of rays.[2]

In the laser community the matrix approach is referred to as the ABCD-matrix method [15], and typically these matrix elements are used to derive the stable Rayleigh range and wavefront curvature for Hermite–Gaussian cavity modes. The lowest-order mode is analogous to a Gaussian particle beam with emittance $\lambda/4\pi$, while its Rayleigh length at the waist $Z_R = \sigma_r^2/(\lambda/4\pi) = (\lambda/4\pi)\sigma_{r'}^2$ is equivalent to the Courant–Snyder beta function from particle optics.

[2] Non-ideal elements, apertures, and nonlinear transformations introduce interference effects that may not be well described by the methods presented here.

In order to understand the transverse X-ray profile, we first consider the simple two-mirror resonator shown in Figure 7.2. We model this optical cavity as containing one ideal mirror of focal length f, such that the round trip distance in the cavity is L_c. We restrict our discussion to the 2D phase space (x, ϕ), and recall from Section 1.1.4 that the matrices associated with a drift length ℓ and a focusing mirror are, respectively,

$$\mathsf{L}(\ell) = \begin{bmatrix} 1 & \ell \\ 0 & 1 \end{bmatrix} \qquad \mathsf{F}(f) = \begin{bmatrix} 1 & 0 \\ -1/f & 1 \end{bmatrix}. \tag{7.18}$$

Stable resonator modes exist when the RMS size, divergence, and correlation are periodic over one round trip through the cavity. These mode sizes can be determined from the the matrix map that starts and ends in the middle of the undulator; for the two-mirror resonator this is given by $\mathsf{M}_{2\text{res}} = \mathsf{L}(L_c/2)\mathsf{F}(f)\mathsf{L}(L_c/2)$. The matrix $M_{2\text{res}}$ maps (x, ϕ) from and back to the center of the cavity such that

$$\begin{bmatrix} x \\ \phi \end{bmatrix}_{\text{out}} = \mathsf{M}_{1\text{res}} \begin{bmatrix} x \\ \phi \end{bmatrix}_{\text{in}}, \tag{7.19}$$

while the second-order moment matrix Σ_{out} at the output plane is related to the initial Σ_{in} via

$$\begin{aligned} \Sigma_{\text{out}} \equiv \begin{bmatrix} \langle x^2 \rangle & \langle x\phi \rangle \\ \langle \phi x \rangle & \langle \phi^2 \rangle \end{bmatrix}_{\text{out}} &= \mathsf{M}_{2\text{res}} \begin{bmatrix} \langle x^2 \rangle & \langle x\phi \rangle \\ \langle \phi x \rangle & \langle \phi^2 \rangle \end{bmatrix}_{\text{in}} \mathsf{M}_{2\text{res}}^T \\ &= \mathsf{M}_{2\text{res}} \Sigma_{\text{in}} \mathsf{M}_{2\text{res}}^T. \end{aligned} \tag{7.20}$$

Equating Σ_{out} and Σ_{in} implies that at the cavity middle the correlation vanishes (the radiation has a waist), and that the cavity round trip length L_c and mirror focal length f are related to the trapped mode Rayleigh length through the following relation:

$$f = \frac{L_c}{4} + \frac{1}{L_c} \frac{\langle x^2 \rangle_{\text{in}}}{\langle \phi^2 \rangle_{\text{in}}} = \frac{L_c}{4} + \frac{Z_R^2}{L_c}. \tag{7.21}$$

Note that stable operation requires $f > L_c/4$, which in terms of the mirror's radius of curvature is $2f = r > L_c/2$. This inequality can be violated if there is sufficient FEL amplification, but for low-gain devices it provides a good starting point for optical cavity design.

In the next section we will describe cavity designs suitable for X-ray FEL oscillators (XFELOs). In addition, we will extend the cavity analysis described here to include a more general configuration and tolerance analysis.

7.2 X-Ray Cavity Configurations

FEL oscillators are composed of two basic components: the gain medium, consisting of an undulator and electron beam, and a closed loop of optical elements that trap and focus the radiation to be amplified. Having already explained the essential physical principles behind the XFELO, here we describe how a suitable X-ray optical cavity could be formed. Although no such cavity yet exists, the various optical elements required for an

XFELO have been developed over the past few decades, largely due to the demands of synchrotron light sources. For example, the X-ray focusing required for an XFELO cavity can be done with high-quality grazing incidence mirrors or with compound refractive lenses (CRLs). However, neither of these focusing elements can work alone: grazing incidence mirrors rely on total external reflection that is appreciable only at very small grazing incidence angles, while CRLs work in transmission. Hence, another optical element that has high reflectivity at near-normal angles of incidence is required to efficiently return the radiation to the undulator for amplification. This role can be filled by crystal mirrors based on Bragg reflection from the (nearly) perfectly aligned crystal planes.

X-ray optics based on Bragg reflection work via the coherent scattering of light whose wavelength approximately satisfies the Bragg condition $\lambda = 2a \sin\Theta$, where Θ is the angle from grazing incidence and a is the atomic spacing between the crystalline planes. The spectral width of large reflectivity, called the Darwin width, is inversely proportional to the number of crystal planes that contribute to the reflection. For hard X-rays more than 10^5 crystal planes may contribute, so that typically $\sigma_{\text{refl}}/\omega_1 \lesssim 10^{-5}$ at 5 keV and $\sigma_{\text{refl}}/\omega_1 \lesssim 10^{-7}$ at 20 keV. Within the Darwin width the reflectivity can approach 100 percent, meaning that Bragg crystals may enable a low-loss optical cavity for X-rays that efficiently filters the FEL output.

The simplest cavity configuration is illustrated in Figure 7.2, which uses two Bragg crystals to trap the X-rays and a single grazing incidence mirror to shape and focus the transverse mode. This cavity is a realization of the two-mirror resonator described previously. In order to have a large reflectivity, the grazing angle of incidence at the curved mirror must be less than the critical angle of total external reflection, which is typically a few mrad. The radiation in turn must be approximately normal to the Bragg crystal, $\Theta = \pi/2$ within a few mrad; converting this angular extent into a wavelength range using Bragg's law $\lambda = 2a \sin\Theta$ implies that the two-crystal cavity supports FEL lasing over a severely limited spectral/tuning range.

7.2.1 Four-Crystal, Wavelength-Tunable XFELO Cavity

Tunable XFELOs require cavities with more than two Bragg crystals, and we show an example of a four-crystal configuration in Figure 7.3. This particular design wraps up the four-bounce monochromator invented by Cotterill [17] into a closed, bow-tie shaped cavity. The wavelength can be changed by adjusting the four crystals' Bragg angles in unison while keeping a constant round-trip path length by a coordinated translation

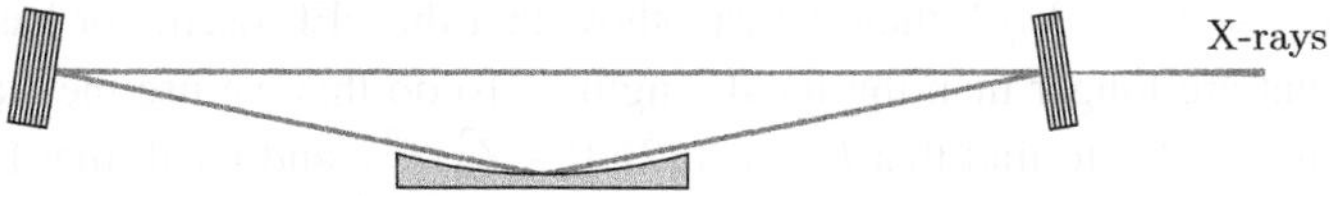

Figure 7.2 A basic, two-crystal scheme for an XFELO optical cavity that is not tunable. Here, the focusing is provided by a grazing incidence mirror, although other options are possible. Adapted from Ref. [16].

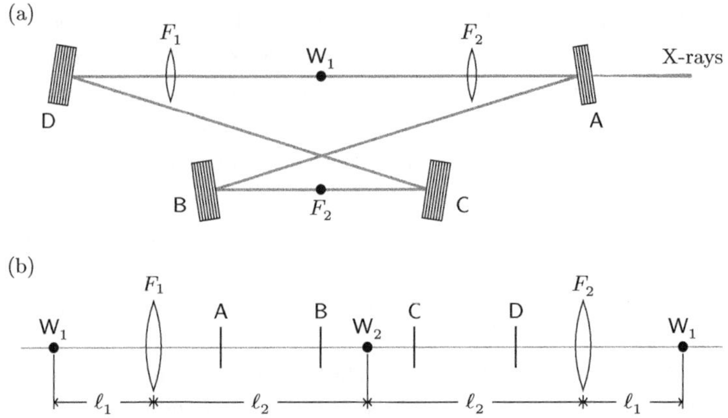

Figure 7.3 (a) A diagram of a tunable cavity configuration using four crystals. The focusing elements F_1 and F_2 could represent grazing incidence mirrors or compound refractive lenses (CRLs). (b) Unfolded optical lattice equivalent to the four-crystal cavity. A, B, C, and D label the crystal positions, while W_1 (W_2) indicate the position of the radiation waist inside (outside) the undulator. Adapted from Ref. [16].

of the crystals. The four-crystal scheme also allows one to use a single crystal material (with different reflection planes) over the entire 5–20 keV spectral range of the XFELO. This is an important advantage, since XFELOs operating at any wavelength can then employ diamond crystals which, as we will discuss shortly, have excellent thermo-mechanical properties.

To study the four-crystal cavity of Figure 7.3(a), we assume that the role of the crystals is solely to deflect the ray, in which case this configuration is equivalent to the linear periodic system shown in Figure 7.3(b). We take the focusing elements F_1 and F_2 to have focal length f, and label the relevant cavity distances as shown. Note that W_1 is the position of the waist at the middle of the undulator, while the other waist is located at W_2. Using the same notation for the free space and focusing transport matrices as in Equation (7.18), one period through the cavity is given by

$$\mathsf{M}_{4\text{cr}} = \mathsf{L}(\ell_1)\,\mathsf{F}(f)\,\mathsf{L}(2\ell_2)\,\mathsf{F}(f)\,\mathsf{L}(\ell_1). \tag{7.22}$$

Stability requires that $|\text{Tr}(\mathsf{M})| < 2$, which can be shown to be equivalent to $f > 0$ and

$$0 < \left(\frac{\ell_1}{f} - 1\right)\left(\frac{\ell_2}{f} - 1\right) < 1. \tag{7.23}$$

Consequently, a stable cavity either has both $\ell_{1,2} > f$ or $\ell_{1,2} < f$. If we assume that this four-crystal cavity reduces to the two-mirror cavity discussed previously in the limit $\ell_2 \to \ell_1 = L_c/2$, then we can show that the FEL oscillator has drift half-lengths $\ell_{1,2}$ that are longer than the focal length f. To do this we use the stability condition equation (7.21) to find that $f = (\ell_1/2)(1 + Z_R^2/\ell_1^2)$, and recall that FEL gain is maximized when $Z_R \sim L_u/2\pi$, so that $Z_R^2/\ell_1 < 1$ and $\ell_1 > f$.

As we just mentioned, the Rayleigh range inside the undulator $Z_{R,1}$ is chosen so as to maximize the FEL gain. On the other hand, there is considerably more freedom in selecting $Z_{R,2}$, which, as shown in Figure 7.3, governs the divergence of the radiation seen by the crystal mirrors. Since the Bragg crystal angular acceptance is related to the bandwidth of large reflectivity through Bragg's law $\lambda = 2a \sin \Theta$, we see that it is advantageous to decrease the radiation divergence at the crystal by increasing $Z_{R,2}$. To determine the stable Rayleigh lengths, we equate the second-order moment matrices via $\Sigma_{\text{out}} = \mathsf{M}_{4\text{cr}} \Sigma_{\text{in}} \mathsf{M}_{4\text{cr}}^T$, to find that

$$Z_{R,1} = \sqrt{\frac{\ell_1 - f}{\ell_2 - f}} [f(\ell_1 + \ell_2) - \ell_1 \ell_2]^{1/2} \qquad Z_{R,2} = \frac{\ell_2 - f}{\ell_1 - f} Z_{R,1}. \tag{7.24}$$

Hence, we can extend the Rayleigh range $Z_{R,2}$ and decrease the X-ray divergence on the crystals by making the length ℓ_2 longer than ℓ_1.

We can determine the tolerance in the orientation of the crystals by using a standard method from accelerator physics. Suppose the crystal A is oriented at an angle $\Delta\Theta$ from the ideal angle, so that an on-axis ray is deflected by an angle $2\Delta\Theta$. The coordinates of the displaced optical axis at W_1 can be found by requiring that the displaced reference trajectory is periodic with the period of the optical system:

$$\begin{bmatrix} \Delta x \\ \Delta \phi \end{bmatrix} = \mathsf{M}_{4\text{cr}} \begin{bmatrix} \Delta x \\ \Delta \phi \end{bmatrix} + \mathsf{L}(\ell_1)\,\mathsf{F}(f)\mathsf{L}(2\ell_2 - d_A) \begin{bmatrix} 0 \\ 2\Delta\Theta \end{bmatrix}, \tag{7.25}$$

where d_A is the distance from F_1 to crystal A. The first term in Equation (7.25) is the round-trip transformation from the waist W_1 at the undulator center, while the second includes the angular deviation due to an error at crystal A.

Solving (7.25) will give the displaced ray position Δx and angle $\Delta\phi$ as a function of the angular error $\Delta\Theta$. The tolerance is set by requiring the resulting ray displacements in position and angle to be much less than mode size and angular divergence, $|\Delta x| \ll \sigma_r$ and $|\Delta\phi| \ll \sigma_{r'}$. For typical XFELO parameters $\sigma_r \approx 10$ μm and $\sigma_{r'} \approx 1$ μrad, and the resulting angular tolerance $\Delta\Theta \lesssim 10$ nrad. It appears that the null-detection feedback technique employed at the Laser Interferometer Gravitational-Wave Observatory (LIGO) may be able to achieve this high level of stability for multiple optical axes with a single detector, and therefore appears to be a promising stabilization approach.

Finally, preserving the wavefronts through the optical cavity places additional tolerances on the smoothness of the optical elements. For diamond crystals, the surface error height δh should be a fraction of the X-ray wavelength times the difference of the index of refraction from unity. Since for hard X-rays the refractive index differs from one by an amount of order 10^{-6}, the tolerance on δh is about a micron, which should be achievable; similar considerations for CRLs give a similar constraint on its surface. On the other hand, the surface requirements of the grazing incidence mirror can be divided up into a height error, which contributes to diffuse scattering and an effective reduction in reflectivity, and figure error, which contributes to mode distortion. The requirement on the height error for a grazing incidence mirror is about 1 nm, while the latter figure error is about 0.1 μrad; both tolerances are demanding but within the current state of the art.

7.2.2 Diamond Crystals for XFELO

Diamond is a material whose superb physical qualities are well suited to an XFELO cavity: high mechanical hardness, high thermal conductivity, high radiation hardness, low thermal expansion, and chemical inertness. An exceptionally high $\geq$ 99 percent reflectivity is predicted in X-ray Bragg diffraction, higher than that from any other crystal. This is because the distance over which X-rays are reflected from diamond (the so-called "extinction length") is much shorter than the characteristic absorption length; in other words, diamond has a uniquely small ratio of the extinction length to the absorption length that leads to a near perfect reflectivity. Hence, the primary source of loss come from imperfections and impurities in the crystal structure. Recent developments in the manufacture of synthetic diamond crystals have shown that nearly defect-free crystals of suitable size for an XFELO can be produced, while experiments with 13.9 keV and 23.7 keV X-ray photons have established that the predicted reflectivity greater than 99 percent at near normal incidence can be achieved [18].

In addition to its high reflectivity, diamond is also well suited to handling the thermal heat load from an XFELO. Specifically, since Bragg crystals rely on the periodic lattice spacing, one must determine if temperature gradients from radiation heating could lead to strain in the material and gradients in the lattice spacing. Fortunately, diamond has a very high thermal diffusivity and an extremely small coefficient of thermal expansion for $T < 100$ K, so that the expansion of cryogenically cooled diamond crystals due to heating can be neglected.

7.3 XFELO Parameters and Performance

The major parameters of an example XFELO system from Ref. [16] are listed in Table 7.1. The electron beam parameters considered here are relatively conservative. XFELO parameters with higher beam qualities, lower bunch charge, and lower electron beam energy may also be feasible [19]. Operating an XFELO with Bragg reflectors will be difficult below 5 keV due to enhanced photo-absorption in the crystal at low photon energies, and limited to a maximum photon energy of around 20 keV because of the decreasing crystal bandwidth at high energy. Although in principle the four-crystal configuration can be adjusted over a very wide range of wavelengths, the practical tuning range for a specific Bragg plane is limited to 2–6 percent because the angular acceptance can become smaller than the X-ray beam divergence at lower Bragg angles. We note that a few percent is in fact a huge tuning range in comparison to the very narrow bandwidth of $\sim 10^{-7}$.

The profiles of the radiation output near the energy of 14.4 keV are shown in Figure 7.4 [16]. In panel (a) the output radiation power as a function of time is indicated as the solid line, with the electron beam current envelope as a dashed line for reference. In panel (b) we plot the corresponding output spectrum with a solid line, showing that the spectral FWHM is approximately 1.8 meV, corresponding to a relative FWHM of $\sim 1.3 \times 10^{-7}$. Note that this bandwidth is much narrower than the reflectivity width of the Bragg crystals (the dashed line), and instead approximately equals c/σ_z. This is

Table 7.1 Major XFELO parameters

Electron Beam	
Energy	5–7 GeV
Bunch charge	25–50 pC
Bunch length (RMS)	0.1–1 ps
Normalized RMS emittance	$\leq$ 0.2–0.3 mm-mrad
Energy spread (RMS)	$\lesssim 2 \times 10^{-4}$
Bunch repetition rate	$\sim$ 1 MHz (constant)
Undulator	
Period length	$\sim$ 2 cm
Deflection parameter K	1.0–1.5
Length	30–60 m
Optical Cavity	
Configuration	2–4 diamond crystals + focusing mirrors
Total roundtrip reflectivity	> 85% (50% for 100 A peak current)
Total length	$\sim$ 100 m
XFELO Output	
Photon energy coverage	5–25 keV (plus the third harmonic)
Spectral purity	1–10 meV (10^{-6}–10^{-7} in relative BW)
Coherence	Full transverse and temporal coherence
X-ray pulse length	0.1–1.0 ps
Tuning range	2–6%
Number of photons/pulse	$\sim 10^9$
Pulse repetition rate	$\sim$ 1 MHz
Peak spectral brightness	10^{32}–10^{34} ph/[s·mm^2·mrad^2(0.1% BW)]
Average spectral brightness	10^{26}–10^{28} ph/[s·mm^2·mrad^2(0.1% BW)]

because, after saturation, the radiation envelope follows that of the electron beam current, so that radiation spectral profile is roughly given by the Fourier transform of the electron current profile. These steady-state pulses are reached after $\sim$ 1000 passes.

7.4 X-Ray Frequency Combs from a Mode-Locked FEL Oscillator

The output of an FEL oscillator is a train of radiation pulses, each of which is essentially a copy of the single trapped pulse that is circulating in the cavity. Hence, if the cavity and accelerator are sufficiently stable, the radiation coherence can be extended across multiple pulses, in which case the output spectrum becomes a comb of narrow spectral lines. This is the same basic physics at play as in frequency stabilized mode locked lasers (see, e.g., [20]).

To understand how these spectral lines emerge, consider the electric field of the pulse train,

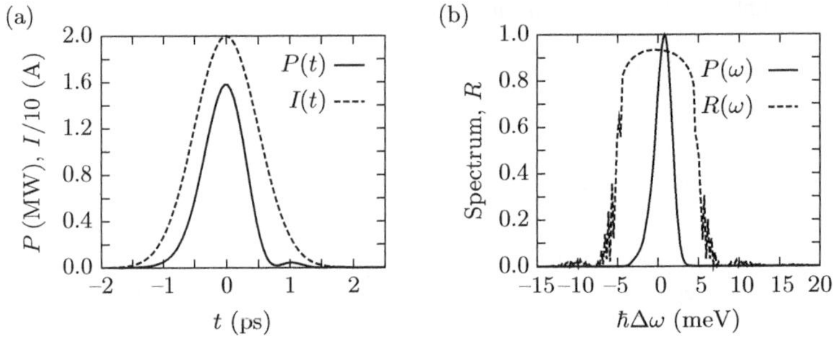

Figure 7.4 (a) The solid line plots the temporal power profile of the XFELO output at 14.4 keV, showing peak powers $\sim$ 1.5 MW; the electron beam current with $\sigma_t = 0.5$ ps is shown as a dashed line. (b) Spectrum of the same XFELO output as a solid line. The FWHM bandwdith of $\sim$ 2.4 meV is narrower than that of the crystal reflectivity, shown as a dashed line. Adapted from Ref. [16].

$$E(t) = \sum_n e^{-i\omega_{\text{FEL}}(t-nT_c)} A(t - nT_e). \tag{7.26}$$

Here $A(t)$ is the radiation envelope, $\omega_{\text{FEL}} \approx \omega_1$ is the FEL frequency in the middle of its bandwidth, T_c is the round-trip time of the EM field in the cavity, and T_e is the electron bunch spacing. The above equation is in accord with the discussion in section 7.1.2, namely, that the phase advance per period is T_c while the envelope advance per period is T_e (we are neglecting here the small index of refraction during the FEL interaction). Fourier transforming to obtain the spectrum, we find that

$$\left|\tilde{E}(\omega)\right|^2 = \left|\tilde{A}(\omega - \omega_{\text{FEL}})\right|^2 \frac{\sin^2\{N[(\omega - \omega_{\text{FEL}})T_e + \omega_{\text{FEL}} T_c]\}}{\sin^2[(\omega - \omega_{\text{FEL}})T_e + \omega_{\text{FEL}} T_c]}. \tag{7.27}$$

First, note that the overall spectral shape is determined by the single pulse envelope $\tilde{A}(\omega)$, so that the total extent in frequency will be of order the e-beam bandwidth $\sim c/\sigma_z$. Within this envelope are a sequence of evenly spaced spectral lines that comprise the frequency comb; the frequency location of each "tooth" of the comb is given when the denominator vanishes, namely, at the frequencies

$$\omega_m = \omega_{\text{FEL}} + \frac{\pi}{T_e}\left(m - \frac{\omega_{\text{FEL}} T_c}{\pi}\right) \tag{7.28}$$

for integer m. Hence, the line spacing is inversely proportional to the time between electron bunches, $\Delta\omega_{\text{line}} = \pi/T_e$. Since the frequency comb extends over the envelope bandwidth $\sim c/\sigma_z$, there are of order cT_e/σ_z comb lines. Additionally, the numerator of (7.27) implies that the width of each line scales inversely with the number of pulses in the train, so that each line's bandwidth $\sim \Delta\omega_{\text{line}}/N = \pi/NT_e$.

The previous discussion assumes that the cavity is perfectly stable, so that both the bunch spacing T_e and round-trip time in the cavity T_c are precisely fixed. If these times fluctuate or otherwise vary, however, the comb lines can be effectively broadened to the point where their width exceeds the spacing $\Delta\omega_{\text{line}}$. In this case the comb structure vanishes and the coherence between different pulses is lost. Looking at the formula for the pulse train (7.26), we see that the cavity round-trip time T_c appears in the phase, while the dependence on bunch spacing T_e is in the field amplitude. Since the comb spectrum relies on careful phase cancellation, we expect that T_c must be held constant to a fraction of a wavelength, while the tolerance associated with T_e will be significantly more relaxed.

To show how variations in T_c and T_e affect the comb structure, we first write the product $\omega_{\text{FEL}} T_c/\pi$ in terms of its integer and fractional part, so that the cavity time is

$$T_c = \frac{\pi M}{\omega_{\text{FEL}}} + t_c, \tag{7.29}$$

with the integer M chosen such that $0 \leq \omega_{\text{FEL}} t_c/\pi < 1$. Then, setting $m - M \to m$ in Equation (7.28), we see that the comb lines are located at

$$\omega_m = \omega_{\text{FEL}} + \frac{\pi}{T_e}\left(m - \frac{\omega_{\text{FEL}} t_c}{\pi}\right). \tag{7.30}$$

The comb lines will persist provided the fluctuations in the fractional cavity time t_c and the bunch spacing T_e move ω_m a small part of the distance between the spectral lines $\Delta\omega_{\text{line}} = \pi/T_e$. If we first consider variations in the electron bunch spacing δT_e, we find that

$$|\delta T_e| \ll \left|\frac{T_e}{m - \omega_{\text{FEL}} t_c/\pi}\right| \lesssim \frac{\sigma_z}{c}, \tag{7.31}$$

where we have used the fact that there are of order cT_e/σ_z comb lines. Variations in the bunch spacing must therefore be much less than the e-beam width, a requirement that we previously found must be met by any oscillator in order to preserve FEL gain. Note that the radiation wavelength does not enter the requirement here, which is a consequence of the dependence on the bunch spacing T_e appearing in the field amplitude.

On the other hand, requiring that variations in the cavity round-trip time δT_c do not adversely affect the comb leads to

$$|c\delta T_c| \ll \frac{c\pi}{\omega_{\text{FEL}}} \approx \frac{\lambda_1}{2}, \tag{7.32}$$

so that the cavity length must be held constant to within a fraction of the FEL wavelength. This can lead to very strict requirements on the stability of the cavity length, which is a consequence of the fact that T_c dictates the radiation phase from pass to pass.

Stabilizing the cavity length to a fraction of a wavelength requires an extremely sensitive feedback system capable of measuring and correcting tiny errors. This is particularly true at X-ray wavelengths, where fluctuations must be less than the diameter of an atom. Although very stringent, the desired XFELO stabilization appears to be feasible by a feedback system referenced by, for example, the ^{57}Fe nuclear resonance [21], since the resonance width and the comb spacing is comparable ($\sim$ 10 neV). The feedback principle is illustrated in Figure 7.5 and works as follows. The comb line moves one comb spacing if the cavity length changes by one FEL wavelength. With one of the comb lines overlapping with the ^{57}Fe resonance well inside the FEL bandwidth, the cavity length is scanned by one wavelength ($\sim$ 0.8 Å) and held fixed where the fluorescence signal is maximum. Note that it is not necessary to identify which comb line overlaps with the

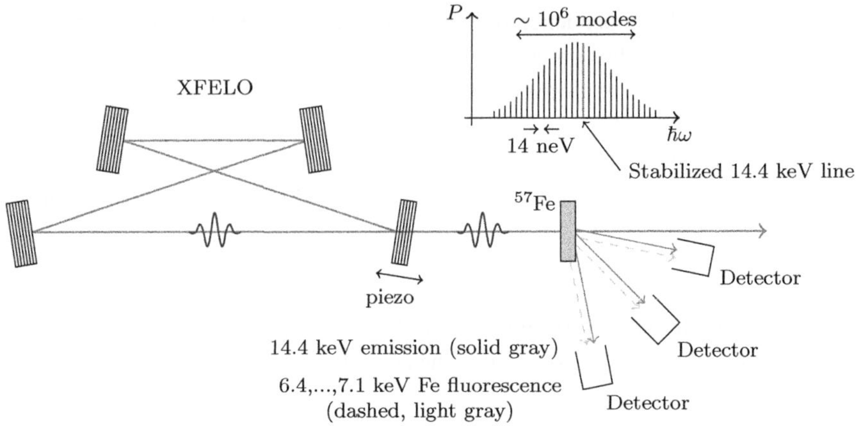

Figure 7.5 Schematic of the cavity-stabilization scheme. A nuclear-resonant sample (here ^{57}Fe) is placed into the XFELO output, and the nuclear-resonant and K-shell electronic fluorescence are monitored as functions of cavity tuning with a piezoelectric actuator. A feedback loop keeps one of the ca. 10^6 longitudinal modes of the XFELO on resonance with the sample. Adapted from Ref. [21].

nuclear resonance. It is only necessary that one of the lines does overlap, and we are keeping that line at that position.

Realizing such a nuclear-resonance-stabilized XFELO (NRS-XFELO) appears to be within the current state of the art. Extending the stabilization scheme to resonances that are an order of magnitude narrower than ^{57}Fe may also be feasible. NRS-XFELOs for even narrower resonances will be quite challenging but worthwhile, as they provide new scientific techniques hitherto not available in hard X-ray wavelengths.

7.5 A Hard X-Ray Master–Oscillator–Power Amplifier (MOPA)

The output of an XFELO can serve as an input to a high-gain amplifier to form an X-ray master–oscillator–power amplifier (MOPA) configuration. Such a facility would bring much of the flexibility of today's optical laser systems to hard X-rays. Compared to similar pulses produced via self-amplified spontaneous emission, those from an X-ray MOPA have better coherence and will be much more stable in both intensity and wavelength. Consequently, one can significantly increase the MOPA X-ray energy by strongly tapering the undulator strength after linear gain saturation, since undulator tapering can be very efficient for coherent FEL pulses (see in Section 4.5). In addition, an X-ray MOPA can potentially reach the comparatively high photon energies of tens to maybe even 100 keV by using the XFELO output as the modulating "laser" in a harmonic generation scheme such as HGHG. Such high-energy photons may have an important niche in certain applications [22].

A possible layout of an X-ray MOPA facility is schematically illustrated in Figure 7.6. Two streams of electron bunches, one for the XFELO and one for the high-gain amplifier, are produced separately by two electron guns and then merged into one stream of

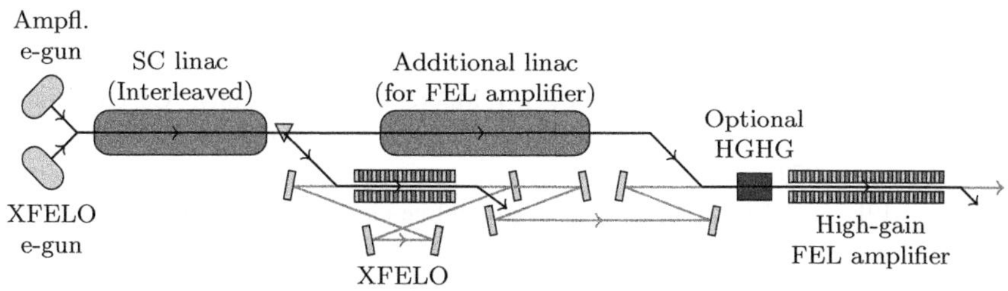

Figure 7.6 A schematic of a hard X-ray MOPA that uses one low-current stream of electron bunches to drive an XFELO that then seeds a high-gain FEL amplifier driven by a high-current electron stream.

interleaved bunches. The characteristics of the XFELO electron bunches can be similar to those listed in Table 7.1, while the high-gain bunches have a high peak current and bunch length that can be tailored to the chosen application. After acceleration, the bunches are separated back into two streams, one to drive the XFELO and the other headed for the amplifier. The X-ray output from the XFELO is delayed and routed as shown by a sequence of Bragg reflections identical to that in the XFELO cavity, so that the X-ray path length can be maintained when the wavelength is tuned by changing the Bragg angles. In this layout the electron path along the high-gain amplifier line can be held fixed. Furthermore, only that portion of the XFELO radiation pulse that overlaps with the high-current e-beam will be amplified in the second undulator. In this way, high intensity X-ray pulses of sub-femtosecond duration can be generated.

Figure 7.6 also shows an additional accelerating stage for the high-gain electron beam. With this addition the fundamental wavelength high-gain amplifier can be shorter than that of the XFELO. This permits one to operate the XFELO at a higher harmonic of the fundamendal, thereby lowering the energy of the first accelerator and reducing the cryogenic requirements for the second superconducting structure. In addition (or alternatively) to this, one can add a harmonic generation stage such as HGHG to reach even shorter wavelengths as shown in Figure 7.6.

References

[1] D. A. G. Deacon, L. R. Elias, J. M. J. Madey, G. J. Ramian, H. A. Schwettman, and T. I. Smith, "First operation of a free-electron laser," *Phys. Rev. Lett.*, vol. 38, p. 892, 1977.

[2] J. M. J. Madey, "Stimulated emission of bremsstrahlung in a periodic magnetic field," *J. Appl. Phys.*, vol. 42, p. 1906, 1971.

[3] C. Brau, *Free Electron Lasers*. Academic Press, 1990.

[4] G. R. Neil, "FEL oscillators," in *Proceedings of the 2003 Particle Accelerator Conference*, Portland, OR, p. 181, 2003.

[5] R. Colella and A. Luccio, "Proposal for a free electron laser in the X-ray region," *Opt. Commun.*, vol. 50, p. 41, 1984.

[6] R. Bonifacio, C. Pellegrini, and L. M. Narducci, "Collective instabilities and high-gain regime in a free electron laser," *Opt. Commun.*, vol. 50, p. 373, 1984.

[7] K.-J. Kim, Y. Shvyd'ko, and S. Reiche, "A proposal for an X-ray free-electron laser with an energy-recovery linac," *Phys. Rev. Lett.*, vol. 100, p. 244802, 2008.

[8] H. Al-Abawi, F. A. Hoff, G. T. Moore, and M. O. Scully, "Coherent transients in the free-electron laser: Laser lethargy and coherence brightening," *Opt. Comm.*, vol. 30, p. 235, 1979.

[9] K.-J. Kim, "Spectral bandwidth in free-electron laser oscillators," *Phys. Rev. Lett.*, vol. 66, p. 2746, 1991.

[10] G. Dattoli, G. Marino, A. Renieri, and F. Romanelli, "Progress in the hamiltonian picture of the free-electron laser," *IEEE J. Quantum Electron.*, vol. 17, p. 1371, 1981.

[11] P. Elleaume, "Microtemporal and spectral structure of storage ring free-electron lasers," *IEEE J. Quantum Electron.*, vol. 21, p. 1012, 1985.

[12] T. M. Antonsen and B. Levush, "Mode competition and suppression in free-electron laser oscillators," *Phys. Fluids B: Plasma Phys.*, vol. 1, p. 1097, 1989.

[13] R. W. Warren, J. E. Sollid, D. W. Feldman, W. E. Stein, W. J. Johnson, A. H. Lumpkin, and J. C. Goldstein, "Near-ideal lasing with a uniform wiggler," *Nucl. Instrum. Methods Phys. Res., Sect. A*, vol. 285, p. 1, 1989.

[14] R. Hajima, N. Nishimori, R. Nagai, and E. J. Minehara, "Analyses of superradiance and spiking-mode lasing observed at JAERI-FEL," *Nucl. Instrum. Methods Phys. Res., Sect. A*, vol. 475, p. 270, 2001.

[15] A. E. Siegman, *Lasers*. Sausalito, CA: University Science Book, 1986.

[16] R. R. Lindberg, K.-J. Kim, Y. Shvydko, and W. M. Fawley, "Performance of the free-electron laser oscillator with crystal cavity," *Phys. Rev. ST Accel. Beams*, vol. 14, p. 010701, 2011.

[17] R. M. J. Cotterill, "A universal planar X-ray resonator," *Appl. Phys. Lett.*, vol. 12, p. 403, 1968.

[18] Y. Shvyd'ko, S. Stoupin, V. Blank, and S. Terentyev, "Near-100% Bragg reflectivity of X-rays," *Nature Photonics*, vol. 5, p. 539, 2011.

[19] R. Hajima and N. Nishimori, "Simulation of an X-ray FEL oscillator for the multi-GeV ERL in Japan," in *Proceedings of the 2009 Free Electron Laser Conference*, Liverpool, UK, 2009.

[20] S. T. Cundiff and J. Ye, "Femtosecond optical frequency combs," *Rev. Mod. Phys.*, vol. 75, p. 325, 2003.

[21] B. Adams and K.-J. Kim, "X-ray comb generation from nuclear-resonance-stabilized X-ray free-electron laser oscillator for fundamental physics and precision metrology," *Phys. Rev. ST Accel. Beams*, vol. 18, p. 030711, 2015.

[22] (2016) Matter-radiation interactions in extremes (MaRIE). [Online]. `www.lanl.gov/science-innovation/science-facilities/marie/index.php`

8 Practical Considerations and Experimental Results for High-Gain FELs

The primary goal of this book is to acquaint the reader with the essential physics of synchrotron radiation and free-electron lasers. However, even with an in-depth understanding of this material, there are many other technical and physics issues that must be addressed when designing an actual device. The first half of this chapter covers "practical considerations" for FEL design, including the effects of and tolerances to beam trajectory and undulator errors, the physical consequences of beam-pipe induced wakefields, and the utility of tapering the undulator strength and/or wavelength. In general, practical considerations required for designing and realizing an FEL facility encompass a wide range of knowledge and components including electron beam production, compression, acceleration, and transport to and through the undulator; optical transport, focusing, and detection of the radiation; and diagnostics of both the electron and photon beams. Thus, we only discuss a small subset of this list that can be analyzed using the FEL theory developed in the previous chapters.

The second half of this chapter gives a very brief tour of certain experimental results that have helped promote the development of single pass X-ray FELs. Our intention is to give a sense of how experimental progress has complemented FEL theory, and give a flavor of the extent of progress made on realizing and improving intense X-ray FEL devices. This chapter primarily covers single-pass, high-gain FELs, since at the time of its writing X-ray FEL oscillators are still at the conceptual stage of development.

8.1 Undulator Tolerances and Wakefields

The design of a typical X-ray FEL calls for a small-gap undulator system that is tens to hundreds of meters in length. This system consists of many individual undulator sections that are separated by a suite of beam focusing, steering, and diagnostic stations. Each undulator section is typically a few meters in length both for ease in construction and installation and to allow for the appropriate placement of e-beam optical elements. Up to this point we have assumed that the electron beam centroid propagates exactly along the optical axis in a perfect undulator field. However, errors in both the undulator magnetic field and in the electron beam steering can degrade the FEL performance. In addition, wakefields induced by a high-current beam in the small-gap vacuum chamber can also interfere with the FEL gain process. In this section we illustrate how FEL theory can be extended to study these effects.

8.1.1 Undulator Errors and Tolerances

Magnetic field errors in an undulator can cause the electron beam to be deflected from its nominal (straight) trajectory, while variations in the magnetic field strength will lead to variations in the resonance condition and a subsequent reduction in gain. Here, we will assume that each undulator segment has been appropriately shimmed to have negligibly small first and second magnetic field integrals, which implies that the net steering errors are similarly negligible. In this case, the 1D analysis of Yu *et al.* [1] is sufficient to determine how variations in the undulator parameter K due to magnetic field errors or transverse misalignments among segments affect FEL performance. For a segment whose undulator deflection parameter has an error of ΔK, the total field strength is given by $K = K_0 + \Delta K$. We consider the effect of this error on the particle phase evolution, writing

$$\frac{d\theta}{dz} = k_u - k_1 \frac{1 + (K_0 + \Delta K)^2/2}{2\gamma^2} \approx 2k_u\eta - k_u \frac{K_0 \Delta K(z)}{1 + K_0^2/2}. \tag{8.1}$$

Here, the first term represents the ideal motion, and the second term describes the phase variation due to the small deviations in $K(z)$. As a concrete, analytic model for the undulator errors, we take ΔK to be comprised of a series of piecewise constant phase kicks, whose length is given by the magnetic correlation length $L_{\text{mag}} = N_{\text{mag}}\lambda_u$:

$$\Delta K(z) = \Delta K_n \quad \text{for } (n-1)L_{\text{mag}} < z < nL_{\text{mag}} \quad (n = 1, 2, 3, \ldots). \tag{8.2}$$

We assume that ΔK_n is a random quantity with vanishing ensemble average $\langle \Delta K_n \rangle = 0$, and that the magnetic correlation length is much shorter than the FEL gain length, i.e., $L_{\text{mag}} \ll L_G \approx \lambda_u/4\pi\rho$. Using the phase equation (8.1), the net phase shift per gain length for errors described by (8.2) is

$$\Delta\theta = 2\pi N_{\text{mag}} \sum_{n=1}^{L_G/L_{\text{mag}}} \frac{K_0 \Delta K_n}{1 + K_0^2/2}. \tag{8.3}$$

Under our assumption that $2L_G/L_{\text{mag}} \gg 1$, the mean phase deviation over an amplitude gain length vanishes, $\overline{\Delta\theta} = 0$, while its variance is given by

$$\overline{(\Delta\theta)^2} = \frac{L_G}{L_{\text{mag}}} \left(2\pi N_{\text{mag}} \frac{K_0 \sigma_K}{1 + K_0^2/2} \right)^2 \approx \frac{\pi N_{\text{mag}} K_0^4}{(1 + K_0^2/2)^2} \frac{(\sigma_K/K_0)^2}{\rho}$$
$$\approx 4\pi\rho N_{\text{mag}} \left(\frac{\sigma_K}{\rho K_0} \right)^2, \tag{8.4}$$

where σ_K is the RMS value of ΔK_n. This RMS drift in phase due to undulator errors is sometimes referred to as "phase shake." Physically, the phase variation (8.4) leads to a diffusive-type spreading of the resonant phase that grows $\sim \overline{(\Delta\theta)^2}(z/L_G)$. Thus, we might expect that these magnetic field errors would lead to an increase in gain length similar to the quadratic dependence on energy spread given in Equation (4.62), so that the power in the presence of undulator errors grows $\sim e^{z/L_G[1+q\overline{(\Delta\theta)^2}]}$, with q some

$O(1)$ constant. Indeed, a more rigorous perturbation analysis agrees with this heuristic argument at lowest order, finding that the radiation power up to saturation is given by [1]

$$P \approx P_0(z)\exp\left[-\frac{z}{L_G}\frac{\overline{(\Delta\theta)^2}}{9}\right], \tag{8.5}$$

where $P_0(z)$ is the power along the undulator without any error (i.e., in the linear regime $P_0 \propto e^{z/L_G}$).

To ensure negligible power degradation when the FEL gain saturates near $z \approx 20L_G$ requires the mean square of the ponderomotive phase shift per gain length $\overline{\Delta\theta^2} \ll 1$. For those errors in the peak on-axis magnetic field B_0 that are uncorrelated from undulator period to period, $N_{\text{mag}} \sim 1$, and the condition becomes [1]

$$\frac{\sigma_K}{K} = \frac{\sigma_B}{B_0} \ll \sqrt{\frac{\rho}{4\pi}} \text{ if } N_{\text{mag}} \sim 1. \tag{8.6}$$

Hence, the pole field error tolerance between adjacent magnets is quite relaxed because it scales as $\sqrt{\rho}$ instead of ρ; for typical parameters $\sqrt{\rho} \gtrsim 10^{-2}$ and the pole-to-pole field strength should vary by no more than a fraction of a percent.

On the other hand, if the length of the undulator segment is a significant fraction of L_G (as is the case at the LCLS), the average error in K over the entire segment has a correlation length $L_{\text{mag}} \to L_G$, implying that $N_{\text{mag}} \to 1/(4\pi\rho)$. In this case neither the model used to obtain (8.4) nor the perturbation analysis leading to (8.5) is strictly valid; nevertheless, Equation (8.4) suggests that the magnetic error tolerance over an entire undulator segment whose length is of order the gain length is given by

$$\frac{\sigma_K}{K} \ll \rho \text{ if } N_{\text{mag}} \sim 1/(4\pi\rho). \tag{8.7}$$

As an example of the utility of (8.7) even for long magnetic correlation lengths, we consider the LCLS which has an FEL Pierce parameter $\rho \approx 4.5 \times 10^{-4}$ and is built using 33 undulator segments, each of which is 3.4 m $\approx L_G$ in length [3]. In Figure 8.1 we plot the normalized output power P/P_0 as a function of the undulator error σ_K/K obtained from a series of GENESIS SASE simulations. We find that the tolerances are in agreement with the requirement of Equation (8.7), with the Gaussian fit of the power having an RMS width of $4.2 \times 10^{-4} \approx \rho$. Note that if we took (8.5) at face value with $N_{\text{mag}} = (4\pi\rho)^{-1}$ and $z = 18L_G$ we would expect an RMS width of $\rho/2$.

It is important to recognize that these magnetic error tolerances are only one contribution to the entire "error budget" for FEL design. Hence, one typically requires that (8.7) be well-satisfied to allow for other errors/misalignments. For example, the degree to which the undulator gap can be held at a fixed and constant distance, and accuracy of the undulator hall temperature control will also contribute to the error budget. Design of a facility must devise methods to control these and other errors/uncertainties so that the total errors are less than what the FEL will allow [4].

8.1.2 Beam Trajectory Errors

FEL performance can also be diminished by trajectory errors of the electron beam as it is transported down the undulator line. The effects of a non-straight beam trajectory may

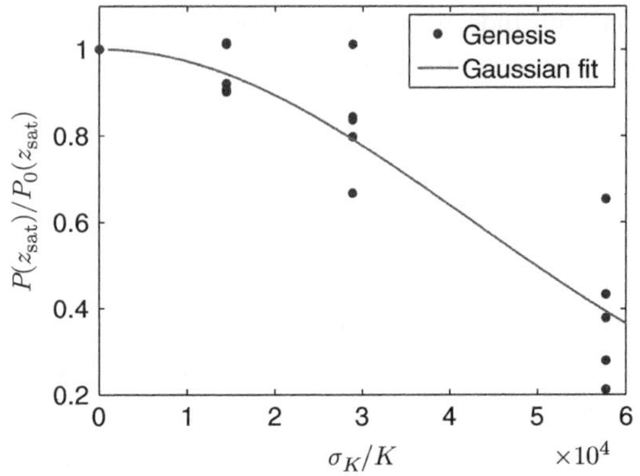

Figure 8.1 Power degradation factor P/P_0 at FEL saturation versus σ_K/K in the LCLS 33 undulator segments. Here, σ_K is the RMS value of a uniform segment K error distribution. Five random error distributions are used for a given σ_K. The RMS width of the Gaussian fit is 4.2×10^{-4}. Reproduced from Ref. [2].

be illustrated using a heuristic 3D model in which we assume that the microbunched beam gets a kick (an instantaneous change in $\boldsymbol{x}'$ at constant $\boldsymbol{x}$) by a single error dipole field [5]; this can be caused, for example, by a misaligned quadrupole. While the kick changes the direction of the beam trajectory by a deflecting angle ψ, the wavefront orientation remains parallel to the plane defined by the microbunching. This discrepancy results in two mechanisms for gain degradation: a decrease in coherent radiation power and an increase in the smearing of the microbunching due to the induced angular deviation. For a high gain FEL, both mechanisms are characterized by the critical angle [5]

$$\psi_{\rm crit} = \sqrt{\frac{\lambda_1}{L_G}} \approx \sqrt{4\pi}\,\sigma_{r'}, \tag{8.8}$$

which we see is proportional to the characteristic radiation divergence in the high-gain limit. Again, we expect that trajectory errors will result in significant gain degradation if $\psi \gtrsim \psi_{\rm crit}$, while FEL performance is maintained if the field tolerances are such that the induced angular deviations $\psi \ll \psi_{\rm crit}$. In fact, after the kick ψ the gain length is increased approximately to $L_G/(1 - \psi^2/\psi_{\rm crit}^2)$ [5]. As an example of the required tolerances at X-ray wavelengths, the LCLS has $\psi_{\rm crit} \approx 6$ μrad at $\lambda_1 = 1.5$ Å for the typical $L_G \approx 4$ m.

While the previous conditions apply to a single angular error, a physical undulator line will induce many small random errors that must be periodically corrected by steering elements located at beam position monitor stations between the undulator sections. To determine the effects of these errors, Ref. [1] developed a statistical analysis based on the phase error model mentioned in the previous section covering magnetic field tolerances. Before quoting the obtained result, we will again try to estimate the relevant scalings using simple physical ideas. If we assume that the correction stations are spaced a distance L_s apart and that $x_{\rm rms}$ specifies the RMS spatial deviation of the e-beam, than the characteristic angular deviation of electrons $\psi \sim x_{\text{rms}}/L_s$. Recalling that an electron with angular deviation ψ leads to a variation in phase according to (5.49), we find, in analogy with (8.1) and (8.4), that angular errors typified by ψ lead to a phase spread whose variance

$$\overline{(\Delta\theta)^2} = \frac{L_G}{L_{\text{traj}}}\left(\frac{2\pi L_{\text{traj}}}{\lambda_1}\psi^2\right)^2 = 4\pi^2\frac{L_G L_{\text{traj}}}{\lambda_1^2}\frac{x_{\text{rms}}^4}{L_s^4}. \tag{8.9}$$

Here, L_{traj} is the analogous "coherence length" of the trajectory errors. If the distance between correction stations is much larger than the FEL gain length, $L_s \gg L_G$, then gain guiding will effectively compensate for trajectory errors over the length $L_{\text{traj}} \sim L_G$. In the opposite limit, $L_{\text{traj}} \to L_s$ because the errors only accumulate over the distance between the correction stations.

Thus, we expect that the phase spread (8.9) will again result in a reduction of the output power given by $P(z) \sim P_0(z)e^{-\overline{(\Delta\theta)^2}(z/L_G)}$ with $\overline{(\Delta\theta)^2}$ given by (8.9) and L_{traj} being proportional to the smaller of either L_s or L_G. The rigorous calculation performed in Ref. [1] corroborates the scalings determined here, showing that the radiation power for an RMS trajectory deviation x_{rms} can be written as

$$P \approx P_0 \exp\left(-\frac{x_{\text{rms}}^4}{x_{\text{tol}}^4}\right), \quad x_{\text{tol}} = \begin{cases} 0.145\left(\dfrac{L_s^4\lambda_1^2}{zL_G}\right)^{1/4} & \text{if } L_s \gg L_G \\ 0.266\left(\dfrac{L_s^3\lambda_1^2}{z}\right)^{1/4} & \text{if } L_s \ll L_G \end{cases}. \tag{8.10}$$

As a practical example, the LCLS has diagnostic and steering stations every $L_s =$ 3.4 m, while its gain length $L_G \approx 4$ m. Using the latter version of x_{tol} from (8.10) and the fact that the undulator saturation distance $z/L_G \approx 20$ at $\lambda_1 = 1.5$ Å, then $x_{\text{tol}} \approx 3$ μm, and the RMS trajectory angle should be controlled to within 1 μrad in order to guarantee a small power degradation.

The fact that a large trajectory distortion destroys the FEL interaction can also be used to facilitate z-dependent FEL power measurements: by kicking the beam at selected undulator locations one can effectively "turn-off" the FEL gain and measure the power at that specific location. This technique is especially useful if intra-undulator radiation diagnostics are not practical to install and only a single FEL measurement station is available at the end of the undulator beamline.

The z-dependent measurements can be easily done by displacing, for example, a focusing quadrupole along the undulator line; if the quadrupole focal length is f, than a horizontal offset Q_x results in kicking the beam by the angle $\psi_x = Q_x/f$, which will interrupt the FEL gain if $\psi_x \gtrsim \psi_{\text{crit}}$. For example, a typical quad at the LCLS has $f = 10$ m, and the horizontal offset that corresponds to the critical angle $\psi_{\text{crit}} = Q_x/f$ is $Q_x = 60$ μm, namely, $\psi_{\text{crit}} \approx 6$ μrad. Figure 8.2 shows that a quadrupole at $z = 40$ m with a horizontal offset $Q_x \geq 60$ μm (i.e., a kick angle $\psi \geq \psi_{\text{crit}}$) inhibits further growth of the FEL fundamental mode. The on-axis radiation intensity remains approximately constant over the rest of the FEL, which may then may be detected by a far-field X-ray

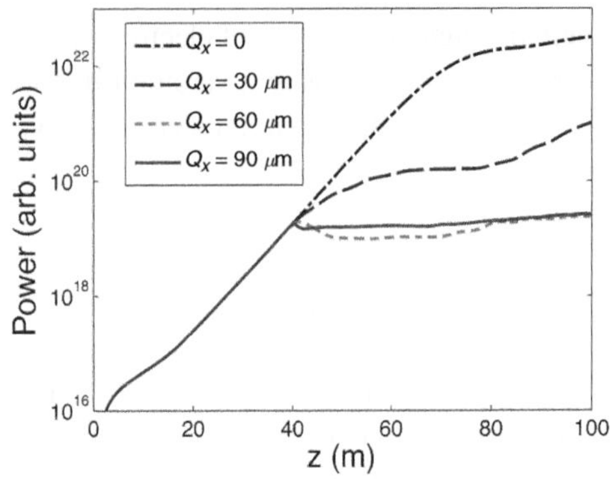

Figure 8.2 GENESIS simulation of the LCLS far-field power for various quadrupole offsets Q_x at $z = 40$ m. Reproduced from Ref. [2].

diagnostic station after the undulator. Similar conclusions hold at other undulator locations in the exponential growth regime, from which one can measure the energy $U(z)$ and experimentally determine, for example, the FEL gain length.

8.1.3 Wakefield Effects, Energy Loss, and Undulator Tapering

As we have seen, any effect that leads to a change in the ponderomotive phase disrupts the FEL gain and results in lower power at saturation. Here, we consider how energy loss due to physics other than the FEL interaction affects FEL performance. Specifically, we will investigate the influence of wakefields from the vacuum pipe, which we include in the 1D approximation by dividing the energy loss into the part due to the FEL interaction and that due to other factors. We continue to define the ponderomotive phase with respect to the initial resonant wavelength as we have done throughout this book, but rewrite the phase equation (3.76) as

$$\frac{d\theta}{dz} = 2k_u \frac{\gamma(z) - \gamma_r}{\gamma_r} = 2k_u \left[\frac{\gamma(z) - \gamma_c(z)}{\gamma_r} + \frac{\gamma_c(z) - \gamma_r}{\gamma_r} \right]$$
$$\equiv 2k_u \big[\eta(z) + \zeta(z)\big], \tag{8.11}$$

where $\eta(z)$ is now the normalized energy deviation from the slice central energy $\gamma_c mc^2$. In turn, $\gamma_c(z)$ differs from the resonant Lorentz factor γ_r because of the externally generated energy deviation $\zeta(z)$, which for our purposes will be typically wakefield-induced. Note that the evolution of η is still dictated by the FEL interaction, so that its equation of motion continues to be Equation (3.77). Additionally, the external energy change $\zeta(z)$ is analogous to a variation of the undulator strength as can be seen by comparing Equations (8.11) and (8.1); the two effects are equal when

$$\zeta(z) = -\frac{K\Delta K(z)}{(2 + K^2)} \approx -\frac{\Delta K(z)}{K} \quad \text{for } K^2 \gg 2. \tag{8.12}$$

Note that the energy loss ζ and hence, the ΔK in (8.12), is not a randomly fluctuating quantity as was the case for undulator errors. Instead, $\zeta(z)$ is a prescribed function of z whose precise form depends on its source. For example, if it is to represent the energy loss due to spontaneous radiation, than ζ is a linearly increasing function of the propagation distance that on average acts on all particles identically.[1] Thus, the energy loss due to spontaneous emission can be compensated for in the phase equation by a linear taper of the undulator strength that satisfies Equation (8.12).

Spontaneous emission is not the only external mechanism for energy loss. In addition, a high-current electron bunch induces a short-range wakefield that changes the beam properties as it propagates along the undulator vacuum chamber. For a typical X-ray FEL, like the LCLS, the dominant (longitudinal) wakefield is caused by the resistive wall of the vacuum pipe [7], which changes the mean beam energy both as a function of the undulator distance as well as the position along the bunch. The distance over which the longitudinal wakefield varies is of order the bunch length, which implies that the energy variation over an X-ray coherence time t_{coh} can be neglected if the electron beam length $T \gg t_{\text{coh}}$. Thus, for long electron beams the dominant resistive wall effect is to decrease the central energy of an FEL slice and consequently shift its resonant wavelength along the undulator distance. Here, we use the term "FEL slice" to denote a section of the electron beam whose temporal length is of order $t_{\text{coh}} \sim \lambda/4\pi\rho$. The energy loss due to the resistive wall wakefield also increases linearly with z, so that for any particular FEL slice one can identify an equivalent linear taper of the undulator via (8.12). However, a single undulator taper cannot perfectly compensate for the energy change due to wakefields for all FEL slices in the bunch, since the energy loss varies from slice to slice.

In general, the energy loss $\zeta(z)$ is not small, but it does typically vary slowly in the sense that the fractional energy change per field gain length is less than ρ. Under this assumption, one can apply WKB theory in the small signal regime before saturation to approximately solve the 1D FEL equations [8]. For negligible energy spread the SASE power is found to be

$$P(z) \approx P_{\text{max}}(z) \exp\left\{-\frac{1}{2}\left[\frac{\zeta(z) - \zeta_{\text{max}}(z)}{\sqrt{3}\sigma_\omega(z)/\omega_1}\right]^2\right\}, \tag{8.13}$$

where P_{max} is the maximum power at the optimal energy change $\zeta_{\text{max}} > 0$. First, we note that because the high-gain bandwidth $\sigma_\omega(z_{\text{sat}})/\omega_1$ is close to ρ, Equation (8.13) indicates that the SASE power has an FWHM in ζ of about 4ρ at saturation. In addition, if we again denote the radiation power for no external energy change [$\zeta(z) \equiv 0$] as P_0, we find that $P_{\text{max}} > P_0$, and that one can therefore improve the SASE output by increasing the electron energy by a small amount during the FEL interaction. To understand why this is true, we note that the frequency of maximum gain decreases slightly as the electrons are modulated and bunched in the ponderomotive potential, so that the electron beam preferentially radiates into longer wavelengths as it propagates down the undulator. Increasing the e-beam energy by an amount that compensates for

[1] In addition to the mean energy loss, the stochastic nature of spontaneous emission also results in energy diffusion. For $K \approx 1$, this diffusive increase of the energy spread grows as $\sigma_\eta \sim \left(\lambda_e/\sqrt{4\lambda_1\lambda_u}\right)\sqrt{z/\lambda_u}$ (λ_e is the Compton wavelength). This is typically much less than ρ except for very short radiation and undulator wavelengths. The expression for the diffusion coefficient at arbitrary K was derived in [6].

this frequency decrease results in maximal gain at the same frequency throughout the FEL interaction.

To reiterate, maximum energy exchange can be obtained by slightly accelerating the electron beam in the FEL, which is typically not practical, or by decreasing the undulator parameter K by an equivalent amount. If there are other mechanisms that cause electrons to lose energy, then the output power is maximized when the undulator taper is adjusted to be slightly larger than that required to cancel the average energy loss. For the LCLS, simulations show that an equivalent fractional energy increase of 2ρ over the saturation distance $z_{sat} \approx 90$ m improves the saturation power by about a factor of two as compared to the nominal saturation power for a uniform undulator with zero external/wakefield-induced energy loss.

For realistic systems, the combined energy loss due to wakefields, spontaneous radiation, and the like is a function of both the propagation distance z and the bunch coordinate θ. Once this energy variation is computed and the undulator taper is specified, Equation (8.13) can be used to estimate the FEL power at each FEL slice along the bunch and to find the average SASE power. As a numerical example, Ref. [8] studied the case for a sinusoidal energy oscillation that resembles the variation of the resistive wall wakefield in the core part of a 1-nC bunch at the LCLS [7]. Figure 8.3 shows how the output power is degraded from its maximum P_{max} as a function of the fractional energy oscillation amplitude ζ_A. The dashed line plots the power degradation factor if we assume that the undulator is tapered to exactly compensate the average energy loss, so that $\zeta(z) = 0$ at the bunch center. This should be compared to the solid line that plots the case when the undulator taper is increased so as to yield a total equivalent energy gain of $\zeta(z_{sat}) = 2\rho$, which delivers twice the power when the energy loss is independent of θ (i.e., when $\zeta_A = 0$). Assuming a round vacuum pipe that is 5 mm in diameter, at $z_{sat} = 90$ m the wakefield amplitude $\zeta_A \approx 6\rho$ if the pipe is made of copper, while $\zeta_A \approx 3\rho$ for an aluminum pipe. For these large wakefield amplitudes, the average radiation power produced by the 1 nC bunch is about 25 percent (50 percent) of P_{max}

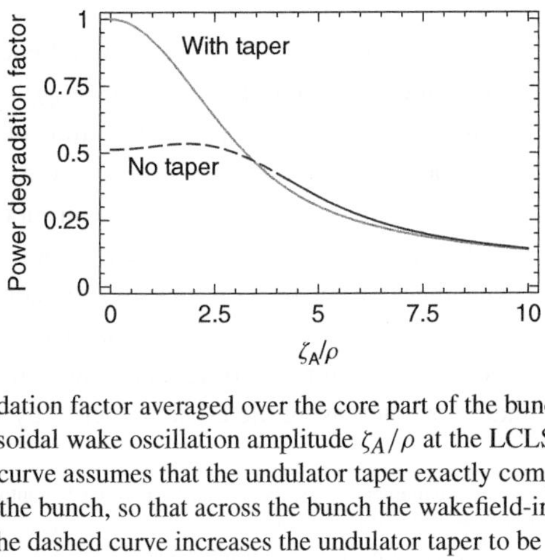

Figure 8.3 Power degradation factor averaged over the core part of the bunch (about 30 μm in length) versus the sinusoidal wake oscillation amplitude ζ_A/ρ at the LCLS saturation distance $z = 90$ m. The dashed curve assumes that the undulator taper exactly compensates for the average energy loss of the bunch, so that across the bunch the wakefield-induced energy loss $\zeta = \zeta_A \sin(t/\tau_{wake})$. The dashed curve increases the undulator taper to be equivalent to an average energy gain $\approx 2\rho$ at the bunch center. Reproduced from Ref. [8].

with the Cu (Al) vacuum pipe, and Figure 8.3 indicates that the average power is nearly insensitive to the precise undulator taper. This is because the wakefield produces a large energy variation across the bunch, and the tapering is only effective for some fraction of the bunch charge.

In order to reduce the wakefield effects in both the undulator and the accelerator, Ref. [9] proposed reducing the bunch charge from 1 nC to 200 pC. In this case, the equivalent energy modulation amplitude ζ_A is small, and one expects from Figure 8.3 that an undulator taper that more than compensates the average energy loss may improve the FEL performance. In fact, start-to-end LCLS simulations [10] show that using an undulator taper designed to be equivalent to an average energy gain of 2ρ increases the saturation power by about a factor of two, which makes the total number of photons produced by the 200 pC bunch comparable to the output of a 1 nC bunch that suffers stronger wakefield effects.

8.2 FEL Experimental Results

It was hoped that electron bunches from a storage ring might be used to drive a high-gain FEL amplifier for SASE by directing the beam down a dedicated bypass line [11, 12]. Unfortunately, storage ring-based FEL amplifiers have only proven feasible for wavelengths in the UV range and longer. This limitation is due to rather basic properties of a storage ring, in which collective instabilities limit the peak current, while the e-beam energy spread is determined by the balance between quantum excitation and radiation damping [13]. A linear accelerator is better suited for driving X-ray FELs, since the electron beam properties can, with proper care, be preserved from the gun to exit. With the invention of photocathode gun [14] and its subsequent development (see, e.g., [15, 16]), a suitable high-brightness electron source became available, and an X-ray FEL was finally technologically feasible.

A typical, high-gain FEL amplifier facility can be schematically represented as in Figure 8.4. An electron beam is initially produced by a photocathode gun, and then conditioned by an emittance corrector that compensates for the growth in correlated emittance due to space charge [16], and with a laser heater that adds a small amount of energy spread to combat longitudinal instabilities in the linac and bunch compressor [17, 18, 19]. It is then accelerated and compressed to become a bright, intense bunch suitable for a high-gain FEL producing SASE. The basic design shown here was the result of many years of progress that was largely inspired by Claudio Pellegrini's 1992 proposal to use the SLAC linac to drive an X-ray FEL [20]. That proposal relied on the 3D FEL theory that we described previously, and also provided the impetus for further theoretical and experimental X-ray FEL developments. In addition, years of experience designing, building, and operating linacs for high-energy collider physics, in conjunction with tremendous progress in generating, preserving, and controlling high-brightness electron beams, paved the way for the eventual realization of X-ray FELs.

Short-wavelength FELs with the general layout shown in Figure 8.4 are now beginning to revolutionize ultrafast X-ray science. Currently, the Free-Electron LASer in

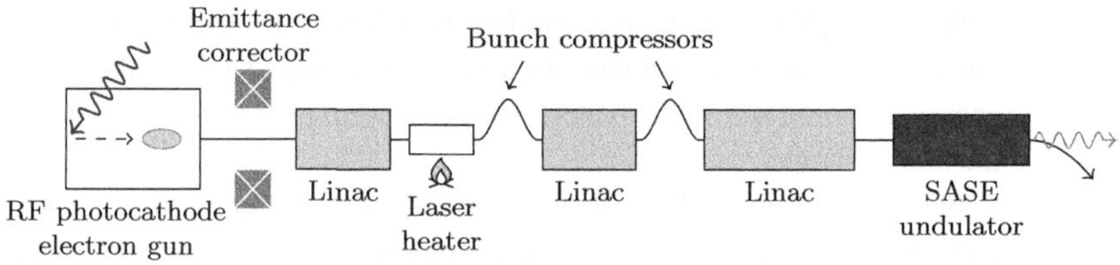

Figure 8.4 Schematic of a SASE-based X-ray FEL facility, including the electron gun, linac, bunch compressors, laser heater, and undulators.

Hamburg (FLASH) [21] produces intense soft X-rays for science, while the Linac Coherent Light Source (LCLS) [22] and the SPring-8 Angstrom Compact free-electron LAser (SACLA) [23] are two operational hard X-ray facilities with a large user community; in addition, there are several other machines in various stages of development and construction around the world (e.g., the European XFEL, Korean XFEL, and Swiss XFEL, among others). Although these facilities employ different accelerator and undulator technologies, they are all based on self-amplified spontaneous emission (SASE) to generate intense X-ray photons. One important improvement over SASE that is also being pursued is the increase of temporal coherence by various seeding techniques. To this end, the FERMI FEL [24] is the first operating facility employing seeding in the EUV to soft X-ray wavelength range.

In this section, we first review some experimental achievements that have helped validate the SASE theory and guide the development of X-ray FELs. Next, we discuss some of the experimental progress made on seeded FELs that improve longitudinal coherence, and conclude with a few examples of short-pulse FEL techniques.

8.2.1 SASE FELs

Although there were early experimental results that observed SASE FEL gain at millimeter wavelengths [25], the most convincing measurements relevant to X-ray FELs were obtained in the infrared, visible, and ultraviolet spectral regions. We review here some of the more significant demonstration experiments that verified our understanding of SASE and indicated that it could be extended to X-rays. The typical parameters of these experiments are summarized in Table 8.1.

In the UCLA/LANL experiment [26], radiation intensity at 12 μm was increased by more than 10^4 when the electron charge was changed by a factor of seven (from 0.3–2.2 nC) as shown in Figure 8.5. Using the measured beam parameters at 2.2 nC, a power gain of 3×10^5 over the spontaneous radiation was deduced. In addition, the characteristic SASE statistics were confirmed by measuring the pulse energy over many instances and fitting to the Gamma function distribution as depicted in Figure 8.5b. The fit indicated that there were $M \approx 8.8$ modes for these parameters.

One of the first large-scale experiments that measured SASE at optical wavelengths was performed in the Low Energy Undulator Test Line (LEUTL) tunnel at Argonne's Advanced Photon Source (APS). In the APS LEUTL experiment [27], the FEL output

Table 8.1 Representative parameters of some FEL experiments and facilities

Parameters	UCLA/LANL	VISA	LEUTL	TTF	FERMI	LCLS	SACLA
γmc^2 [GeV]	0.018	0.071	0.22	0.233	1.2	13.6	6.14
σ_γ/γ [%]	0.25	0.18	0.2	0.15	0.015	0.01	0.01
I [kA]	0.17	0.25	0.2	0.4	0.8	3	3
$\varepsilon_{x,n}$ [μm]	4	2	7	6	0.8	0.4	0.85
$2\pi\bar{\beta}_x$ [m]	1.2	1.8	9	6	50	140	185
λ_u [cm]	2.05	1.8	3.3	2.7	3.5	3	1.5
K	1.2	1.04	3.1	1.2	$\sim$1	3.5	1.36
L_u [m]	2	4	20	27	20	110	100
λ_1 [nm]	12000	800	385	100	4	0.15	0.1

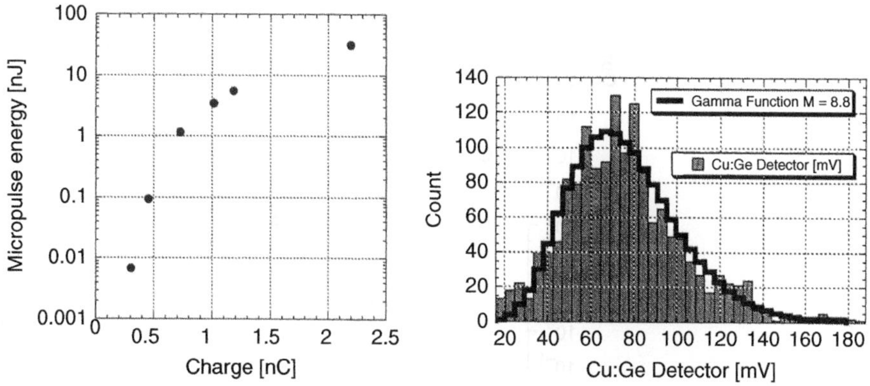

(a) FEL energy vs. charge. (b) FEL energy fluctuation at 2.2 nC.

Figure 8.5 In the UCLA/LANL experiment, the average FEL output energy (nJ) were measured for different electron bunch charges (nC) and individual FEL pulse energy fluctuation (at 2.2 nC) was compared to the predicted gamma distribution function. Reproduced from Ref. [26].

at 530 nm and 385 nm was observed along the nine undulator sections using extensive electron and radiation beam diagnostics capable of measuring SASE evolution as a function of undulator distance z. The experimentally determined gain length, near and far-field mode sizes, and radiation spectrum were found to agree with SASE theories for the measured beam parameters. An intensity amplification of more than 10^5 was measured when the electron beam was compressed to high current using a magnetic chicane. Additionally, saturation at both wavelengths was achieved within the $\sim$ 22 m long undulator as can be seen in Figure 8.6.

Around the same time, the TESLA Test Facility (TTF) at DESY accelerated electron beams using a superconducting rf linac that then drove a high-gain FEL at wavelengths down to about 100 nm [28]. The observed energy gain and saturation power at this wavelength compared quite well with simulations, as shown with the solid lines and circles in Figure 8.7. In addition, the energy fluctuations were measured as a function of the undulator distance (dotted lines and open circles in Figure 8.7); the linear stage

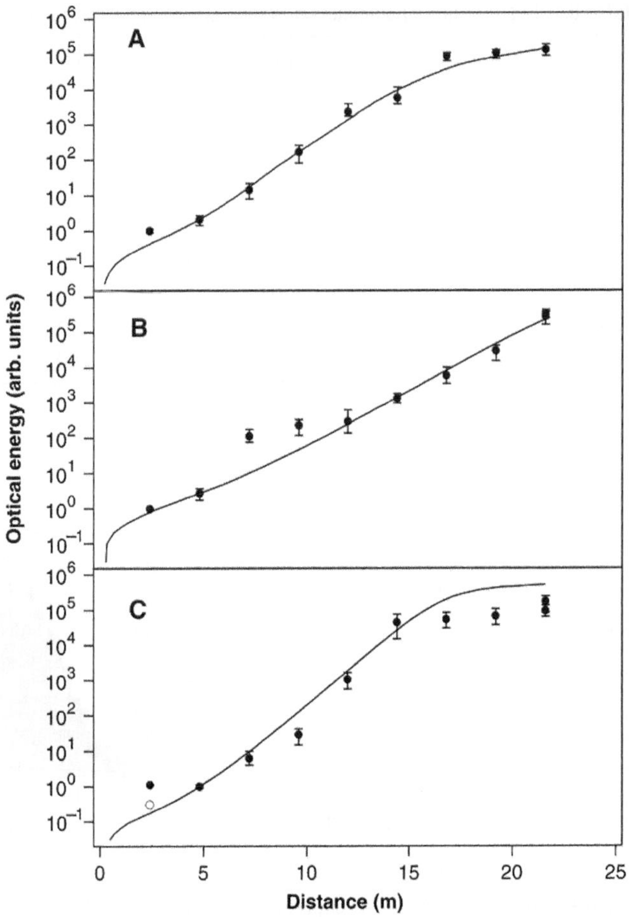

Figure 8.6 Measured LEUTL FEL energy as a function of the undulator distance for (A) 530 nm radiation wavelength, saturated condition, (B) 530 nm radiation wavelength, unsaturated condition and (C) 385 nm radiation wavelength, saturated condition. Reproduced from Ref. [27].

of FEL amplification exhibits rather large fluctuation levels that reduce markedly after saturation, as predicted by the simulations. The TTF FEL has since upgraded its electron beam energy to more than 1 GeV to push the radiation wavelengths down to the so-called water window around $\sim$ 4 nm. With its improved performance, the facility was renamed FLASH in 2005 and subsequently opened to X-ray users. As discussed before, SASE radiation possesses excellent transverse coherence due to gain guiding, and the degree of transverse coherence can be measured using Young's double slit experiment. We show a measurement of the interference fringes indicative of transverse coherence in Figure 8.8, where the distance between the slits is varied at the fixed FEL wavelength of 13.7 nm.

The VISA (Visible to Infrared SASE Amplifier) collaboration conducted another SASE experiment at Brookhaven National Lab. VISA employed distributed strong focusing quadrupoles and operated at a fundamental wavelength of 800 nm. In addition to the normal SASE saturation, nonlinear harmonic radiation was observed together with

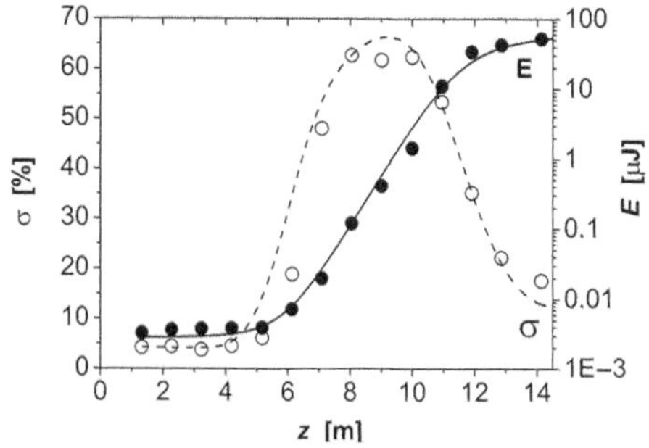

Figure 8.7 Average energy in the radiation pulse (solid circles) and the RMS energy fluctuation in the radiation pulse (empty circles) as a function of the active undulator length in the TTF experiment. The wavelength is 98 nm. Circles denote experimental results while the curves are from numerical simulations. Reproduced from Ref. [28].

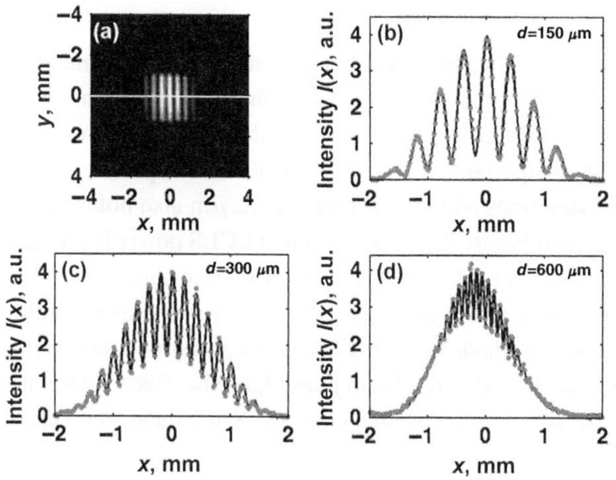

Figure 8.8 Results of a double-slit experiment at FLASH. (a) A typical data set measured on detector with the vertical slits of 150 μm separation. The next panels compare theory (solid lines) to experimental data (points) for different slit separation in the horizontal direction: $d = 150$ μm (b), $d = 300$ μm (c), and $d = 600$ μm (d). The radiation wavelength is 13.7 nm. Reproduced from Ref. [29].

the fundamental radiation mode. The gain lengths and energy evolution of the three lowest FEL harmonics were experimentally characterized using a spectrometer as shown in Figure 8.9. The measured nonlinear harmonic gain lengths and central spectral wavelengths decrease with harmonic number h, which is consistent with nonlinear harmonic theory [30].

These demonstration experiments clearly validated FEL theory and simulation at ultraviolet and longer wavelengths, and pointed the way toward even shorter FEL wavelengths. Nevertheless, it still took a huge leap of faith to build the world's first hard X-ray FEL – the Linac Coherent Light Source (LCLS) at SLAC. This is particularly true when one considers that the LCLS design called for a significant increase in electron beam energy and brightness over that used in the test experiments, as indicated by Table 8.1.

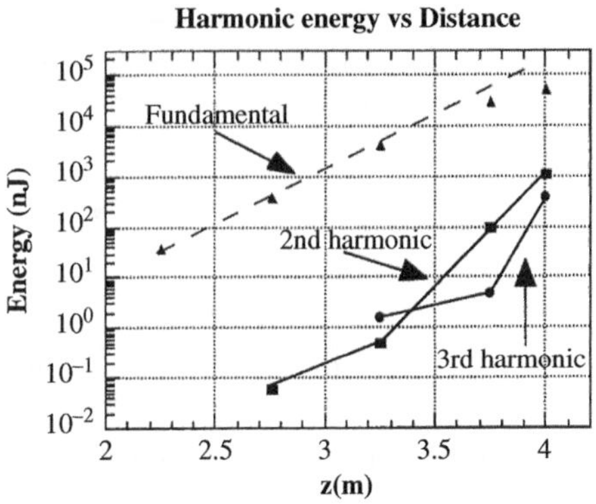

Figure 8.9 FEL energy vs. distance for the fundamental, second and third harmonics in the VISA experiment. The gain lengths for these modes are 19, 9.8, and 6.0 cm, respectively. Reproduced from Ref. [30].

The various technical challenges that were overcome to make the LCLS a success are beyond the scope of this book; rather, we will mention only a few FEL achievements and measurements that resulted in these pioneering experiments.

One important fact that the initial LCLS experiments proved was that the same theory that describes an FEL operating at 12 μm also holds at wavelengths down to 1.5 Å. As shown in Figure 8.10, the measured LCLS power level agrees quite well with GENESIS simulations [22], and the transverse beam profile is Gaussian-like with good spatial coherence. Other radiation properties have shown similar levels of agreement to the predictions made with the FEL theory and simulation techniques that we have described, provided the electron beam phase space has been sufficiently well characterized. In fact, modern experiments often combine X-ray measurements with FEL simulations to infer electron beam properties.

Another interesting FEL measurement at the LCLS controlled the slice energy spread using the so-called laser heater [31]. A laser heater uses the ponderomotive force of high-power laser and wiggler to first imprint an energy modulation on the electron beam. This modulation is then converted to incoherent energy spread using a large dispersion followed by phase mixing and/or Landau damping in the accelerator. The degree of heating is a monotonic function of the modulation amplitude determined by the laser power. Without the controlled heating provided by the laser heater, a microbunching instability in the accelerator can result in a significant increase in the slice energy spread and reduced FEL performance [32]. The right amount of laser heating increases the slice energy spread to 0.01 percent, hindering the instability and maximizing FEL gain; heating the beam further increases the slice energy spread beyond this value and reduces gain. Figure 8.11 shows how the measured FEL gain length at 1.5 Å depends on the laser-heater induced energy spread. The curves plot Ming Xie's fitting formula equation (5.157) including only the heater-induced energy spread for two different emittances that are consistent with other measurements. The fitting formula matches the experiment quite well except when the heater is off, in which case the microbunching instability induces a large additional energy spread.

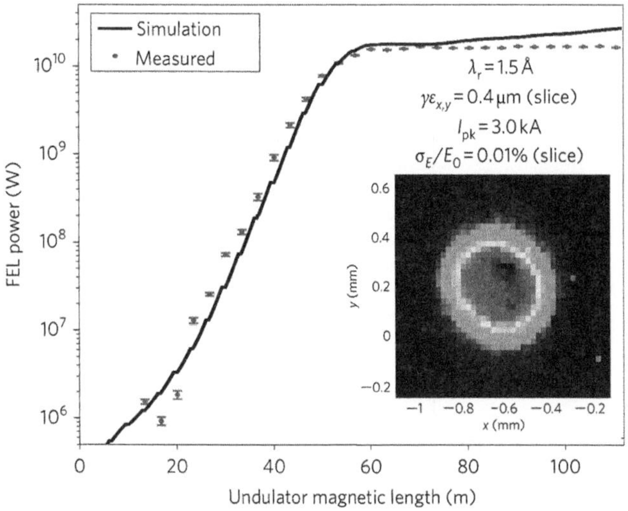

Figure 8.10 LCLS FEL power gain length measurement at 1.5 Å made by kicking the beam after each undulator sequentially. The measured gain length is 3.3 m, and a Genesis simulation prediction is overlaid using the e-beam parameters shown. There are twenty-five 3.35 m-long undulators installed here. Reproduced from Ref. [22].

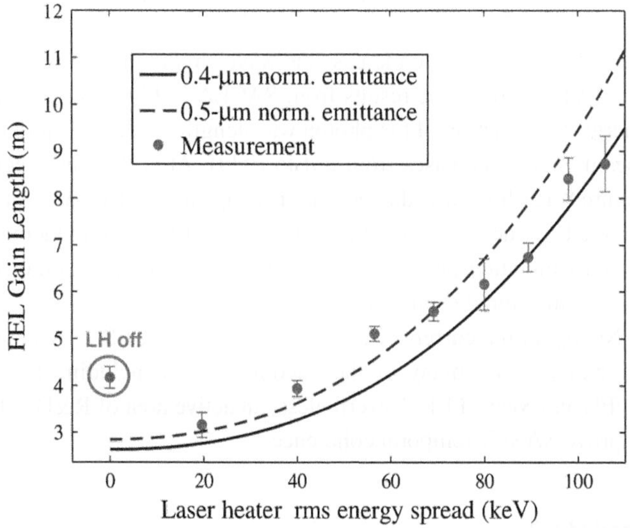

Figure 8.11 Measured LCLS FEL gain length at 1.5 Å vs. laser heater-induced energy spread in keV. Reproduced from Ref. [31].

The SACLA X-ray FEL began operation in 2011 at the SPring-8 facility in Japan [23]. One notable difference in the SACLA design was its goal to be comparatively compact: the lower beam energy of 8.5 GeV required a significantly shorter accelerating linac than that of the LCLS, and the undulator period was almost halved to 1.8 mm. To increase the *K* parameter of this short-period device, SACLA reduced the magnetic gap by employing in-vacuum undulators, and was also the first X-ray FEL facility to use

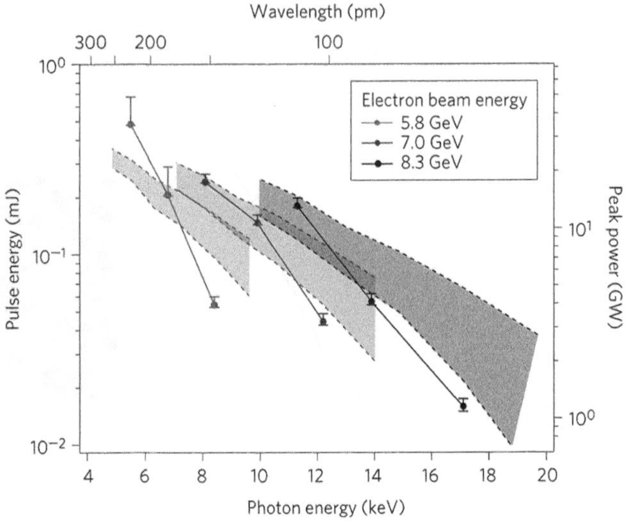

Figure 8.12 Pulse energy as a function of photon energy at the SACLA facility. Lines show experimental results at three different e-beam energies that increase from left to right, while the shaded areas bound simulation predictions assuming a peak current $I = 3.5$ kA and an emittance $\varepsilon_x = 0.6$ mm-mrad (top) and $\varepsilon_x = 0.8$ mm-mrad (bottom). The right shows the predicted peak power from the simulation. Reproduced from Ref. [23].

variable-gap devices to increase the wavelength tuning flexibility. We show some of the first FEL performance results from SACLA in Figure 8.12, which plots the FEL pulse energy as a function of the photon wavelength for three different electron beam energies (beam energy increases from left to right). At each fixed electron energy, the lines in Figure 8.12 show how the measured X-ray energy decreases as smaller wavelengths are targeted by increasing the magnetic gap and thereby reducing K. The shaded regions bound simulation predictions assuming a normalized emittance of 0.6 mm-mrad (top) and 0.8 mm-mrad (bottom).

Nearly all the current X-ray FELs (FLASH, LCLS, SACLA) are based on the SASE principle, as are many of the uncoming ones including the European XFEL, PAL-XFEL and Swiss-FEL. Nevertheless, an active area of R&D is FEL seeding, which will improve SASE's temporal coherence.

8.2.2 Seeded FEL

Seeding techniques can be loosely separated into two categories: external seeding and self-seeding. External seeding begins with a laser pulse that is used to generate electron microbunching at the laser wavelength and higher harmonics, followed by an FEL radiator tuned to selectively amplify one of the higher harmonics of the laser; we discussed the theory behind two external seeding methods in Chapter 6. Self-seeding [33], on the other hand, begins with a relatively short SASE undulator that produces an intense signal at the wavelength of interest. Next, the SASE pulse goes through a monochromator to narrow its bandwidth and improve its temporal coherence. Finally, the filtered radiation seeds a radiator which amplifies the signal to FEL saturation. The advantages of seeded FELs over SASE include a more stable central wavelength, a narrower bandwidth with increased spectral brightness, and the potential of longitudinally coherent X-rays. Novel seeding techniques are still being actively developed and we will only review two representative ones that have been successfully implemented in short-wavelength FEL operation.

The most mature external seeding technique is HGHG (see Section 6.2). HGHG begins with an external seed laser that is used to coherently modulate the electron energy in a wiggler. A dispersive section turns the electron energy modulation into a density modulation whose spectrum contains components at the laser frequency and its higher harmonics. Finally, shorter-wavelength coherent radiation is generated in an FEL radiator whose resonance condition is tuned to a harmonic of the initial seed laser. This scheme was first demonstrated at the BNL Source Development Lab (SDL). The SDL experiment employed an 800 nm Ti:Saphire laser to generate temporally coherent FEL radiation at the third harmonic, $\lambda = 266$ nm. Single-shot spectra at 266 nm were recorded and compared to the SASE spectrum as shown in Figure 8.13, clearly demonstrating the narrower bandwidth of a seeded FEL. More recently, the FERMI FEL has successfully used the HGHG mechanism to convert a 266 nm laser signal into coherent FEL radiation down to 20 nm in a single stage ($h = 13$) [24]. Figure 8.14 shows highly stable and clean radiation spectra at 32.5 nm for FERMI FEL-1. FERMI-2 is a cascaded harmonic generation scheme that uses two stages of HGHG to reach wavelengths below 10 nm [34]. The great success of the FERMI project may encourage other facilities to employ HGHG or EEHG-based schemes in the near future.

Because self-seeding does not rely on any external laser signal, it is particularly suitable to apply to the very short X-ray wavelengths. LCLS has recently demonstrated a novel hard X-ray self-seeding scheme and implemented it for user operation [36]. This self-seeding technique was proposed in Ref. [37], and uses a single diamond crystal as a monochromator and introduces a relatively small path length delay between

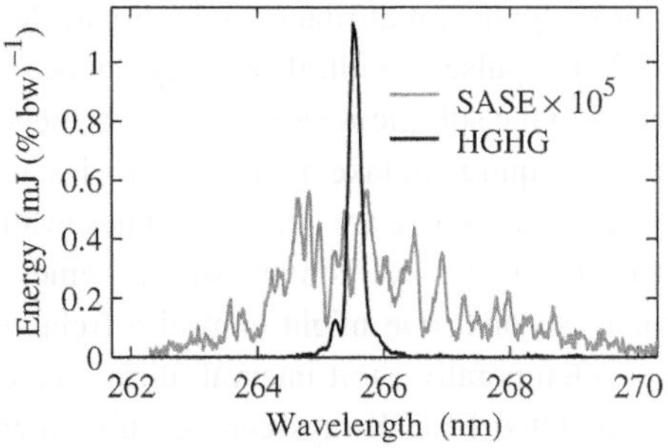

Figure 8.13 Single-shot HGHG spectrum measured at BNL SDL experiment for 30 MW seed laser at 800 nm, exhibiting a 0.1 percent FWHM bandwidth. The gray line is the single-shot SASE spectrum far from saturation when the 30 MW seed was removed. This spectrum serves as the background of the HGHG output. Reproduced from Ref. [35].

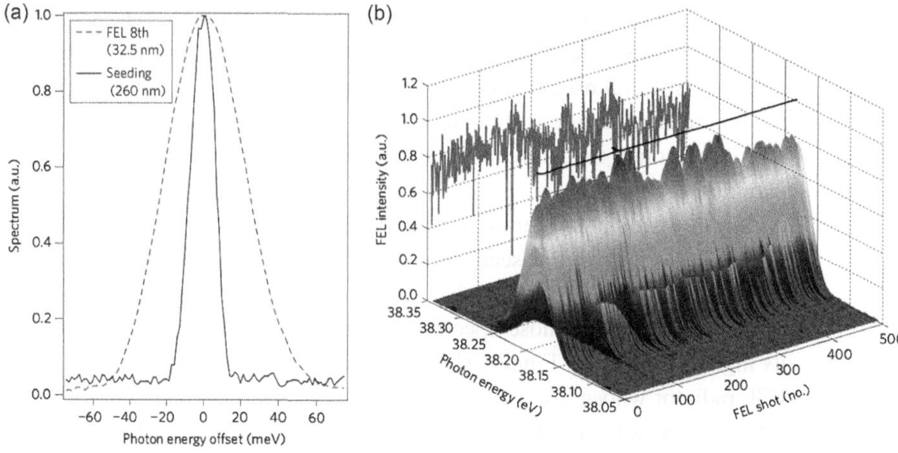

Figure 8.14 Spectra at 32.5 nm for the FERMI FEL. (a) Measured single-shot FEL spectrum (dashed line) and seed laser spectrum (solid line). (b) Acquisition of 500 consecutive FEL spectra. Reproduced with permission from Ref. [24].

the filtered X-ray pulse and the electron bunch. Thus, a relatively compact chicane can be used to delay the electron bunch and to wash out the noisy SASE microbunching. Since the filtered radiation is originated from the SASE process, its intensity will fluctuate according to the Gamma distribution illustrated in earlier chapters. In the second part of the undulator, the filtered radiation is amplified by the delayed electron bunch to reach power saturation and to decrease intensity fluctuation. Measurements of single-shot and average spectra of the seeded LCLS pulses vs. the SASE pulses at 1.5 Å (8.3 keV photon energy) are shown in Figure 8.15. We may note that an XFELO discussed in the previous chapter may be regarded as the ultimate self-seeded X-ray FEL.

8.2.3 Short-Pulse Generation

One of the most significant attributes of high-gain X-ray FELs is the ability to produce femtosecond X-ray pulses for ultrafast X-ray science. Although typical electron bunches are compressed to the sub-picosecond level to achieve a high-peak current, even shorter X-ray pulses are required to take molecular movies and to realize the diffraction-before-destruction technique. There are several avenues available to produce such short pulses. First, one can use aggressive compression schemes to further shorten a (low-charge) electron bunch; second, one might control a (relatively long) electron bunch to radiate light over a temporally short interval; third, one can manipulate the electron bunch so that the generated FEL X-rays can be subsequently tailored in time. In response to wide-ranging scientific demands, many novel short-pulse schemes have been proposed to generate femtosecond and attosecond X-ray pulses. We will review some recent experimental progress in this section.

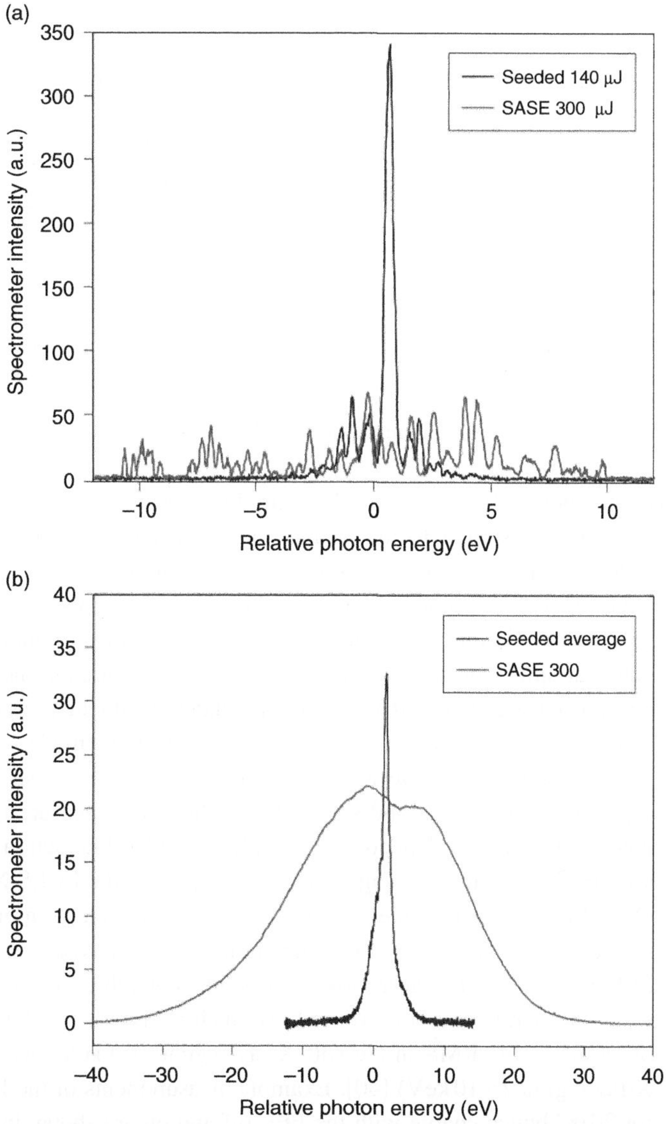

Figure 8.15 Comparison of the single-shot (a) and averaged (b) LCLS X-ray spectrum at 1.5 Å (8.3 keV) in self-seeded mode (narrow peak) with that in SASE mode. The FWHM single-shot seeded bandwidth is 0.4 eV, whereas the SASE FWHM bandwidth is ~ 20 eV. Reproduced from Ref. [36].

Low-Charge Operation

To reach exceptionally short-pulse durations, the LCLS has developed a low-charge operating mode. The reduced bunch charge (20 pC) provides even better transverse emittance from the gun and also mitigates collective effects in the accelerator, allowing for extreme bunch compression [38]. The compressed electron bunch length has an estimated FWHM bunch length < 5 fs. Stable, saturated FEL operation is routinely achieved

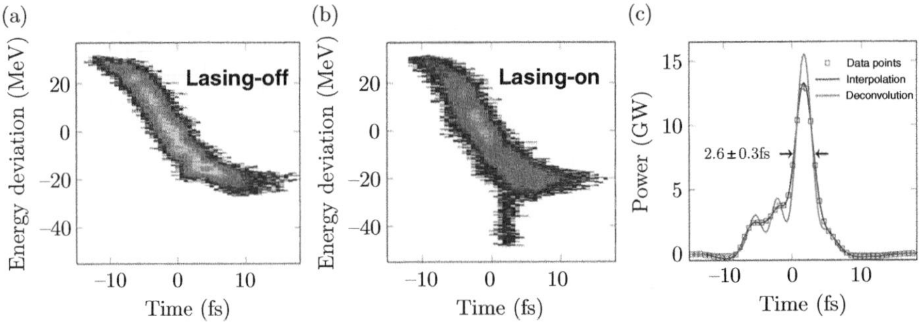

Figure 8.16 X-band deflecting cavity (XTCAV) measurements of sub-5 fs, 1 keV X-ray pulse. The electron bunch charge is 20 pC with electron beam energy of 4.7 GeV. (a) One single-shot lasing-off image; (b) One single-shot lasing-on image; (c) Reconstructed X-ray profile with a calculated 2.6 fs fwhm duration. Adapted from Ref. [40].

in this short-pulse mode over the entire LCLS wavelength range, and the estimated output power is similar to that at the nominal charge of 150–250 pC. The X-ray pulse energy and photon number, however, is nearly an order of magnitude lower in the short-pulse mode than in nominal operation, in proportion to its reduced bunch length/charge. In addition, the low-charge mode produces FEL pulses comprised of only one or two coherent spikes of radiation in the soft X-ray regime, and hence has better temporal coherence (*cf.* Figure 8.16). On the other hand, the fluctuations of the total pulse energy is increased compared to that from the nominal charge mode.

Novel techniques have to be developed to characterize such brief electron and X-ray pulses. One robust and successful method is to measure the electron longitudinal phase space using a deflecting cavity after the FEL interaction on a screen in a dispersive location. Without going into technical details, the LCLS X-band deflecting cavity (XTCAV) provides a time-resolved measure of the electron energy distribution along the bunch, and can be used to compare the time-resolved energy with and without the FEL interaction. The energy loss distribution along the bunch can then be used to reconstruct the X-ray temporal profile [39]. Such measurements have achieved a resolution of less than 1 fs RMS in the soft X-ray regime ($\sim$1 keV) and $\sim$4 fs RMS in the hard X-ray regime ($\sim$10 keV) [40]. Example measurements of the longitudinal phase space for 20 pC bunch charge with the FEL off and on are shown in Figure 8.16 (a) and (b) respectively, while the reconstructed 1 keV X-ray pulse profile with 2.6 fs fwhm duration is shown in Figure 8.16 (c).

Slotted Foil

Another method for femtosecond pulse generation is to use an emittance-spoiling slotted foil, which was first proposed in 2004 and has been used at LCLS since 2010 [41]. When the dispersed electron beam in the middle of a bunch compressor passes through a foil with single or double slots, most of the beam emittance is spoiled by Coulomb collisions in the foil material. Only the very short section(s) of the beam that pass(es) through the slot(s) undisturbed has sufficient brightness to generate X-rays in the FEL. By outfitting a 3 micron thick aluminum foil with an array of different slots, LCLS was

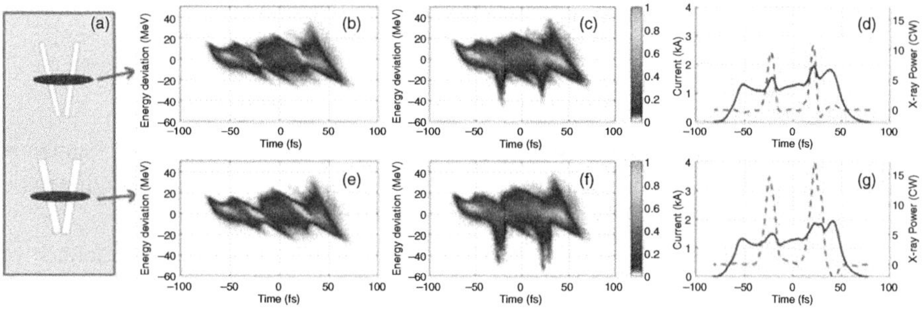

Figure 8.17 X-band deflecting cavity (XTCAV) measurements of sub-5 fs, 1 keV X-ray pulse. The electron bunch charge is 20 pC with electron beam energy of 4.7 GeV. (a) Two V-shaped double-slotted foils of different slot thickness; (b) One single-shot lasing-off image; (c) One single-shot lasing-on image; (d) Reconstructed X-ray profile with a calculated 2.6 fs fwhm duration measurement examples for the double-slotted foil. The examples are measured at the middle of the thin and thick double-slotted geometries. The lasing-off images are shown in (b) and (e), and the lasing-on images in (c) and (f). The current profile and reconstructed X-ray profile are shown in (d) and (g). The double-pulse separation (45 fs) is the same for the two sets, but the pulse duration is different. Adapted from Ref. [42].

able to produce X-ray pulses with various durations and separations. Depending on the bunch charge and the final current, a single slot with variable slot width can control the soft X-ray duration from 50 fs down to 6 fs, while V-shaped double slots with different slot separation can provide two short soft X-ray pulses separated by about 10 fs to 80 fs for pump-probe experiments [42] (see Figure 8.17).

Chirped SASE and Undulator Taper

As we mentioned in the beginning of this section, it can also be possible to produce short pulses by manipulating the radiation field after its production. For example, if the mean frequency of the radiation varies linearly in time (i.e., is "chirped"), one can perform X-ray pulse compression [43] or X-ray slicing [44]. Frequency-chirped SASE may be produced using an electron beam that is chirped in energy through the FEL resonance condition. It can be shown [45] that the SASE coherence time is independent of the frequency chirp $u \equiv \Delta\omega/\Delta t$ as long as the frequency variation over the time t_{coh} is much smaller than the FEL bandwidth, $|u| \ll \sigma_\omega^2$.

The physics of X-ray pulse compression is identical to that used for optical lasers (and to chicane-based electron beam bunch compression): a dispersive medium induces an energy-dependent path length designed to remove the frequency chirp and combine the frequency components into a shorter temporal duration. For X-rays, a grating can be used to provide the dispersion and resulting compression. If such a high-quality grating at the frequency of interest is not available, one can instead use a monochromator to select (or slice out) a temporally short section of the chirped X-ray pulse. The output length is roughly the inverse of the convolved spectral widths of the chirped SASE and the monochromator; specifically, if σ_m is the RMS bandwidth of the monochromator, the sliced RMS X-ray pulse duration is [45]

$$\sigma_t = \sqrt{\frac{\sigma_\omega^2 + \sigma_m^2}{u^2} + \frac{1}{4\sigma_m^2}}\,. \tag{8.14}$$

The minimum pulse duration for an optimized monochromator bandwidth is $(\sigma_t)_{\min} = (\sigma_\omega/|u|)\sqrt{1 + u^2/\sigma_\omega^2}$; for $|u| \sim \sigma_\omega^2$ we have $\sigma_t \sim t_{\text{coh}}$, and a single temporal mode of a few hundred attoseconds duration may be selected.

The energy variation that can be produced by the rf accelerator over the entire electron bunch is usually much smaller than that required to select a single SASE spike, with the typical rf-induced chirp resulting in a pulse duration after slicing on the order of 10 fs [45]. A sufficiently large energy chirp can be produced over a small fraction of the bunch through the resonant interaction of the electrons with a high-power optical laser in an undulator. While the local energy chirp can be sufficiently large to select a single temporal mode, with $|u| \sim \sigma_\omega^2$, the large variation in mean energy over the coherence time disrupts the resonance condition and reduces the FEL gain.

On the other hand, one can overcome this limitation by varying either the undulator period or strength along its length to effectively counteract the energy chirp [46]. For simplicity we will only consider variations in $K(z)$, as the extension to the undulator period is straightforward. The basic physics can be understood by considering radiation that is initially generated in the middle of the bunch and satisfies the local FEL resonance condition at $z = 0$ given by $\lambda_1 = \{[1 + K(0)^2/2]\lambda_u\}/2\gamma_r^2$. During subsequent evolution, the radiation slips forward toward the head of the bunch, where it interacts with electrons whose energy is γ_h. Thus, to maintain resonance we require $\lambda_1 = \{[1 + K(z)^2/2]\lambda_u\}/2\gamma_h^2$, meaning that we must increase (decrease) the magnetic field strength K as a function of z if $\gamma_h > \gamma_r$ ($\gamma_h < \gamma_r$). For a linear chirp, we have $\gamma_h = \gamma_r + (d\gamma/dt)dt$, so that using $K(z) = K(0) + (dK/dz)dz$ and $dt/dz = -\lambda_1/c\lambda_u$ (time decreases toward the bunch head), we find that resonance is maintained if

$$\frac{\lambda_u K(0)}{1 + K(0)^2/2}\frac{dK}{dz} = -\frac{2\lambda_1}{c\gamma_r}\frac{d\gamma}{dt}. \tag{8.15}$$

Thus, a tapered undulator can automatically "select" a small fraction of an energy-modulated bunch that has the right chirp with a pulse duration of about 200 attoseconds [46]. The compensation of undulator taper to an energy-chirped electron beam has been demonstrated at the SPARC FEL test facility at a visible radiation wavelength [47].

Finally, we note that the energy modulation generated by a high-power optical laser in an undulator can be converted to the density modulation by a magnetic chicane, leading to very large peak currents spikes (each with a width of a sub-optical period) and typical sub-femtosecond X-ray pulses [48]. Various laser manipulations of electron beams for generation of extremely short pulses are reviewed in Ref. [49].

References

[1] L.-H. Yu, S. Krinsky, R. L. Gluckstern, and J. B. J. van Zeijts, "Effect of wiggler errors on free-electron laser gain," *Phys. Rev. A*, vol. 45, p. 1163, 1992.

[2] Z. Huang and K.-J. Kim, "Review of X-ray free-electron laser theory," *Phys. Rev. ST Accel. Beams*, vol. 10, p. 034801, 2007.

[3] J. Galayda *et al.*, "Linac Coherent Light Source (LCLS) Conceptual Design Report," *SLAC, Report* SLAC-R-593, 2002.

[4] H.-D. Nuhn *et al.*, "LCLS undulator commissioning, alignment, and performance," in *Proceedings of the 2009 FEL Conference*, p. 714, 2009.

[5] T. Tanaka, H. Kitamura, and T. Shintake, "Consideration on the BPM alignment tolerance in X-ray FELs," *Nucl. Instrum. Methods Phys. Res. A*, vol. 528, p. 172, 2004.

[6] E. L. Saldin, E. A. Schneidmiller, and M. V. Yurkov, "Calculation of energy diffusion in an electron beam due to quantum fluctuations of undulator radiation," *Nucl. Instrum. Methods Phys. Res., Sect. A*, vol. 381, p. 545, 1996.

[7] K. Bane and G. Stupakov, "Resistive wall wakefield in the LCLS undulator beam pipe," *SLAC, Report* SLAC-PUB-10707, 2004.

[8] Z. Huang and G. Stupakov, "Free electron lasers with slowly varying beam and undulator parameters," *Phys. Rev. ST Accel. Beams*, vol. 8, p. 040702, 2005.

[9] P. Emma, Z. Huang, C. Limborg, J. Wu, W. M. Fawley, M. Zolotorev, and S. Reiche, "An optimized low-charge configuration of the Linac Coherent Light Source," in *Proceedings of the 2005 Particle Accelerator Conference*. Piscataway, NJ: IEEE, 2005.

[10] W. Fawley, K. Bane, P. Emma, Z. Huang, H.-D. Nuhn, S. Reiche, and G. Stupakov, "LCLS X-ray FEL output performance in the presence of highly time-dependent undulator wakefields," in *Proceedings of the 2005 Free Electron Laser Conference*, Stanford, CA, USA, 2005.

[11] J. B. Murphy and C. Pellegrini, "Free electron lasers for the xuv spectral region," *Nucl. Instrum. Methods Phys. Res., Sect. A*, vol. 237, p. 159, 1985.

[12] K.-J. Kim, J. J. Bisognano, A. A. Garren, K. Halbach, and J. M. Peterson, "Issues in storage ring design for operation of high-gain FEL," *Nucl. Instrum. Methods Phys. Res., Sect. A*, vol. 239, p. 54, 1985.

[13] S. Krinsky, "Some comments on the design of electron storage rings for free electron lasers," in *Free electron generation of extreme ultraviolet coherent radiation*, J. M. J. Madey and C. Pellegrini, Eds., no. 118. SPIE, 1984, p. 44.

[14] J. S. Fraser, R. L. Sheffield, and E. R. Gray, "A new high brightness electron injector for free-electron lasers driven by RF linacs," *Nucl. Instrum. Methods Phys. Res., Sect. A*, vol. 250, p. 71, 1986.

[15] K.-J. Kim, "Rf and space-charge effects in laser-driven rf electron guns," *Nucl. Instrum. Methods Phys. Res., Sect. A*, vol. 275, p. 201, 1988.

[16] B. E. Carlsten, "New photoelectron injector design for the Los Alamos national laboratory XUV FEL accelerator," *Nucl. Instrum. Methods Phys. Res., Sect. A*, vol. 285, p. 313, 1989.

[17] M. Borland, "Coherent synchrotron radiation and microbunching in bunch compressors," in *Proceedings of LINAC 2002*, p. 11, 2002.

[18] E. Saldin, E. Schneidmiller, and M. Yurkov, "Longitudinal space charge-driven microbunching instability in the TESLA test facility linac," *Nucl. Instrum. Methods Phys. Res., Sect. A*, vol. 528, p. 355, 2004.

[19] Z. Huang, M. Borland, P. Emma, J. Wu, C. Limborg, G. Stupakov, and J. Welch, "Suppression of microbunching instability in the linac coherent light source," *Phys. Rev. ST Accel. Beams*, vol. 7, p. 074401, 2004.

[20] C. Pelligrini, "A 4 to 0.1 nm FEL based on the SLAC linac," in *Proceedings of the Workshop on Fourth Generation Light Sources*, M. Cornacchia and H. Winick, Eds., p. 364, 1992.

[21] W. Ackermann *et al.*, "Operation of a free-electron laser from the extreme ultraviolet to the water window," *Nature Photonics*, vol. 1, p. 336, 2007.

[22] P. Emma *et al.*, "First lasing and operation of an ångstrom-wavelength free-electron laser," *Nature Photonics*, vol. 4, p. 641, 2010.

[23] T. Ishikawa *et al.*, "A compact X-ray free-electron laser emitting in the sub-ångström region," *Nature Photonics*, vol. 6, p. 540, 2012.

[24] E. Allaria *et al.*, "Highly coherent and stable pulses from the FERMI seeded free-electron laser in the extreme ultraviolet," *Nature Photonics*, vol. 6, p. 699, 2012.

[25] T. J. Orzechowski, B. Anderson, W. M. Fawley, D. Prosnitz, E. T. Scharlemann, S. Yarema, D. Hopkins, A. C. Paul, A. M. Sessler, and J. Wurtele, "Microwave radiation from a high-gain free-electron laser amplifier," *Phys. Rev. Lett.*, vol. 54, p. 889, 1985.

[26] M. J. Hogan, C. Pellegrini, J. Rosenzweig, S. Anderson, P. Frigola, A. Tremaine, C. Fortgang, D. C. Nguyen, R. L. Sheffield, J. Kinross-Wright, A. Varfolomeev, A. A. Varfolomeev, S. Tolmachev, and R. Carr, "Measurements of gain larger than 10^5 at 12 μm in a self-amplified spontaneous-emission free-electron laser," *Phys. Rev. Lett.*, vol. 81, p. 4867, 1998.

[27] S. Milton *et al.*, "Exponential gain and saturation of a self-amplified spontaneous emission free-electron laser," *Science*, vol. 292, p. 2037, 2001.

[28] V. Ayvazyan *et al.*, "Generation of GW radiation pulses from a VUV free-electron laser operating in the femtosecond regime," *Phys. Rev. Lett.*, vol. 88, p. 104802, 2002.

[29] A. Singer, I. A. Vartanyants, M. Kuhlmann, S. Duesterer, R. Treusch, and J. Feldhaus, "Transverse-coherence properties of the free-electron laser FLASH at DESY," *Phys. Rev. Lett.*, vol. 101, p. 254801, 2008.

[30] A. Tremaine, X. J. Wang, M. Babzien, I. Ben-Zvi, M. Cornacchia, H.-D. Nuhn, R. Malone, A. Murokh, C. Pellegrini, S. Reiche, J. Rosenzweig, and V. Yakimenko, "Experimental characterization of nonlinear harmonic radiation from a visible self-amplified spontaneous emission free-electron laser at saturation," *Phys. Rev. Lett.*, vol. 88, p. 204801, 2002.

[31] Z. Huang *et al.*, "Measurements of the linac coherent light source laser heater and its impact on the X-ray free-electron laser performance," *Phys. Rev. ST Accel. Beams*, vol. 13, p. 020703, 2010.

[32] M. Borland, Y.-C. Chae, P. Emma, J. W. Lewellen, V. Bharadwaj, W. M. Fawley, P. Krejcik, C. Limborg, S. V. Milton, H.-D. Nuhn, R. Soliday, and M. Woodley, "Start-to-end simulation of self-amplified spontaneous emission free electron lasers from the gun through the undulator," *Nucl. Instrum. Methods Phys. Res., Sect. A*, vol. 483, p. 268, 2002.

[33] J. Feldhaus, E. L. Saldin, J. R. Schneider, E. A. Schneidmiller, and M. V. Yurkov, "Possible application of X-ray optical elements for reducing the spectral bandwidth of an X-ray SASE FEL," *Optics Comm.*, vol. 140, p. 341, 1997.

[34] E. Allaria, D. Castronovo, P. Cinquegrana, P. Craievich, M. Dal Forno, M. B. Danailov, G. D'Auria, A. Demidovich, G. De Ninno, S. Di Mitri, B. Diviacco, W. M. Fawley, M. Ferianis, E. Ferrari, L. Froehlich, G. Gaio, D. Gauthier, L. Giannessi, R. Ivanov, B. Mahieu, N. Mahne, I. Nikolov, F. Parmigiani, G. Penco, L. Raimondi, C. Scafuri, C. Serpico, P. Sigalotti, S. Spampinati, C. Spezzani, M. Svandrlik, C. Svetina, M. Trovo,

M. Veronese, D. Zangrando, and M. Zangrando, "Two-stage seeded soft-X-ray free-electron laser," *Nature Photonics*, vol. 7, p. 913, 2013.

[35] L. H. Yu *et al.*, "First ultraviolet high-gain harmonic-generation free-electron laser," *Phys. Rev. Lett.*, vol. 91, p. 074801, 2003.

[36] J. Amann *et al.*, "Demonstration of self-seeding in a hard X-ray free-electron laser," *Nature Photonics*, vol. 6, p. 693, 2012.

[37] G. Geloni, V. Kocharyan, and E. Saldin, "A novel self-seeding scheme for hard X-ray FELs," *J. Modern Optics*, vol. 58, p. 1391, 2011.

[38] Y. Ding, A. Brachmann, F.-J. Decker, D. Dowell, P. Emma, J. Frisch, S. Gilevich, G. Hays, P. Hering, Z. Huang, R. Iverson, H. Loos, A. Miahnahri, H.-D. Nuhn, D. Ratner, J. Turner, J. Welch, W. White, and J. Wu, "Measurements and simulations of ultralow emittance and ultrashort electron beams in the linac coherent light source," *Phys. Rev. Lett.*, vol. 102, p. 254801, Jun 2009.

[39] W. E. Stein and R. L. Sheffield, "Electron micropulse diagnostics and results for the Los Alamos free-electron laser," *Nucl. Instrum. Methods Phys. Res., Sect. A*, vol. 250, p. 12, 1986.

[40] C. Behrens *et al.*, "Few-femtosecond time-resolved measurements of X-ray free-electron lasers," *Nature Commun.*, vol. 5, 2014.

[41] P. Emma, K. Bane, M. Cornacchia, Z. Huang, H. Schlarb, G. Stupakov, and D. Walz, "Femtosecond and subfemtosecond X-ray pulses from a self-amplified spontaneous-emission-based free-electron laser," *Phys. Rev. Lett.*, vol. 92, p. 074801, 2004.

[42] Y. Ding *et al.*, "Generating femtosecond X-ray pulses using an emittance-spoiling foil in free-electron lasers," *Appl. Phys. Lett.*, vol. 107, no. 19, p. 191104, 2015.

[43] C. Pellegrini, "High power femtosecond pulses from an X-ray SASE-FEL," *Nucl. Instrum. Methods Phys. Res., Sect. A*, vol. 445, no. 1-3, p. 124, 2000.

[44] C. Schroeder, C. Pellegrini, S. Reiche, J. Arthur, and P. Emma, "Chirped-beam two-stage SASE-FEL for high power femtosecond X-ray pulse generation," *Nucl. Instrum. Methods Phys. Res., Sect. A*, vol. 483, no. 1-2, p. 89, 2002, proceedings of the 23rd International Free Electron Laser Conference and 8th FEL Users Workshop.

[45] S. Krinsky and Z. Huang, "Frequency chirped self-amplified spontaneous-emission free-electron lasers," *Phys. Rev. ST Accel. Beams*, vol. 6, p. 050702, May 2003.

[46] E. L. Saldin, E. A. Schneidmiller, and M. V. Yurkov, "Self-amplified spontaneous emission FEL with energy-chirped electron beam and its application for generation of attosecond X-ray pulses," *Phys. Rev. ST Accel. Beams*, vol. 9, p. 050702, May 2006.

[47] L. Giannessi *et al.*, "Self-amplified spontaneous emission free-electron laser with an energy-chirped electron beam and undulator tapering," *Phys. Rev. Lett.*, vol. 106, p. 144801, Apr 2011.

[48] A. A. Zholents, "Method of an enhanced self-amplified spontaneous emission for X-ray free electron lasers," *Phys. Rev. ST Accel. Beams*, vol. 8, p. 040701, Apr 2005.

[49] E. Hemsing, G. Stupakov, D. Xiang, and A. Zholents, "Beam by design: Laser manipulation of electrons in modern accelerators," *Rev. Mod. Phys.*, vol. 86, pp. 897–941, Jul 2014.

Appendix A **Hamilton's Equations of Motion on Phase Space**

Traditional Hamiltonian mechanics considers the electron motion to occur in a six-dimensional phase space defined by the particle positions $\boldsymbol{q}$ and canonical momenta $\boldsymbol{p}$, with particle trajectories given by curves in the $(\boldsymbol{q},\boldsymbol{p})$ manifold that are parametrized by the time t. As we have discussed, however, we find it convenient to use the position along the undulator z as the independent evolution parameter, with t one of the particle coordinates. While changing the independent variable from t to z can be done in several different ways, we will choose the route that requires the least additional formalism. We begin with the single particle Hamiltonian, which is given by the sum of the kinetic energy γmc^2 and the potential energy $-e\Phi$ expressed in terms of the canonical coordinates:

$$\mathscr{H} = \sqrt{m^2c^4 + c^2(\boldsymbol{P} + e\boldsymbol{A})^2 + c^2(P_z + eA_z)^2} - e\Phi. \tag{A.1}$$

The canonical momentum in the $\boldsymbol{x}$ and z directions are given by $\boldsymbol{P}$ and P_z, respectively, which are related to the kinetic momentum and the vector potential $(\boldsymbol{A}, A_z)$ via the usual relations $\boldsymbol{P} = \gamma m\boldsymbol{v} - e\boldsymbol{A}$ and $P_z = \gamma mv_z - eA_z$. We introduce dimensionless coordinates by scaling the transverse momenta by mc and vector potential by (mc/e), defining

$$\boldsymbol{p} \equiv \frac{\boldsymbol{P}}{mc} \equiv \frac{\gamma}{c}\frac{d\boldsymbol{x}}{dt} - \frac{e}{mc}\boldsymbol{A} \equiv \frac{\gamma}{c}\frac{d\boldsymbol{x}}{dt} - \boldsymbol{a} \tag{A.2}$$

$$p_z \equiv \frac{P_z}{mc} \equiv \frac{\gamma}{c}\frac{dz}{dt} - \frac{e}{mc}A_z \equiv \frac{\gamma}{c}\frac{dz}{dt} - a_z. \tag{A.3}$$

If we additionally define the scaled electrostatic potential $\phi \equiv (e/mc^2)\Phi$, the dimensionless version of the Hamiltonian (A.1) is

$$\mathscr{H} = \sqrt{1 + \left[\boldsymbol{p} + \boldsymbol{a}(\boldsymbol{x}, z; t)\right]^2 + \left[p_z + a_z(\boldsymbol{x}, z; t)\right]^2} - \phi(\boldsymbol{x}, z; t), \tag{A.4}$$

where the canonical coordinates are $(\boldsymbol{x}, z)$, whose conjugate momenta are $(\boldsymbol{p}, p_z)$. To change the independent variable from t to z requires the Jacobian dt/dz, which can be obtained using Hamilton's equation for z:

$$\frac{dt}{dz} = \left(\frac{\partial\mathscr{H}}{\partial p_z}\right)^{-1} = \frac{H + \phi}{c(p_z + a_z)} = \frac{H + \phi}{c\sqrt{(H + \phi)^2 - 1 - (\boldsymbol{p} - \boldsymbol{a})^2}}. \tag{A.5}$$

To arrive at (A.5), we have eliminated p_z using (A.4), and defined the scaled particle energy $H = \gamma - \phi$, which equals the value of the Hamiltonian function $\mathscr{H}$ on a physical orbit. Using Hamilton's equations of motion and the chain rule, the particle equations of motion along z are

$$\boldsymbol{x}' = \frac{d\boldsymbol{x}}{dz} = \frac{\partial \mathscr{H}}{\partial \boldsymbol{p}}\frac{dt}{dz} = \frac{\boldsymbol{p}+\boldsymbol{a}}{\sqrt{(H+\phi)^2 - 1 - (\boldsymbol{p}+\boldsymbol{a})^2}} \tag{A.6}$$

$$\boldsymbol{p}' = \frac{d\boldsymbol{p}}{dz} = -\frac{\partial \mathcal{H}}{\partial \boldsymbol{x}}\frac{dt}{dz} = -\frac{\nabla\left[(p+a)^2 - (H+\phi)^2\right]}{2\sqrt{(H+\phi)^2 - 1 - (p+a)^2}} - \nabla a_z \tag{A.7}$$

$$H' = \frac{dH}{dz} = \frac{\partial \mathcal{H}}{\partial t}\frac{dt}{dz} = \frac{\frac{\partial}{\partial t}\left[(\boldsymbol{p}+\boldsymbol{a})^2 - (H+\phi)^2\right]}{2\sqrt{(H+\phi)^2 - 1 - (\boldsymbol{p}+\boldsymbol{a})^2}} + \frac{1}{c}\frac{\partial a_z}{\partial t}. \tag{A.8}$$

Inspection of the transverse equations (A.6)–(A.7) reveals that they can be derived from the new Hamiltonian $\mathcal{H} = -\sqrt{(H+\phi)^2 - 1 - (\boldsymbol{p}+\boldsymbol{a})^2} + a_z$, with $(\boldsymbol{x},\boldsymbol{p})$ canonical coordinates. The time–energy equations (A.5) and (A.8) nearly obtain from the same Hamiltonian, with only a sign difference and a factor of c. If we introduce the position variable $\tau \equiv -ct$ the coordinates (τ, H) form a canonical pair whose equations of motion are also derivable from H. Thus, the single particle dynamics are governed by the Hamiltonian

$$\mathcal{H} = -\sqrt{[H+\phi(x,\tau;z)]^2 - 1 - \left[\boldsymbol{p}+\boldsymbol{a}(\boldsymbol{x},\tau;z)\right]^2} + a_z(\boldsymbol{x},\tau;x). \tag{A.9}$$

We recognize the new Hamiltonian $\mathcal{H} = -p_z$ under the assumption that $p_z > 0$. Thus, $\mathcal{H}$ is proportional to the generator of z-translations, and the canonical formalism yields evolution equations whose independent parameter is z. This is in contrast to the Hamiltonian, which equals the energy (the generator of time translations) and produces dynamical equations with respect to t. We can make two additional simplifications to the Hamiltonian (A.9) for paraxial beams relevant to X-ray generation. First, the electrostatic force due to ϕ is typically negligible with respect to the forces due to $\boldsymbol{a}$. For this reason, we neglect $\phi(\boldsymbol{x},\tau;z)$ from (A.9), in which case the canonical energy $H = \gamma$. Second, in the paraxial approximation we have $|\boldsymbol{p}+\boldsymbol{a}|/\gamma \sim |\boldsymbol{x}'| \ll 1$, so that we can expand the square root in (A.9). Thus, the paraxial Hamiltonian through $O(1/\gamma^2)$ is

$$\mathcal{H}(x,\tau,p,\gamma;z) = \frac{1}{2\gamma}\left[p + a(x,\tau;z)\right]^2 - \left(\gamma - \frac{1}{2\gamma}\right) + a_z(x,\tau;z). \tag{A.10}$$

The equations of motion are given by the canonical formulas

$$\boldsymbol{q}' = \frac{\partial \mathcal{H}}{\partial \boldsymbol{p}} \qquad \boldsymbol{p}' = -\frac{\partial \mathcal{H}}{\partial \boldsymbol{q}}, \tag{A.11}$$

and we recall that the coordinates and conjugate momenta are $\boldsymbol{q} = (x,y,\tau)$ and $\boldsymbol{p} = (p_x, p_y, \gamma)$, respectively.

In addition to providing the equations of motion, the Hamiltonian formulation also implies that the resulting dynamics are symplectic, which has many important consequences including Liouville's theorem (phase space volume conservation) and the preservation of Poincaré invariants (which generalizes Liouville's theorem to certain projected areas in phase space). Hence, the Hamiltonian/symplectic structure strongly restricts the possible dynamics; for example, it precludes attracting fixed points. We can insure that any approximate description derivable from $\mathcal{H}$ inherits the Hamiltonian structure by using Hamiltonian perturbation/approximation techniques. In the next section we briefly show how this can be done for the FEL, and conclude this appendix with a simple Hamiltonian description of a test electron in a high-gain FEL.

A.1 FEL Particle Equations from Transformation Theory

Here, we derive the "wiggle-averaged" FEL equations of motion from the Hamiltonian, including slow variations in the undulator parameters λ_u and K. The result will be similar to the early work in Ref. [1], but will be fully 3D. We begin with the dimensionless FEL vector potential associated with the undulator and radiation,

$$\begin{aligned} \boldsymbol{a} &= \hat{x}\left\{K(z)\cosh[k_u(z)y]\cos\left[\textstyle\int dz'\, k_u(z')\right] + a(\boldsymbol{x},\tau;z)\right\} \\ &= \hat{x}\left[K\cosh(k_u y)\cos\varphi_u + a(\boldsymbol{x},\tau;z)\right], \end{aligned} \tag{A.12}$$

where a describes the radiation and for notational simplicity we have defined the undulator phase $\varphi_u \equiv \int^z dz'\, k_u(z')$ as shown and dropped the explicit dependence of the undulator parameters K and k_u on z. Inserting (A.12) into the Hamiltonian (A.10) and using $\cos^2\varphi_u = [1+\cos(2\varphi_u)]/2$ leads to

$$\begin{aligned} \mathcal{H} \approx\ & \frac{1}{2\gamma} + \frac{K^2}{4\gamma}\cosh^2(k_u y) - \gamma + \frac{p_x^2 + p_y^2}{2\gamma} + \frac{Ka}{\gamma}\cosh(k_u y)\cos\varphi_u \\ & + \frac{K^2}{4\gamma}\cosh^2(k_u y)\cos(2\varphi_u) + \frac{Kp_x}{\gamma}\cosh(k_u y)\cos\varphi_u, \end{aligned} \tag{A.13}$$

where we have dropped the higher-order terms $\sim p_x a/\gamma$ and $\sim a^2/\gamma$.

The fast oscillations in the undulator field give rise to the terms $\sim \cos\varphi_u$ and $\sim \sin(2\varphi_u)$. In our previous derivation we argued that the effect of these terms can be averaged away from the dynamics. From the Hamiltonian perspective we perform this "wiggle averaging" by making a canonical coordinate transformation. The basic idea is that if we can find a coordinate transformation that approximately removes the oscillating terms from $\mathcal{H}$, the remaining slowly varying Hamiltonian will yield the average motion of these new averaged or "oscillation center" coordinates.

There are systematic methods to remove oscillating terms in the Hamiltonian order by order; Lie transformation techniques (see, e.g., [2]) provide one elegant way to asymptotically eliminate the fast variation in $\mathcal{H}$. However, we will take a more simplistic view both because we are only interested in the lowest-order equations and since using more sophisticated methods would require developing a fair bit of additional mathematical machinery.

We will perform the canonical transformation using the familiar mixed variable generating function. We denote the new/averaged coordinates and Hamiltonian by bars, so that the type-two generating function (see, e.g., [3, 4]) that depends on the old coordinates and new momenta is $F_2(\boldsymbol{q},\bar{\boldsymbol{p}};z)$. F_2 connects the old and new coordinates, momenta, and Hamiltonian via

$$\bar{\boldsymbol{q}} = \frac{\partial F_2}{\partial \bar{\boldsymbol{p}}} \qquad \boldsymbol{p} = \frac{\partial F_2}{\partial \boldsymbol{q}} \qquad \bar{\mathcal{H}} = \mathcal{H} + \frac{\partial F_2}{\partial z}. \tag{A.14}$$

First, we wish to transform away the fast oscillations from the FEL Hamiltonian that are associated with the motion in the undulator. The transformation equation (A.14) for $\bar{\mathcal{H}}$ suggests that this might be accomplished if we choose $\partial F_2/\partial z$ to cancel the last line in (A.13), so that

$$\begin{aligned} F_2(\boldsymbol{q},\bar{\boldsymbol{p}};z) &= \bar{p}_x x + \bar{p}_y y + \bar{\gamma}\tau \\ &\quad - \frac{K^2}{8k_u\bar{\gamma}}\cosh^2(k_u y)\sin(2\varphi_u) - \frac{Kp_x}{k_u\bar{\gamma}}\cosh(k_u y)\sin\varphi_u. \end{aligned} \tag{A.15}$$

The first line is the identity transformation; the rest of F_2 eliminates from the new $\bar{\mathcal{H}}$ the second, oscillating line of (A.13), and leads to the following nontrivial coordinate transformations:

$$x = \bar{x} + \frac{K}{k_u\bar{\gamma}}\cosh(k_u y)\sin\varphi_u \tag{A.16}$$

$$\tau = \bar{\tau} - \frac{K^2}{8k_u\bar{\gamma}^2}\cosh^2(k_u y)\sin(2\varphi_u) - \frac{Kp_x}{k_u\bar{\gamma}^2}\cosh(k_u y)\sin\varphi_u \tag{A.17}$$

$$p_y = \bar{p}_y - \frac{K^2}{8\bar{\gamma}}\sinh(2k_u y)\sin(2\varphi_u) - \frac{K\bar{p}_x}{\bar{\gamma}}\sinh(k_u y)\sin\varphi_u. \tag{A.18}$$

Since $y = \bar{y}$, $p_x = \bar{p}_x$, and $\gamma = \bar{\gamma}$ the right-hand side of (A.16)–(A.18) can be written entirely in terms of the new coordinates. The first two averaged coordinates are familiar from the previous derivation: they define the oscillation center $\bar{x}$ and $\bar{\tau}$ by subtracting off the figure-eight motion in the undulator. The introduction of $\bar{p}_y$ is new and due to motion off-axis. The oscillation center/wiggle-averaged Hamiltonian is now

$$\begin{aligned} \bar{\mathcal{H}} &\approx \frac{1}{2\bar{\gamma}} + \frac{K^2}{4\bar{\gamma}}\cosh^2(k_u\bar{y}) - \bar{\gamma} + \frac{\bar{p}_x^2 + \bar{p}_y^2}{2\bar{\gamma}} + O\left[\frac{K'}{k_u\bar{\gamma}}, \frac{k_u'}{k_u^2\bar{\gamma}}\right] \\ &\quad + \frac{K}{\bar{\gamma}} a[\boldsymbol{q}(\bar{\boldsymbol{q}},\bar{\boldsymbol{p}});z]\cosh(k_u\bar{y})\cos\varphi_u + O\left[\frac{\bar{p}_y k_u\bar{y}}{\bar{\gamma}^2}, \frac{\bar{p}_y\bar{p}_x}{\bar{\gamma}^2}, \frac{(k_u\bar{y})^2}{\bar{\gamma}^3}\right]. \end{aligned} \tag{A.19}$$

The neglected terms on the second line are all small corrections, while those of the first line embody the slowly varying approximation: they can be neglected only if the undulator parameters vary slowly over one wavelength, meaning that $k_u' \ll k_u^2$ and $K' \ll k_u K$. Within the slowly varying approximation, the undulator motion is obtained from the first line in (A.19).

We complete the FEL equations by assuming that the radiation can be described by a slowly-varying envelope and phase, so that

$$\begin{aligned} a[\boldsymbol{q}(\bar{\boldsymbol{q}},\bar{\boldsymbol{p}});z] &= \tilde{a}[\boldsymbol{q}(\bar{\boldsymbol{q}},\bar{\boldsymbol{p}});z]e^{ik_1[z+\tau(\bar{\tau},\bar{y},\bar{p}_x,\bar{\gamma};z)]} + c.c. \\ &= \left[\tilde{a}(\bar{\boldsymbol{q}};z) + O\left(\frac{K}{k_u\bar{\gamma}}\frac{\partial a}{\partial\bar{x}}, \frac{K^2}{k_u\bar{\gamma}^2}\frac{\partial a}{\partial\bar{\tau}}\right)\right]e^{ik_1(z+\bar{\tau})} \end{aligned}$$

$$\times \sum_{m,\ell} J_m\left[\frac{k_1 K^2 \cosh^2(k_u y)}{8k_u \bar{\gamma}^2}\right] J_\ell\left[\frac{k_1 K \bar{p}_x \cosh(k_u y)}{k_u \bar{\gamma}^2}\right] e^{-i(2m+\ell)\varphi_u} + c.c.$$

$$\approx \tilde{a}(\bar{\boldsymbol{q}}; z) e^{ik_1(z+\bar{\tau})} \sum_m J_m\left(\frac{k_1 K^2}{8k_u \bar{\gamma}^2}\right) e^{-2im\varphi_u} + c.c.. \tag{A.20}$$

To go from the second to third line we have assumed that the beam coordinates are small, as before, and that the radiation field varies slowly over one wavelength, with $(1/k_u\bar{\gamma}^2)\,|\partial a/\partial \bar{\tau}| \sim (1/k_1)\,|\partial a/\partial \bar{\tau}| \sim |\partial a/\partial \theta| \ll |a|$.

When we insert the expression (A.20) for a into $\bar{\mathcal{H}}$, we find that the FEL coupling can be simplified via

$$\frac{K}{\bar{\gamma}} a[\boldsymbol{q}(\bar{\boldsymbol{q}}, \bar{\boldsymbol{p}}); z] \cosh(k_u \bar{y}) \cos\varphi_u$$

$$\approx \frac{K}{2\bar{\gamma}}\left[\tilde{a}(\bar{\boldsymbol{q}}; z) e^{ik_1(z+\bar{\tau})} + c.c.\right](e^{i\varphi_u} + e^{-i\varphi_u}) \sum_m J_m\left(\frac{K^2}{4+K^2}\right) e^{-2im\varphi_u}$$

$$= \frac{K[\mathrm{JJ}]}{2\bar{\gamma}}\left[\tilde{a} e^{i(k_1 z + \varphi_u + k_1 \bar{\tau})} + c.c.\right]$$

$$+ \left[K\tilde{a} e^{ik_1(z+\bar{\tau})} \frac{e^{i\varphi_u}}{2\bar{\gamma}} \sum_{m\neq 0} J_m\left(\frac{K^2}{4+K^2}\right) e^{-2im\varphi_u}\right. \tag{A.21}$$

$$\left. + K\tilde{a} e^{ik_1(z+\bar{\tau})} \frac{e^{-i\varphi_u}}{2\bar{\gamma}} \sum_{m\neq -1} J_m\left(\frac{K^2}{4+K^2}\right) e^{-2im\varphi_u} + c.c.\right].$$

Assuming that the ponderomotive phase $\theta \equiv k_1(z+\bar{\tau}) + \varphi_u$ varies slowly, we have divided the coupling into its fast and slow components in (A.21). The fast variations can be eliminated with another coordinate transformation derivable from an appropriate generating function. For example, $\partial F_2/\partial z$ would approximately eliminate the second to last line if $F_2 \sim \sum_{m\neq 0}[e^{i(1-2m)\varphi_u}/(1-2m)]$. The rest of the transformation is analogous, long, and not particularly illuminating. We skip its explicit form, drop the bars from the new coordinates, and expand the hyperbolic functions to write the FEL Hamiltonian as

$$\begin{aligned}\mathcal{H}_{\mathrm{FEL}}(\boldsymbol{q}, \boldsymbol{p}; z) \approx\ & \frac{1+K^2/2}{2\gamma} - \gamma + \frac{p_x^2 + p_y^2}{2\gamma} + \frac{K^2 k_u^2}{4\gamma} y^2 \\ & + \frac{K[\mathrm{JJ}]}{2\gamma}\left[\tilde{a}(x, y, \tau; z) e^{i(k_1 z + \varphi_u + k_1 \tau)} + c.c.\right].\end{aligned} \tag{A.22}$$

It is straightforward to show that the equations of motion implied by $\mathcal{H}_{\mathrm{FEL}}$ give rise to the 3D FEL equations previously derived, plus a few additional terms $\sim \partial\tilde{a}/\partial x$, etc. that are small.

A.2 Motion of a Test Electron in a High-Gain FEL

In this section we describe the motion of a test electron in a high-gain FEL with the goal of giving some additional insight into the dynamics. We being with the 1D particle equations for a monochromatic field

$$\frac{d\theta}{d\hat{z}} = \hat{\eta} \qquad \frac{d\hat{\eta}}{d\hat{z}} = a(\hat{z})e^{i\theta} + a(\hat{z})^* e^{-i\theta}. \tag{A.23}$$

We will assume that the field $a(\hat{z})$ is given by the exponentially growing solution of the cold, 1D FEL theory:

$$a(\hat{z}) = -\frac{a_0}{2i}e^{-i\mu_3\hat{z}} = -\frac{a_0}{2i}e^{(\sqrt{3}+i)\hat{z}/2}, \tag{A.24}$$

where the a_0 was chosen such that

$$\frac{d\theta}{d\hat{z}} = \hat{\eta} \qquad \frac{d\hat{\eta}}{d\hat{z}} = -a_0 e^{\sqrt{3}\hat{z}/2}\sin(\theta + \hat{z}/2). \tag{A.25}$$

The equations of motion (A.25) can be derived from the Hamiltonian

$$\mathcal{H}(\theta, \hat{\eta}) = \frac{1}{2}\hat{\eta}^2 - a_0 e^{\sqrt{3}\hat{z}/2}\left[\cos(\theta + \hat{z}/2) - 1\right], \tag{A.26}$$

which describes a particle moving in a growing pendulum-type potential with scaled phase velocity equal to $-1/2$. We transform to coordinates (Ψ, p) that are moving with the wave using the generating function $F_2(p, \theta) = p(\theta + \hat{z}/2)$. Applying Equations (A.14) implies that the new and old coordinates are related via $\Psi = \partial F_2/\partial p = \theta + \hat{z}/2$ and $p = \partial F_2/\partial p = \hat{\eta}$, while the new Hamiltonian

$$\mathcal{K}(\Psi, p) = \mathcal{H}(\Psi, p) + \frac{\partial F_2}{\partial z} = \frac{1}{2}\left(p + \tfrac{1}{2}\right)^2 - a_0 e^{\sqrt{3}\hat{z}/2}(\cos\Psi - 1). \tag{A.27}$$

This is a pendulum Hamiltonian with exponentially growing amplitude whose equilibrium point is at $(\Psi, p) = (0, -1/2)$. As the field grows, a particle whose initial energy deviation $\hat{\eta} = p = 0$ will eventually become trapped by the wave and oscillate about this equilibrium point. These particles move with the phase velocity of the wave, having lost on average an energy $\Delta\gamma = \rho\gamma_r\hat{\eta} = \rho\gamma_r/2$.

References

[1] N. M. Kroll, P. L. Morton, and M. N. Rosenbluth, "Free-electron lasers with variable parameter wigglers," *IEEE J. Quantum Electron.*, vol. 17, p. 1436, 1981.

[2] A. J. Lichtenberg and M. A. Lieberman, *Regular and Stochastic Motion*. New York: Springer-Verlag, 1983.

[3] J. C. Goldstein, "Theory of the sideband instability in free electron lasers," *Nucl. Instrum. Methods Phys. Res., Sect. A*, vol. 237, p. 27, 1985.

[4] I. Percival and D. Richards, *Introduction to Dynamics*. Cambridge, UK: Cambridge University Press, 1982.

Appendix B Simulation Methods for FELs

Numerical simulation has become an integral tool of FEL research, and in this appendix we discuss certain aspects of FEL simulations. We will only briefly mention the integration algorithms *per se*, and instead try to give a flavor for the numerical techniques and approximations that are commonly employed to increase computational speed and efficiency.

The 3D FEL equations that are typically solved on a computer are a bit different from those that we have written down thus far, in that they dispense with several simplifications that we made for analytic tractability. First, it is straightforward to solve the 3D particle equations with natural focusing in the undulator, and separately apply the transverse FODO focusing as necessary. Thus, the smooth focusing approximations discussed in Section 5.2.2 are not required. Second, nearly all codes are written in the time domain, because it is much more convenient when solving the fully nonlinear problem associated with saturation. In the time domain the paraxial current is comprised of an FEL slice average over the particle phases, rather than a sum over the entire beam as dictated in the frequency representation. This averaged current is accurate if the FEL slice is chosen to be some integral number of wavelengths much less than the coherence time t_{coh}, and if the e-beam parameters are nearly constant over this time interval. Third, there is typically no compelling reason to assume that the energy deviation is small, so that one can dispense with the expansion in $\eta \ll 1$ that we have used throughout this book. Fourth, we permit the undulator parameters K and λ_u to slowly vary over the length of the FEL. For N_Δ particles in an FEL slice, the generalized time domain representation of the dimensionless field equation (5.96) is

$$\left(\frac{\partial}{\partial \hat{z}} + \frac{k_u}{2\rho k_{u,0}}\frac{\partial}{\partial \theta} - \frac{i}{2}\frac{\partial^2}{\partial \hat{\boldsymbol{x}}^2}\right) a(\theta, \hat{\boldsymbol{x}}; z) = -\frac{1}{N_\Delta}\sum_{j\in\Delta}\frac{K[\text{JJ}]}{K_0[\text{JJ}]_0}\frac{e^{-i\theta_j}}{1+\rho\hat{\eta}_j} \times 2\pi\hat{\sigma}_x^2\delta(\hat{\boldsymbol{x}} - \hat{\boldsymbol{x}}_j), \tag{B.1}$$

while the FEL particle equations as typically coded can be written as

$$\frac{d}{d\hat{z}}\theta_j = \frac{1}{2\rho}\left[\frac{k_u(z)}{k_{u,0}} - \frac{1+K^2/2}{1+K_0^2/2}\frac{1}{(1+\rho\hat{\eta}_j)^2}\right] - \frac{\hat{\boldsymbol{p}}_j^2 + \hat{k}_\beta^2\hat{\boldsymbol{x}}_j^2}{2(1+\rho\hat{\eta}_j)^2} - \frac{i\rho}{(1+\rho\hat{\eta}_j)^2}\frac{K[\text{JJ}]}{K_0[\text{JJ}]_0}\left[a(\theta, \hat{\boldsymbol{x}}; \hat{z})e^{i\theta_j} - a(\theta, \hat{\boldsymbol{x}}; \hat{z})^* e^{-i\theta_j}\right] \tag{B.2}$$

$$\frac{d}{d\hat{z}}\hat{\eta}_j = \frac{1}{1+\rho\hat{\eta}_j}\frac{K[\mathrm{JJ}]}{K_0[\mathrm{JJ}]_0}\left[a(\theta,\hat{\boldsymbol{x}};\hat{z})e^{i\theta_j} + a(\theta,\hat{\boldsymbol{x}};\hat{z})^* e^{-i\theta_j}\right] \tag{B.3}$$

$$\frac{d}{d\hat{z}}\hat{\boldsymbol{x}}_j = \frac{\hat{\boldsymbol{p}}_j}{(1+\rho\hat{\eta}_j)} \tag{B.4}$$

$$\frac{d}{d\hat{z}}\hat{\boldsymbol{p}}_j = -\frac{\hat{k}_\beta^2\hat{\boldsymbol{x}}_j}{(1+\rho\hat{\eta}_j)} \tag{B.5}$$

with the scaled natural focusing $\hat{k}_\beta \equiv \sqrt{2}K/(4\gamma_r\rho)$, and $k_{u,0}$, K_0, and $[\mathrm{JJ}]_0$ denote reference values used to express the equations in dimensionless form, while the quantities without subscripts may be functions of z. These equations agree with those implied by (5.97) and (5.98) to lowest order in $|\rho\hat{\eta}_j| \ll 1$ when the undulator parameters are independent of z.

Recall that we have still made a number of approximations to arrive at (B.1)–(B.5). First, we assumed that the electron beam is paraxial, which is a good approximation for relativistic electron beams, and an extremely good approximation for the multi-GeV energy beams that are often considered. Second, the wave equation neglects any backward-going radiation (in general a very good approximation) and assumes that the forward radiation is well characterized by a fast oscillation near the resonant frequency and a slowly varying envelope, so that the second-order longitudinal derivatives may be neglected. Third, the electron-radiation coupling results from taking the wiggle-averaged (oscillation center) particle equations, which assumes that the electromagnetic field envelope is slowly varying in the same sense as was previously assumed. Fourth, we have neglected the space-charge force on the electron motion, which is equivalent to assuming that the FEL gain occurs over a much shorter distance than that required for a space-charge plasma oscillation. We can express this condition in terms of the relativistic plasma frequency $\Omega_p \equiv \sqrt{e^2 n_e/\epsilon_0 mc^2\gamma^3}$ as $\Omega_p/2k_u\rho \ll 1$, which can be rewritten as

$$\frac{\Omega_p}{2k_u\rho} = \frac{\sqrt{\rho}}{2k_u}\frac{\Omega_p}{\rho^{3/2}} = \frac{2\sqrt{2\rho}}{K[\mathrm{JJ}]} \ll 1. \tag{B.6}$$

We see that (B.6) is satisfied for a very wide range of FEL devices, the one exception being when the electron beam is very dense (which increases ρ) and the undulator magnetic field is very weak. If the electrostatic force is small, $\Omega_p < 2k_u\rho$, the dominant effect of space charge is to shift the frequency of maximum gain by an amount proportional to $\Omega_p/2k_u\rho$, while the correction to the FEL growth rate scales $\sim (\Omega_p/2k_u\rho)^2$.

While we have made a number of caveats regarding the applicability of (B.1)–(B.5), there is ample experience indicating that these equations accurately represent the physics. Furthermore, numerical simulations have demonstrated remarkably good agreement with experiments over a wide range of parameters.

B.1 The Design of FEL Simulation Codes

The FEL community has developed a set of commonly employed methods to numerically integrate the FEL equations very efficiently with sufficient accuracy. The basis for many of these methods originates in the fact that one can treat the longitudinal and transverse directions differently. FEL codes typically identify longitudinal "slices" of both the electron beam and the radiation field. An electron beam slice contains all the particles in any given slice average, meaning that at phase θ it includes all the electrons with coordinates $|\theta_j - \theta| \le \Delta\theta/2$. Since the current varies slowly, it is typical to take periodic boundary conditions within a slice, in which case neighboring e-beam slices only communicate via the radiation slippage.

Integrating the particle motion within each electron beam slice is relatively straightforward, as the position of each electron is governed by a set of ordinary differential equations (ODEs) that are amenable to most standard ODE solvers. A commonly used general-purpose solver is the fourth-order Runge–Kutta (RK4) integrator, but for illustration we demonstrate the idea using the simpler second-order RK2. Denoting the particle coordinates by $\mathcal{Z}$ and their evolution equations by $d\mathcal{Z}/dz = \mathcal{F}(\mathcal{Z}; z)$, the RK2 update over a step Δz is given by

$$\begin{aligned} \mathcal{Z}_1 &= \mathcal{F}(\mathcal{Z}; z) \\ \mathcal{Z}_2 &= \mathcal{F}[\mathcal{Z} + (\Delta z/2)\mathcal{Z}_1; z + \Delta z/2] \\ \mathcal{Z}(z + \Delta z) &= \mathcal{Z}(z) + (\Delta z)\mathcal{Z}_2. \end{aligned}$$

At each step the RK2 algorithm above is accurate to $O[(\Delta z)^3]$; obtaining $\mathcal{Z}$ at a macroscopic distance $O(1)$ requires $\sim 1/(\Delta z)$ time steps, for which we expect the RK2 algorithm to converge to the true solution $\sim (\Delta z)^2$.[1]

In addition to the electron beam slices, FEL codes also employ slices of radiation, which are typically taken to interact with one e-beam slice at any given undulator location z. We show a 1D diagram of the particle and electric field slices in Figure B.1; in 3D the particles have four additional phase-space coordinates while the electromagnetic field also depends on (x, y). Within each longitudinal slice the field is numerically approximated by some finite number of values; the most common method defines the

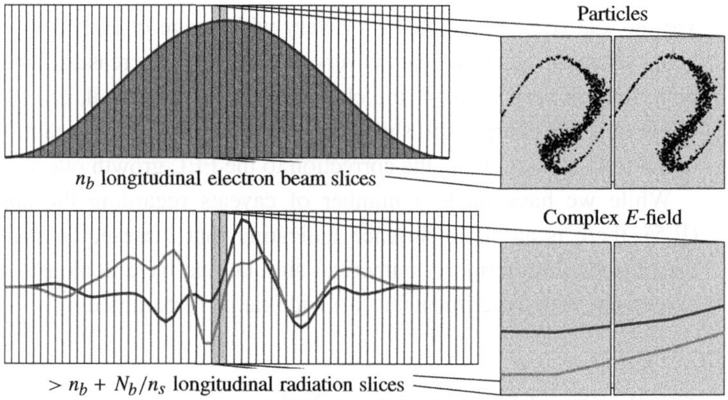

Figure B.1 Schematic of the FEL simulation slices for a 1D code.

[1] The more complicated RK4 algorithm mentioned earlier requires twice as many evaluations per step but converges $\sim (\Delta z)^4$, so that it is often faster for a given accuracy.

field on a discrete transverse grid in the (x, y) plane, and the radiation–particle coupling is obtained by interpolating between the fixed grid and the transverse particle positions. To complete the field evolution, the electromagnetic diffraction can be computed by any one of a number of standard algorithms. For example, the alternating direction implicit method is one such algorithm that is stable and fast.

The collection of radiation slices maps out the discretized electromagnetic field profile as a function of the beam coordinate θ. As implied by Equation (B.1), the radiation slices move at the speed of light and only communicate with each other indirectly through their interaction with the more slowly moving particles. For this reason, any given radiation slice interacts only with those particles that are ahead of it and within the total slippage time defined by the undulator, $N_u\lambda_1/c$; equivalently, a radiation slice only (indirectly) affects those radiation slices whose initial phase is behind it and within $2\pi N_u$.

The division into radiation and electron beam slices that only interact in one direction over a limited distance provides an opportunity for significant computational savings. For example, by starting with the slice at the electron bunch tail, one need only load a single electron beam slice into memory at a time to compute the FEL generated light. We illustrate this in Figure B.2, where the slice of particles representing the electron beam tail is at the top in dark gray, and we have also included the (initially prescribed)

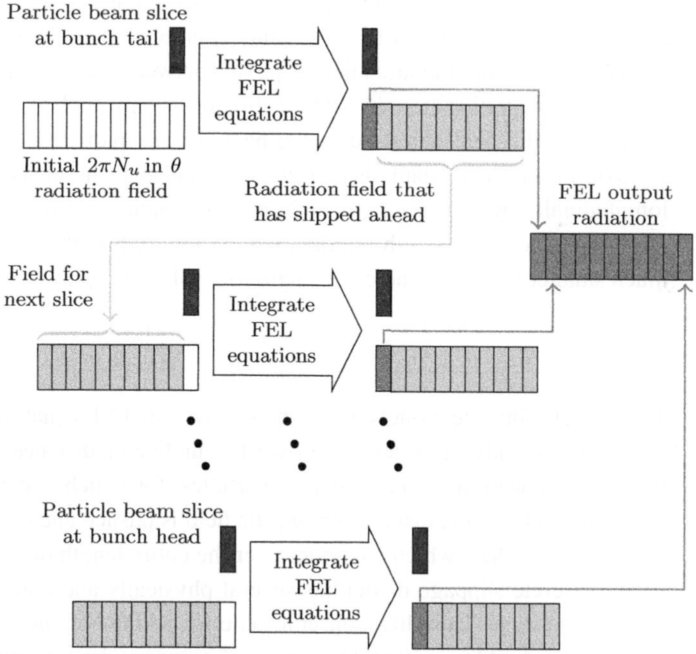

Figure B.2 Numerical integration scheme of the FEL equations that takes advantage of the fact that the communication is limited to the total distance the radiation field slips ahead of the particles in the undulator, which is $2\pi N_u$ in θ. After integrating the tail electron beam slice and obtaining the output FEL radiation, the final field (in gray) is saved while the other radiation slices (in light gray) are used to initialize the field for interaction with the next slice of particles. This serves to communicate information with those electron beam slices that are ahead and within $2\pi N_u$ in phase.

radiation slices with which it will interact over the undulator length, meaning that the discrete radiation slices span the total phase $2\pi N_u$ in θ. A simulation begins by evolving the tail slice through the undulator and computing the generated light. At the end of the FEL, any desired diagnostic information is written out or saved; the macroparticles can then be discarded to free memory while the field is retained. The radiation slice that last interacted with the particles (gray in Figure B.2) represents the amplified output field, while the slices that slipped ahead of the particles (in light gray) initialize the field array that interacts with the next electron beam slice as shown. Now, when the FEL equations are integrated through the undulator, the radiation field generated by the tail slice interacts with the next slice of particles after one slippage distance. Again, after evolution through the undulator the last slice of radiation is added to the (gray) output array, while the rest is saved to interact with the next set of particles via slippage. This procedure continues, with the electron beam slices being sequentially selected and discarded from tail to head, until the entire field has been computed. Alternatively, the slice formulation of the FEL affords a straightforward and efficient implementation on parallel computers: each node can be assigned some small number of electron beam slices, and the comparatively slow communication between nodes is limited to passing the complex field envelope for slippage.

Numerically implementing the radiation advection between electron beam slices is made particularly simple with the commonly employed discrete slippage model. For this algorithm, the entire radiation field is passed between slices after an appropriate discrete number of integration steps in $\hat{z}$. Thus, slippage is only applied after $\Delta\hat{z}/(2\rho\Delta\theta)$ $\hat{z}$-steps, after which the field is passed to the next slice of particles. As an example, suppose we load n_b electron beam slices (buckets), and that the bucket centers are separated longitudinally by $n_s\lambda_1$ (i.e., $\Delta\theta = 2\pi n_s$). To simulate the total radiation slippage over the whole undulator length requires $n_b n_s > N_u$, while the slice separation should be much smaller than the coherence length to resolve the spiky structure of SASE:

$$n_s \ll \frac{c\sigma_\tau}{\lambda_1} \approx \frac{1}{4\pi\rho}. \tag{B.7}$$

The discrete slippage model applies the steady-state FEL equations (namely (B.1)–(B.5) neglecting the advection term ∂_θ) over the undulator distance required for the radiation to slip ahead to the next slice of particles, for which $\hat{z}$ increases by $4\pi\rho n_s$. After every such integration distance $4\pi\rho n_s$ the field is pushed ahead to interact with the next electron beam slice, which continues over the entire length of the undulator.

The discrete slippage model has several physically and numerically desirable properties: it preserves causality, conserves energy, and is both numerically stable and fast. The primary problem is that the numerical accuracy depends on the number of z-steps between the slippage. This problem can in principle be alleviated with more advanced flux conserving algorithms, but to date there has been little perceived need to do so. Note, however, that trying to numerically solve the slippage using more well-known schemes like finite differences almost universally results in a number of unphysical effects, including numerical damping/growth, a-causal behavior, and artificial smoothing of the field. An alternative implementation of slippage that is stable, causal, and accurate uses a discrete formulation of the Green function solution to the driven paraxial equation.

Finally, there is often a prohibitively large number of electrons in the beam for each one to be included in a simulation. Instead, FEL codes typically sub-sample the particle phase space using macroparticles, meaning that simulations of an electron beam with N_{real} particles per slice (or bucket) are performed using $N_{\text{sim}} = N_\Delta \ll N_{\text{real}}$ particles in each slice. This approximation is often described as assuming that each simulation macroparticle represents $N_{\text{real}}/N_{\text{sim}} \gg 1$ "real" electrons, but for the FEL it is more useful and accurate to interpret the macroparticles as being chosen to sample the electron beam distribution function in as representative a manner as possible. The use of macroparticles does not change the form of the FEL equations, but one must take care that using a small number of simulation particles does not change the statistical properties of the particle current, as it is the statistics of shot noise that determines the seeding for SASE FELs. One common way to accomplish this is to initially load the particles uniformly in phase on the interval $[0, 2\pi)$, and then add an appropriately chosen small random deviation to each particle position so that the low-order statistical properties of the particle bunching matches that of the physical beam. According to this scheme, the initial phase of the N_{sim} electrons is assigned according to

$$\theta_j = \frac{2\pi j}{N_{\text{sim}}} + (2\delta)r_j, \tag{B.8}$$

where r_j is an independently chosen random number between 0 and 1 for each particle and δ controls the size of the initial displacement. In order to match the average initial bunching to that of the real shot noise,

$$\left\langle \left| \frac{1}{N_\Delta} \sum_{j\in\Delta} e^{-i\theta_j} \right|^2 \right\rangle = \left\langle \left| \frac{1}{N_{\text{sim}}} \sum_{j=1}^{N_{\text{sim}}} e^{-i\theta_j} \right|^2 \right\rangle = \frac{1}{N_{\text{real}}}, \tag{B.9}$$

Penman and McNeil showed in Ref. [1] that the appropriate choice for δ is given by

$$\text{sinc}^2\delta = 1 - \frac{N_{\text{sim}}}{N_{\text{real}}}, \qquad \text{or} \quad \delta \approx \sqrt{\frac{3N_{\text{sim}}}{N_{\text{real}}}} \quad \text{for} \quad \frac{N_{\text{sim}}}{N_{\text{real}}} \ll 1. \tag{B.10}$$

Note that if $N_{\text{sim}} = N_{\text{real}}$, $\delta = \pi$ and the particles are randomly loaded over the entire phase interval. On the other hand, when the number of simulation macroparticles is much smaller than the number of electrons in the real beam, the random deviation is small with $\delta \ll 1$. This algorithm was extended to yield appropriate bunching statistics at the nonlinear harmonics in Ref. [2]. Although this extension is not necessary to predict the amplification of the fundamental frequency, it is crucial when assessing harmonic generation schemes of the type discussed in Chapter 6.

The particle loading scheme defined by (B.8) and (B.10) initializes the phase in such a way that the initial statistics of the macroparticle bunching matches that associated with the shot noise of the true beam. However, when one considers electron beams with a finite energy spread and/or in multiple dimensions, it alone is not sufficient to correctly predict the noise that effectively seeds SASE. The basic reason for this can be understood by considering the effect of energy spread. First, we assume that the particle phases are initialized as described above, while the starting energies are assigned according to a Gaussian distributed (pseudo)-random number. Over the first few gain lengths where SASE is initialized, the electron phase θ is largely unaffected by the field, which therefore changes by an amount proportional to its energy deviation at $z = 0$. Because $\hat{\eta}_j(\hat{z} = 0)$ was loaded at random, the initial bunching will change by a significant amount if the change in phase is of order the initial deviation δ or larger. Thus, a random Gaussian load in $\hat{\eta}_j$ introduces unphysically large fluctuations in the bunching if $\sigma_{\hat{\eta}} \gtrsim \delta$, which in turn results in errors in the level of effective SASE seeding.

One simple way of addressing this problem is to initially divide the total number of simulation particles in any slice into a number of "beamlets," where each beamlet has an equal number of N_b particles and every particle in a given beamlet has the same energy [2]. The initial energy of these N_{sim}/N_b beamlets is chosen to sample the energy distribution; the particle phase is defined in an analogous manner to the Penman–McNeil scheme, only now the particles in each beam span the full 2π in phase. For integers n and ℓ, where $1 \leq n \leq N_b$ labels the particle within a beamlet and $1 \leq \ell \leq N_{\text{sim}}/N_b$ identifies the beamlet, we write

$$\theta_{n,\ell} = \frac{2\pi}{N_b}\left(n + \frac{N_b^2}{N_{\text{sim}}}\ell\right) + (2\delta) r_{n,\ell}, \tag{B.11}$$

where $r_{n,\ell}$ is a random number between 0 and 1 and δ is given by (B.10). The particles within each beamlet (i.e., at fixed ℓ) have identical energy, which implies that the single particle motion in the absence of an electric field preserves the single slice bunching initialized by (B.11). While this is not quite correct for the "actual" dynamics since the physical bunching level has small fluctuations due to the particle motion, nevertheless the initial bunching level is more faithfully represented and the SASE more accurately modeled. In fact, for initial energy spreads $\sigma_\eta \lesssim 0.5$ the initial SASE seeding is correct to within a factor of two, while we have found that the two lowest-order correlation functions (the coherence function (4.63) and the intensity correlation function) are identical within numerical accuracy over several coherence times, which is sufficient for most applications.

There are several methods for choosing the energies of the individual beamlets; for example, one can assign the energies according to Gaussian random numbers. At the other extreme, one might uniformly sample the energy distribution, which for a Gaussian implies that the spacing in energy is given by the error function erf(x); to be more specific, we use the beamlet index ℓ with $1 \leq \ell \leq N_{\text{sim}}/N_b$, and assign their energy according to

$$\frac{1}{2}\left[\text{erf}(\eta_\ell) - \text{erf}(\eta_{\ell-1})\right] = \frac{1}{N_{\text{sim}}/N_b + 1}, \tag{B.12}$$

with the first energy η_1 obtained using $\eta_0 \equiv -\infty$. A loading scheme that represents a reasonable compromise between these two is based on quasirandom numbers such as the Hammersley or Halton sequence. These quasirandom sequences fill space more uniformly than a set of uniformly distributed random numbers, and are therefore a common tool in numerical simulations; in this context they are often referred to as a "quiet start," although quiet start can also be used to refer to any of the techniques we've discussed here for reducing macroparticle noise. Similar considerations must also be made for the transverse phase space coordinates, as these degrees of freedom also couple to the phase evolution. Thus, a full "quiet start" loading scheme first determines the initial 5D coordinates of each beamlet (i.e., $(\eta_\ell, \boldsymbol{x}_\ell, \boldsymbol{p}_\ell)$ for $1 \le \ell \le N_{\text{sim}}/N_b$), and then within each beamlet loads the N_b phases over the full 2π according to (B.11). Typical simulations may have $2 \le N_b \le 16$ and $32 \le N_{\text{sim}} \lesssim 16384$ or more, depending on the number of dimensions, the size of emittance and energy spread, the need to simulate harmonics, etc. Note that this is still many fewer than the number of particles in a typical FEL slice, which is often $\sim 10^5$ to 10^6 or more.

B.2 Existing FEL Codes

The history of FEL simulation goes back over 30 years, and many codes that are no longer actively being used or developed have contributed to the present suite of codes that are available. While we will try to recognize some of the code genealogy here, we will primarily focus on describing a few of the more commonly employed FEL simulation codes presently in use.

Two of the most widely used multi-D FEL codes include GINGER [3] and GENESIS [4]. Both codes track the electron motion in 3D, and can be run in both single slice (also called "time-independent," "steady-state," or "seeded") or multi-slice mode (alternatively referred to as "time-dependent," "polychromatic," or, less precisely, "SASE"). GINGER presumes that the radiation field is axially symmetric, which is then solved for as a function of the radius on a nonuniform grid. The precursor of GENESIS, known as TDA [5] also solved for an axisymmetric radiation field, but this assumption has been relaxed in GENESIS, which is fully 3D in both the particle motion and field solver. Figures 4.5 and 5.2 show examples of GINGER and GENESIS time-dependent simulations, respectively.

Both GINGER and GENESIS numerically solve for the diffraction using a stable differencing scheme for the derivatives (the alternating direction implicit method being one such method). An alternative method based on a discretized integral solution has been implemented in RON [6] and FAST [7], which can reduce the computation time and aid design optimization. The code MEDUSA [8] uses an entirely different algorithm, known as the source-dependent expansion, to solve for the radiation field. The source-dependent expansion projects the field onto a set of orthogonal modes whose parameters are adjusted according to the equations of motion; in MEDUSA's case, the radiation is represented as a sum of Gauss–Hermite modes whose complex width, which defines the mode RMS beam size and effective Rayleigh range, are slow functions of z determined by the current source.

Finally, since one often also wants to understand, predict, and control FEL emission at harmonics of the fundamental, all the codes listed above can also compute the emission of higher harmonics. One straightforward but approximate way to do this merely computes the emission from the nonlinear components of the induced electron beam bunching without including the back-reaction of the field on the particles. This method requires little additional computational expense and is a reasonable approximation for those applications in which the harmonic power is the typical 1 percent or less of that in the fundamental. Nevertheless, all the FEL codes can also compute the self-consistent interaction of the harmonic fields with the particles, which is essential to compute any harmonic gain.

References

[1] C. Penman and B. W. J. McNeil, "Simulation of input electron noise in the free-electron laser," *Opt. Commun.*, vol. 90, p. 82, 1992.

[2] W. M. Fawley, "Algorithm for loading shot noise microbunching in multidimensional, free-electron laser simulations," *Phys. Rev. ST Accel. Beams*, vol. 5, p. 070701, 2002.

[3] —, "A user manual for GINGER and its post-processor XPLOTGIN," *LBL*, Report LBNL-49625, 2002.

[4] S. Reiche, "GENESIS 1.3: a fully 3D time-dependent FEL simulation code," *Nucl. Instrum. Methods Phys. Res., Sect. A*, vol. 429, p. 243, 1999.

[5] T.-M. Tran and J. Wurtele, "Free-electron laser simulation techniques," *Phys. Report*, vol. 195, p. 1, 1990.

[6] R. J. Dejus, O. A. Shevchenko, and N. V. Vinokurov, "An integral equation based computer code for high-gain free-electron lasers," *Nucl. Instrum. Methods Phys. Res., Sect. A*, vol. 429, p. 225, 1999.

[7] E. L. Saldin, E. A. Schneidmiller, and M. V. Yurkov, "FAST: a three-dimensional time-dependent FEL simulation code," *Nucl. Instrum. Methods Phys. Res., Sect. A*, vol. 429, p. 233, 1999.

[8] H. P. Freund, S. G. Biedron, and S. V. Milton, "Nonlinear harmonic generation in free-electron lasers," *IEEE J. Quantum Electron.*, vol. 36, p. 275, 2000.

Appendix C Quantum Considerations for the FEL

While Madey's first calculation of the FEL gain mechanism was quantum mechanical in nature [1], currently all built and planned FEL facilities are essentially classical devices. In this appendix we will give an introduction to the quantum analysis that will help clarify when quantum effects become important. Our discussion will rely on some fundamentals of quantum mechanics, but will try to introduce all the relevant notation and results in as self-contained a manner as possible. We use the Heisenberg representation that evolves the quantum operators while keeping the states fixed, and we will try to differentiate the classical numbers from quantum operators with a judicious choice of fonts. Quantum effects are typically negligible for FELs, but we will show when quantum noise in the FEL amplifier may play a role. Finally, we conclude this appendix by in some sense going back the beginning and re-deriving Madey's theorem using a quantum approach.

C.1 The Quantum Formulation

The simplest way to investigate quantum effects is to quantize the frequency domain FEL equations assuming that the field is periodic over the time T. This is applicable for long electron beams, and will serve as a quantum generalization of the cold, 1D limit we derived in Section 4.3. We begin by replacing the integrals over frequency with Fourier sums whose frequency spacing $d\omega = \omega_1 d\nu \to 2\pi/T$. This changes the classical equations for the scaled particle energy as follows:

$$\frac{d\hat{\eta}_j}{d\hat{z}} = \frac{\lambda_1}{cT}\sum_{\nu}\left(a_\nu e^{i\nu\theta_j} + c.c\right). \tag{C.1}$$

The phase equation remains $d\theta_j/d\hat{z} = \hat{\eta}_j$, while we write the dimensionless form of the classical Maxwell equation in frequency space as

$$\left(\frac{\partial}{\partial\hat{z}} + \frac{i\Delta\nu}{2\rho}\right)a_\nu = -\frac{1}{N_\lambda}\sum_j e^{-i\nu\theta_j} = -\frac{cT}{\lambda_1}\frac{1}{N_e}\sum_j e^{-i\nu\theta_j}. \tag{C.2}$$

We will find it most convenient to associate with a_ν the quantum field operator $\mathfrak{a}_\nu$ defined such that $\mathfrak{a}_\nu^\dagger\mathfrak{a}_\nu$ is the number operator for photons with scaled frequency ν ($\mathfrak{a}_\nu^\dagger$ denotes the Hermitian conjugate of $\mathfrak{a}_\nu$). Thus $\mathfrak{a}_\nu^\dagger$ and $\mathfrak{a}_\nu$ are respectively the creation and annihilation operators of a single photon with frequency $\nu\omega_1$, and $\hbar\omega\mathfrak{a}_\nu^\dagger\mathfrak{a}_\nu$ is the electromagnetic energy operator of mode ν. To identify the classical field a_ν with the quantum operator $\mathfrak{a}_\nu$, we use the expression (4.38) for $dP/d\omega$ and the definition of the dimensionless a_ν equation (3.84) to write the classical expression for the field energy as

$$U = \int d\omega\, T\frac{dP}{d\omega} \to \frac{2\pi\lambda_1}{T}\sum_\nu \frac{\epsilon_0}{\pi c}(2\pi\sigma_x^2)\left\langle |E_\nu(z)|^2\right\rangle \tag{C.3}$$

$$= \rho\gamma_r mc^2 N_e\left(\frac{\lambda_1}{cT}\right)^2\sum_\nu\left\langle |a_\nu(z)|^2\right\rangle. \tag{C.4}$$

Hence, the appropriate definition makes the quantum association

$$a_\nu \to \frac{cT}{\lambda_1}\sqrt{\frac{\hbar\omega}{N_e\rho\gamma_r mc^2}}\,\mathfrak{a}_\nu \equiv \frac{cT}{\lambda_1}\sqrt{\frac{\nu q}{N_e}}\,\mathfrak{a}_\nu. \tag{C.5}$$

Here we have defined the quantum FEL parameter $q \equiv \hbar\omega_1/\rho\gamma_r mc^2$, which is the ratio of the characteristic photon energy to the energy bandwidth of the FEL [2]. When $q \ll 1$ an electron emits $\sim 1/q \gg 1$ photons before falling out of the FEL bandwidth, and the device can be described classically. On the other hand, quantum effects become important when $q \gtrsim 1$ and the emission of a single resonant photon knocks an electron out of the amplification bandwidth.

The definition (C.5) implies that the equal "time" (i.e., $\hat{z}$) commutation relations for the field operators are

$$\left[\mathfrak{a}_\nu(\hat{z}), \mathfrak{a}^\dagger_{\nu'}(\hat{z})\right] \equiv \mathfrak{a}_\nu(\hat{z})\mathfrak{a}^\dagger_{\nu'}(\hat{z}) - \mathfrak{a}^\dagger_{\nu'}(\hat{z})\mathfrak{a}_\nu(\hat{z}) = \delta_{\nu,\nu'}. \tag{C.6}$$

The quantum version of the Maxwell equation (C.2) is

$$\left(\frac{\partial}{\partial\hat{z}} + \frac{i\Delta\nu}{2\rho}\right)\mathfrak{a}_\nu = -\frac{1}{\sqrt{\nu q N_e}}\sum_j e^{-i\nu\Theta_j}, \tag{C.7}$$

where Θ_j is the quantum operator associated with the particle phase θ_j. To determine the quantum conjugate to Θ, we recall our Hamiltonian discussion from Appendix A in which we showed that $\tau = -ct$ and $mc\gamma$ are canonical position–momentum pairs whose Poisson bracket $\{\tau_j, mc\gamma_\ell\} = \delta_{j,\ell}$. Using the fact that $\theta_j = k_1\tau_j - (k_1 + k_u)z$ and $\hat{\eta}_j \equiv (\gamma_j - \gamma_r)/\rho\gamma_r$, at fixed z we have $\{\theta, \hat{\eta}\} = \omega_1/\rho\gamma_r mc^2$. Identifying the Poisson bracket with the quantum commutator via $\{A, B\} = C \to [A, B] = (i/\hbar)C$ suggests that we should define the quantum canonical momentum via the association

$$\hat{\eta} \to \frac{\hbar\omega_1}{\rho\gamma mc^2}\mathfrak{p} \equiv q\,\mathfrak{p} \quad \text{so that} \quad [\Theta_j, \mathfrak{p}_\ell] = i\delta_{j,\ell}. \tag{C.8}$$

We can now infer the electron's quantum equations of motion from the classical counterparts. We will find it convenient, however, to first write the quantum Hamiltonian for the FEL in 1D:

$$\mathscr{H} = \sum_{j=1}^{N_e}\left[\frac{q}{2}\mathfrak{p}_j^2 + \frac{1}{\sqrt{\nu q N_e}}\sum_\nu\left(i\mathfrak{a}_\nu e^{i\nu\Theta_j} + h.c.\right)\right] + \sum_\nu\frac{\Delta\nu}{2\rho}\mathfrak{a}^\dagger_\nu\mathfrak{a}_\nu, \tag{C.9}$$

where $h.c.$ denotes Hermitian conjugate and we have implicitly assumed that the field and particle operators commute. The Heisenberg equations of motion are obtained from the Hamiltonian via

$$\frac{d}{d\hat{z}}\mathfrak{a}_\nu = i[\mathscr{H},\mathfrak{a}_\nu] = -i\frac{\Delta\nu}{2\rho}\mathfrak{a}_\nu - \frac{1}{\sqrt{\nu q N_e}}\sum_{j=1}^{N_e} e^{-i\nu\Theta_j} \tag{C.10}$$

$$\frac{d}{d\hat{z}}\Theta_j = i[\mathscr{H},\Theta_j] = q\,\mathfrak{p}_j \tag{C.11}$$

$$\frac{d}{d\hat{z}}\mathfrak{p}_j = i[\mathscr{H},\mathfrak{p}_j] = \sqrt{\frac{\nu}{qN_e}}\sum_\nu\left[a_\nu e^{i\nu\theta_j} + a_\nu^\dagger e^{-i\nu\theta_j}\right]; \tag{C.12}$$

note that these are identical to those implied by the classical–quantum correspondence that we have developed. We simplify and linearize the equations (C.10)–(C.12) using collective variables analogous to those used in Section 3.4.4. The collective variables were introduced in [3, 2] where the non-commutivity of Θ_j and $\mathfrak{p}_j$ was neglected; this omission was fixed in [4], although the correct linearized version of the quantum FEL equations were given more recently in [5, 6], etc. We introduce the bunching factor operator

$$\mathscr{B}_\nu \equiv \frac{1}{\sqrt{\nu q N_e}}\sum_{j=1}^{N_e} e^{-i\nu\Theta_j}, \tag{C.13}$$

so that the Maxwell equation is

$$\left(\frac{d}{\partial\hat{z}} + i\frac{\Delta\nu}{2\rho}\right)\mathfrak{a}_\nu = -\mathscr{B}_\nu. \tag{C.14}$$

The evolution of the bunching can be determined using the Hamiltonian (C.9) taking due care of the non-commuting operators. We find that

$$\frac{d}{d\hat{z}}\mathscr{B}_\nu = i[\mathscr{H},\mathscr{B}_\nu] = -i\nu\mathscr{P}_\nu \tag{C.15}$$

where the collective momentum

$$\mathscr{P}_\nu \equiv \frac{1}{2}\sqrt{\frac{q}{\nu N_e}}\sum_{j=1}^{N_e}\left(\mathfrak{p}_j e^{-i\nu\Theta_j} + e^{-i\nu\Theta_j}\mathfrak{p}_j\right). \tag{C.16}$$

This symmetric version of $\mathscr{P}_\nu$ is critical to identify the consistent set of linearized equations [6]. To see this, we compute the equation of motion for $\mathscr{P}$. Ignoring the non-resonant terms $\sim e^{-2i\nu\Theta_j}$, we have

$$\begin{aligned}\frac{d}{d\hat{z}}\mathscr{P}_\nu &= i[\mathscr{H},\mathscr{P}_\nu]\\ &= \mathfrak{a}_\nu - i\frac{\nu q}{4}\frac{1}{\sqrt{\nu q N_e}}\sum_{j=1}^{N_e}\left(\mathfrak{p}_j^2 e^{-i\nu\Theta_j} + e^{-i\nu\Theta_j}\mathfrak{p}_j^2 + 2\mathfrak{p}_j e^{-i\nu\Theta_j}\mathfrak{p}_j\right)\\ &= \mathfrak{a}_\nu - i\frac{\nu q}{4}\frac{1}{\sqrt{\nu q N_e}}\sum_{j=1}^{N_e}\left(\nu^2 e^{-i\nu\Theta_j} + 4\mathfrak{p}_j e^{-i\nu\Theta_j}\mathfrak{p}_j\right).\end{aligned} \tag{C.17}$$

The final term is higher order in the field variables, as can be verified by computing its equation of motion.[1] Neglecting the last term, we have

$$\frac{d}{d\hat{z}}\mathscr{P}_\nu = \mathfrak{a}_\nu - i\frac{\nu^3 q^2}{4}\mathscr{B}_\nu. \tag{C.18}$$

The Equations (C.14), (C.15), (C.18) constitute a closed set of linear equations for the quantum FEL in collective variables. Much like in our previous discussion in Chapter 3, the general solution can be written as a linear sum of modes with exponential dependence $e^{-i\mu\hat{z}}$. Writing $\mathfrak{a}_\nu \sim e^{-i\mu\hat{z}}$, etc., we find that μ must satisfy the cubic equation [5]

$$\left(\mu - \frac{\Delta\nu}{2\rho}\right)\left(\mu^2 - \frac{q^2}{4}\right) = 1, \tag{C.19}$$

where we set $\nu \to 1$ except in the detuning term, since $\Delta\nu \sim \rho \ll 1$. Quantum effects play no role in determining the growth rate when the photon energy is much smaller than the FEL energy bandwidth, i.e., when $q^2 = [\hbar\omega_1/(\rho\gamma_r mc^2)]^2 \ll 4$.

Note that Equation (C.19) is identical to the classical dispersion relation (4.51) for a flat-top energy distribution whose full-width is q; we plotted examples of the growing root's dependence on $\Delta\nu$ for several q in Figure 4.2. Here, we complete that discussion by giving the full solution for the initial value problem when μ satisfies the dispersion relation (C.19). Let the three roots be denoted μ_α with $\alpha = 1, 2, 3$. We further define

$$\Upsilon_\alpha \equiv \frac{\mu_\alpha}{(\mu_\alpha - \mu_\beta)(\mu_\alpha - \mu_\gamma)} \tag{C.20}$$

for (α, β, γ) cyclic permutations of $(1, 2, 3)$, which can be shown to satisfy the relations

$$\sum_\alpha \Upsilon_\alpha \mu_\alpha = 1 \qquad \sum_\alpha \Upsilon_\alpha = \sum_\alpha \frac{\Upsilon_\alpha}{\mu_\alpha} = 0 \tag{C.21}$$

$$\sum_\alpha \Upsilon_\alpha \mu_\alpha^2 = -\frac{\Delta\nu}{2\rho} \qquad \sum_\alpha \Upsilon_\alpha \mu_\alpha^3 = \left(\frac{\Delta\nu}{2\rho}\right)^2 + \frac{q^2}{4}. \tag{C.22}$$

Using (C.19)–(C.22), the solution of the initial value problem of the EM operator is as follows:

$$\mathfrak{a}_\nu(\hat{z}) = g(\hat{z})\mathfrak{a}_\nu(0) + f(\hat{z})\mathscr{B}_\nu(0) + h(\hat{z})\mathscr{P}_\nu(0), \tag{C.23}$$

where $\mathfrak{a}_\nu(0)$, $\mathscr{B}_\nu(0)$, $\mathscr{P}_\nu(0)$ are the initial values of the field, bunching, and collective momentum operators, and the c-number functions g, f, and h give the evolution as a sum over the three independent exponentially growing solutions:

$$g \equiv \sum_\alpha g_\alpha \equiv \sum_\alpha \Upsilon_\alpha \left(\mu_\alpha - \frac{q^2}{4\mu_\alpha}\right) e^{-i\mu_\alpha\hat{z}} \tag{C.24}$$

$$f \equiv \sum_\alpha f_\alpha \equiv -i\sum_\alpha \Upsilon_\alpha e^{-i\mu_\alpha\hat{z}} \tag{C.25}$$

$$h \equiv \sum_\alpha h_\alpha \equiv -i\sum_\alpha \frac{\Upsilon_\alpha}{\mu_\alpha} e^{-i\mu_\alpha\hat{z}}. \tag{C.26}$$

[1] The term $\sum \mathfrak{p}_j^2 e^{-i\nu\Theta_j}$ cannot be neglected since its governing equation involves linear terms; the difference arises since $\mathfrak{p}$ and Θ do not commute.

The result (C.23) is the mathematical statement that the FEL acts as a linear amplifier before saturation, with amplitude gain functions given by g, f, and h. For completeness, we also include the solutions for the collective e-beam operators:

$$\begin{aligned}\mathscr{B}_\nu(\hat{z}) = \sum_\alpha \frac{\Upsilon_\alpha}{\mu_\alpha}\left[i\mathfrak{a}_\nu(0) + \mu_\alpha\left(\mu_\alpha + \frac{\Delta\nu}{2\rho}\right)\mathscr{B}_\nu(0)\right. \\ \left. + \left(\mu_\alpha + \frac{\Delta\nu}{2\rho}\right)\mathscr{P}_\nu(0)\right]e^{-i\mu_\alpha\hat{z}}\end{aligned} \tag{C.27}$$

$$\begin{aligned}\mathscr{P}_\nu(\hat{z}) = \sum_\alpha \Upsilon_\alpha\left\{i\mathfrak{a}_\nu(0) + \frac{1}{\mu_\alpha}\left[1 + \frac{q^2}{4}\left(\mu_\alpha + \frac{\Delta\nu}{2\rho}\right)\right]\mathscr{B}_\nu(0)\right. \\ \left. + \left(\mu_\alpha + \frac{\Delta\nu}{2\rho}\right)\mathscr{P}_\nu(0)\right\}e^{-i\mu_\alpha\hat{z}}\end{aligned} \tag{C.28}$$

As we have mentioned, the quantum description is essential when the parameter $q \equiv \hbar\omega_1/\rho\gamma_r mc^2$ becomes large. In the next section we will show an example where quantum considerations can play a role even if $q \ll 1$. Specifically, we will compute the minimum noise in the FEL amplifier due to quantum effects, and determine the initial state of the beam that leads to the minimum noise.

C.2 Quantum Noise in the FEL Amplifier

In this section we compute the intrinsic FEL amplifier noise, including quantum effects. As shown in the previous section (and throughout this book), a high-gain FEL acts as a linear amplifier before saturation, with each frequency component being independently amplified. Thus, we can consider a single frequency component and drop the subscript ν without any loss of generality; the frequency index can be easily restored at the end.

We begin by presenting some results applicable to a generic linear amplifier that relies heavily on the work by Caves [7]. These noise results will then be applied to the FEL. First, the general field solution (C.23) can be divided as

$$\mathfrak{a}(\hat{z}) = g(\hat{z})\mathfrak{a}(0) + \mathscr{F}(\hat{z}), \qquad \mathscr{F}(\hat{z}) = f(\hat{z})\mathscr{B}(0) + h(\hat{z})\mathscr{P}(0). \tag{C.29}$$

The operator $\mathscr{F}$ describes the dynamics of the electron beam (the amplifying medium), while $g(\hat{z})$ is the amplitude gain. Using the (valid for all $\hat{z}$) EM field commutation relations (C.6) and the fact that the e-beam and radiation operators commute, it follows from the definition (C.29) that

$$[\mathscr{F}, \mathscr{F}^\dagger] = 1 - G(\hat{z}) \qquad G(\hat{z}) \equiv |g(\hat{z})|^2. \tag{C.30}$$

Physically, G is the FEL energy gain of the initial input field $\mathfrak{a}(0)$, and we note that any linear amplifier can be written as (C.29) where $\mathscr{F}$ satisfies (C.30). Our goal is to find the minimum noise in the FEL amplifier, which is equivalent to computing the minimum output photon number when there are no photons initially. Our results will apply to a generic linear amplifier, which mathematically amounts to finding

$$\min_\Psi \left\langle \mathfrak{a}^\dagger(\hat{z})\mathfrak{a}(\hat{z})\right\rangle_\Psi \text{ assuming that } \left\langle \mathfrak{a}^\dagger(\hat{0})\mathfrak{a}(\hat{0})\right\rangle_\Psi = 0. \tag{C.31}$$

Here $\langle\mathcal{O}\rangle_\Psi \equiv \langle\Psi|\mathcal{O}|\Psi\rangle$ denotes the expectation value of the operator $\mathcal{O}$ with respect to the initial quantum state Ψ. In the Heisenberg picture Ψ is fixed, and the above implies that this state is the vacuum of the initial EM field operators with $\mathfrak{a}(0)\big|\Psi\big\rangle = 0$ and $\big\langle\Psi\big|\mathfrak{a}^\dagger(0) = 0$.

To solve (C.31), we will find it convenient to introduce the absolute square of a general non-Hermitian operator $\mathcal{O}$:

$$|\mathcal{O}|^2 = \tfrac{1}{2}\left(\mathcal{O}^\dagger\mathcal{O} + \mathcal{O}\mathcal{O}^\dagger\right). \tag{C.32}$$

To understand why this is called the absolute square, we write

$$e^{i\phi}\mathcal{O} = \mathcal{O}_x + i\mathcal{O}_y \tag{C.33}$$

where the phase ϕ is arbitrary for now but will become important later. The two Hermitian operators $\mathcal{O}_x$ and $\mathcal{O}_y$ can be thought of as the real and imaginary parts of $e^{i\phi}\mathcal{O}$, with

$$\mathcal{O}_x = \tfrac{1}{2}\left(e^{i\phi}\mathcal{O} + e^{-i\phi}\mathcal{O}^\dagger\right) \qquad \mathcal{O}_y = \tfrac{1}{2i}\left(e^{i\phi}\mathcal{O} - e^{-i\phi}\mathcal{O}^\dagger\right). \tag{C.34}$$

It then follows that $|\mathcal{O}|^2 = \mathcal{O}_x^2 + \mathcal{O}_y^2$ as should be for an absolute square.[2] Now, the square of the EM field operator is given by

$$|\mathfrak{a}|^2 \equiv \frac{1}{2}\left(\mathfrak{a}^\dagger\mathfrak{a} + \mathfrak{a}\mathfrak{a}^\dagger\right) = \mathfrak{a}^\dagger\mathfrak{a} + \frac{1}{2}. \tag{C.35}$$

The $1/2$ above can be understood as being associated with quantum fluctuations of the vacuum. Using (C.35) and (C.29), the expectation value of the photon number can be written as

$$\begin{aligned}\left\langle\mathfrak{a}^\dagger(\hat{z})\mathfrak{a}(\hat{z})\right\rangle_\Psi &= \left\langle|\mathfrak{a}(\hat{z})|^2\right\rangle_\Psi - \frac{1}{2} = \left\langle\left|g(\hat{z})\mathfrak{a}(0) + \mathscr{F}(\hat{z})\right|^2\right\rangle_\Psi - \frac{1}{2}\\ &= \frac{1}{2}(G-1) + \left\langle\left|\mathscr{F}(\hat{z})\right|^2\right\rangle_\Psi.\end{aligned} \tag{C.36}$$

The term with the factor of $1/2$ arises due to vacuum fluctuations. Thus, the vacuum provides an effective noise source equal to 1/2 photon per mode that is amplified by the factor $G-1$; this contribution is quite substantial in high-gain devices. The other term involving the expectation value of $|\mathscr{F}|^2$ gives the noise contribution from the amplifier itself. It is well known in the laser community that these two noise sources are at best approximately equal, which gives the rule that the minimum noise in a high-gain amplifier corresponds to one photon per mode, with the vacuum contributing 1/2 photon and the amplifier medium providing the additional half.

We will re-derive this well-known result regarding the minimum noise in an amplifier, paying particular attention to the electron beam quantum state Ψ that gives minimal noise. Ultimately, we wish to compute Ψ and $\mathscr{F}$ that minimizes

$$\left\langle|\mathscr{F}|^2\right\rangle_\Psi = \frac{1}{2}\left\langle\mathscr{F}_x^2 + \mathscr{F}_y^2\right\rangle_\Psi, \tag{C.37}$$

[2] In other terminology, $\mathcal{O}_x$ ($i\mathcal{O}_y$) is the Hermitian (anti-Hermitian) part of $e^{i\phi}\mathcal{O}$.

where the "real" and "imaginary" parts of $\mathscr{F}$ are

$$\mathscr{F}_x = \frac{1}{2}\left(e^{i\phi}\mathscr{F} + e^{-i\phi}\mathscr{F}^\dagger\right) \qquad \mathscr{F}_y = \frac{1}{2i}\left(e^{i\phi}\mathscr{F} - e^{-i\phi}\mathscr{F}^\dagger\right). \tag{C.38}$$

Thus, we write the absolute square of the operator $\mathscr{F}$ as

$$\langle |\mathscr{F}|^2 \rangle_\Psi = \left\langle \mathscr{F}_x^2 + \mathscr{F}_y^2 \right\rangle_\Psi \geq 2\sqrt{\langle \mathscr{F}_x^2 \rangle_\Psi \left\langle \mathscr{F}_y^2 \right\rangle_\Psi} \tag{C.39}$$

$$\geq 2\left|\langle \mathscr{F}_x \mathscr{F}_y \rangle_\Psi\right|, \tag{C.40}$$

where the first line follows since the algebraic mean of two real numbers is greater than or equal to their geometric mean, while the second line applies the Cauchy–Schwarz inequality. The equality in Equation (C.39) applies when

$$\left\langle \mathscr{F}_x^2 \right\rangle_\Psi = \left\langle \mathscr{F}_y^2 \right\rangle_\Psi, \tag{C.41}$$

while equality in (C.40) requires that

$$\mathscr{F}_x|\Psi\rangle = \alpha \mathscr{F}_y|\Psi\rangle \tag{C.42}$$

for some constant α. Since we now have $\langle |\mathscr{F}|^2 \rangle_\Psi \geq 2\left|\langle \mathscr{F}_x \mathscr{F}_y \rangle_\Psi\right|$, we consider

$$\begin{aligned}
\left|\langle \mathscr{F}_x \mathscr{F}_y \rangle_\Psi\right|^2 &= \frac{1}{4}\left|\langle \mathscr{F}_x \mathscr{F}_y + \mathscr{F}_y \mathscr{F}_x + [\mathscr{F}_x, \mathscr{F}_y] \rangle_\Psi\right|^2 \\
&= \frac{1}{4}\left|\langle \mathscr{F}_x \mathscr{F}_y + \mathscr{F}_y \mathscr{F}_x \rangle_\Psi\right|^2 + \frac{1}{4}\left|\langle [\mathscr{F}_x, \mathscr{F}_y] \rangle_\Psi\right|^2 \\
&\geq \frac{1}{4}\left|\langle [\mathscr{F}_x, \mathscr{F}_y] \rangle_\Psi\right|^2,
\end{aligned} \tag{C.43}$$

where equality now requires

$$\langle \mathscr{F}_x \mathscr{F}_y + \mathscr{F}_y \mathscr{F}_x \rangle_\Psi = 0. \tag{C.44}$$

Finally, we have $[\mathscr{F}_x, \mathscr{F}_y] = i(1 - G)/2$ from the commutation relation (C.30), so that in view of (C.40) and (C.43) we derive

$$\left\langle \left|\mathscr{F}^2\right| \right\rangle_\Psi \geq \frac{1}{2}|G - 1|. \tag{C.45}$$

Inserting this into the photon number relationship (C.36) leads to the inequality $\langle \mathfrak{a}^\dagger \mathfrak{a} \rangle_\Psi \geq (G - 1 + |G - 1|)/2$, which for positive gain $G \geq 1$ implies that

$$\left\langle \mathfrak{a}^\dagger(\hat{z})\mathfrak{a}(\hat{z}) \right\rangle_\Psi \geq G - 1. \tag{C.46}$$

This inequality is slightly stronger than that given by Caves [7], showing in particular that the minimum noise vanishes in the limit of no gain $G = 1$. Furthermore, comparing (C.46) and (C.36) proves the statement that we made earlier: in the high-gain limit $G \gg 1$, the amplifier produces a noisy signal with an average of G photons per mode. Said another way, the minimum input noise signal is one photon per mode, half of which comes from the amplifier, and half of which comes from the vacuum.

We have shown that the input state with minimal amplifier noise must satisfy (C.41), (C.42), and (C.44). The first and last of these restricts the constant α in (C.42) such that

$(\mathscr{F}_x \pm i\mathscr{F}_y)|\Psi\rangle = 0$. This in turn implies that either $\mathscr{F}|\Psi\rangle = 0$, which leads to the trivial zero photon result $\langle \mathfrak{a}^\dagger(\hat{z})\mathfrak{a}(\hat{z})\rangle_\Psi = 0$, or that

$$\mathscr{F}^\dagger|\Psi\rangle = 0. \tag{C.47}$$

It is straightforward to show that Ψ satisfying (C.47) leads to the minimum number of output noise photons given by (C.46):

$$\begin{aligned}\left\langle \mathfrak{a}^\dagger(\hat{z})\mathfrak{a}(\hat{z})\right\rangle_\Psi &= \frac{G-1}{2} + \frac{1}{2}\left\langle \mathscr{F}\mathscr{F}^\dagger + \mathscr{F}^\dagger\mathscr{F}\right\rangle_\Psi \\ &= \frac{G-1}{2} + \frac{1}{2}\left\langle 2\mathscr{F}\mathscr{F}^\dagger + [\mathscr{F}^\dagger, \mathscr{F}]\right\rangle_\Psi = G-1\end{aligned} \tag{C.48}$$

if $\mathscr{F}^\dagger|\Psi\rangle = 0$.

Now, we apply these general results to determine the input state Ψ that gives the minimal noise for the FEL; specifically, we compute the initial quantum distribution of the e-beam that satisfies $\mathscr{F}^\dagger|\Psi\rangle = 0$. To simplify the calculation, we assume that quantum effects are small, writing the quantum phase variable at $\hat{z} = 0$ as

$$\Theta_j(0) = \theta_j^c + \tilde{\Theta}_j, \tag{C.49}$$

where θ_j^c is a c-number denoting the initial classical position while $\tilde{\Theta}_j$ is the quantum correction that we will treat as small (we will specify what this means later). Assuming further that the momentum $\mathfrak{p}$ is small, we drop second-order terms in $\tilde{\Theta}$ and $\mathfrak{p}$ from the bunching and collective momentum to find that

$$\mathscr{B}(0) \approx \frac{i}{\sqrt{qN_e}}\sum_{j=1}^{N_e} e^{-i\theta_j^c}\left(1 - i\tilde{\Theta}_j\right) \qquad \mathscr{P}(0) \approx \sqrt{\frac{q}{N_e}}\sum_{j=1}^{N_e} e^{-i\theta_j^c}\mathfrak{p}_j. \tag{C.50}$$

We will see that the expansion (C.50) is valid (and quantum corrections are small) if $\sqrt{q} \ll 1$. Subsequent values of $\mathscr{B}$ and $\mathscr{P}$ are obtained by multiplying the initial values by the gain functions $g(\hat{z})$ and $h(\hat{z})$, respectively. While we gave the full solution in (C.23)–(C.26), here we will significantly simplify the calculation by assuming that quantum recoil can be neglected ($q \ll 1$) and by restricting ourselves to the high-gain limit, $\hat{z} \gg 1$ (or $G \gg 1$). In this case only the growing solution with $\mu_3 = (-1 + i\sqrt{3})/2$ is relevant, so that $f \to f_3 = -ie^{-i\mu_3\hat{z}}/\mu_3$ and $h \to h_3 = -ie^{-i\mu_3\hat{z}}/\mu_3^2$. Then, the electron beam operator is given by

$$\mathscr{F} = \mathscr{F}_C + \mathscr{F}_Q; \qquad \mathscr{F}_C = -\frac{e^{-i\mu_3\hat{z}}}{3\mu_3\sqrt{qN_e}}\sum_{j=1}^{N_e} e^{-i\theta_j^c} \tag{C.51}$$

$$\mathscr{F}_Q = -\frac{e^{-i\mu_3\hat{z}}}{3\mu_3\sqrt{qN_e}}\sum_{j=1}^{N_e} e^{-i\theta_j^c}\left(\tilde{\Theta}_j + \frac{iq}{\mu_3}\mathfrak{p}_j\right), \tag{C.52}$$

where we have divided $\mathscr{F}$ into a sum of a purely classical part $\mathscr{F}_C$ and the part involving quantum effects $\mathscr{F}_Q$. Requiring the quantum part to satisfy the minimum noise condition (C.47) implies that

$$\left(\tilde{\Theta}_j - \frac{iq}{\mu_3^*}\mathfrak{p}_j\right)|\Psi\rangle. \tag{C.53}$$

In the $\tilde{\theta}$-representation the momentum operator $\mathfrak{p}_j = -i\partial/\partial\tilde{\theta}$, so that the solution to (C.53) is the quantum wavefunction

$$\psi(\tilde{\theta}_j) = \frac{1}{\sqrt{2\pi q}} e^{\mu_3^* \tilde{\theta}_j^2/2q} = \frac{1}{\sqrt{2\pi q}} \exp\left(-\frac{1+i\sqrt{3}}{4q}\tilde{\theta}_j^2\right), \tag{C.54}$$

where $\tilde{\theta}$ is a c-number coordinate. Thus, the wavefunction of each particle is a Gaussian function centered about its classical position θ_j^c with RMS width equal to $\sqrt{q}$. To understand the quantum phase space distribution, we compute the Wigner function

$$\begin{aligned} W(\tilde{\theta},p) &= \int d\xi \;\psi^*(\tilde{\theta}+\tfrac{\xi}{2})\psi^*(\tilde{\theta}-\tfrac{\xi}{2})e^{-i\xi p} \\ &= \exp\left(-\frac{2}{q}\tilde{\theta}^2 - 2\sqrt{3}\tilde{\theta}p - 2qp^2\right). \end{aligned} \tag{C.55}$$

The distribution is a correlated Gaussian in phase space, with the 1-σ curve given by the tilted ellipse described by

$$\frac{4}{q}\left(\tilde{\theta}^2 + q\sqrt{3}\tilde{\theta}p + q^2p^2\right) = 1. \tag{C.56}$$

The Gaussian distribution (C.55) has an RMS width in $\tilde{\theta}$ equal to $\sqrt{q}$ and an RMS width in p equal to $1/\sqrt{q}$. Since the operator $\mathscr{F}_Q$ in (C.50) is proportional to $\tilde{\Theta}$ and $q\mathfrak{p}$, we find that quantum effects are small provided $\sqrt{q} \ll 1$.

As a final check, we compute the noise term $\langle \mathscr{F}_Q^\dagger \mathscr{F}_Q \rangle_\Psi$. The wavefunction of electron beam is a product of the single particle wavefunctions, so that the state $|\Psi\rangle$ in the $\tilde{\theta}$ representation is

$$\langle \tilde{\theta}|\Psi\rangle = \prod_{j=1}^{N_e} \psi(\tilde{\theta}_j). \tag{C.57}$$

Using the high-gain $q \ll 1$ result $|g_3|^2 = |h_3|^2 = |f_3|^2 = G$, the quantum part contributes the noise signal

$$\begin{aligned} \frac{1}{2}\left\langle \mathscr{F}_Q^\dagger \mathscr{F}_Q \right\rangle_\Psi &= \frac{G}{2qN_e}\left\langle \Psi \left| \sum_{j,\ell=1}^{N_e} e^{i(\theta_j^c - \theta_\ell^c)}\left(\tilde{\Theta}_j - \frac{iq}{\mu_3^*}\mathfrak{p}_j\right)\left(\tilde{\Theta}_\ell + \frac{iq}{\mu_3}\mathfrak{p}_\ell\right)\right| \Psi \right\rangle \\ &= \frac{G}{2q}\int d\tilde{\theta}\;\psi^*(\tilde{\theta})\left(\tilde{\theta}^2 - q^2\frac{\partial^2}{\partial\tilde{\theta}^2} + \frac{q\tilde{\theta}}{\mu_3}\frac{\partial}{\partial\tilde{\theta}} - \frac{q}{\mu_3^*}\frac{\partial}{\partial\tilde{\theta}}\tilde{\theta}\right)\psi(\tilde{\theta}) \\ &= \frac{G}{2}, \end{aligned} \tag{C.58}$$

where the second line follows since $\psi(\tilde{\theta})$ is an even function so that terms with $j \neq \ell$ vanish. Note that the expectation value of the first term (equal to $\tilde{\Theta}^2$) and second term (given by $q^2\mathfrak{p}^2/|\mu_3|^2$) both equal q for a total of $2q$, while the sum of the third and fourth terms is $-q$. The output noise (C.58) is in agreement with (C.45) when $G \gg 1$.

The amplifier noise will be larger if the initial state differs from (C.54). For example, consider the simple Gaussian function $\psi(\tilde{\theta}) = e^{-\tilde{\theta}^2/2\sigma_\theta^2}/\sqrt{2\pi}\sigma_\theta$. Then, the noise signal becomes

$$\frac{1}{2}\left\langle \mathscr{F}_Q^\dagger \mathscr{F}_Q \right\rangle_\Psi = \frac{G}{2}\left[\frac{1}{q}\left(\sigma_\theta^2 + \frac{q^2}{4\sigma_\theta^2} + \frac{1}{2}\right)\right], \tag{C.59}$$

where the last $1/2$ is the sum of the third and fourth terms in the integral above (C.58). The minimum value has $\frac{1}{2}\langle \mathscr{F}_Q^\dagger \mathscr{F}_Q \rangle_\Psi = 3G/2$, which occurs when $\sigma_\theta^2 = q/2$. Equation (C.59) is the same as that obtained by [4], where it was argued that one should set $\sigma_\theta^2 = 2\pi N_u \rho q$ to minimize the positional variance throughout the undulator. Regardless, for the intrinsic quantum noise contribution to be dominant, we must also require that the classical average $\langle |\mathscr{F}_C|^2 \rangle = 0$ over the frequencies of interest, which can be satisfied, for example, if the classical positions are uniformly spaced.

We conclude this section by computing the full FEL power spectral density, i.e., the quantum extension of the cold 1D formula (4.47). To do this in a more general manner we will restore the frequency labels and include an input radiation signal. The latter implies that the wavefunction Ψ may include some initial number of photons, in which the extension of the photon number (C.36) is straightforward if we note that expectation values involving a single field or amplifier operator vanish. We have

$$\begin{aligned}\left\langle \mathfrak{a}_\nu^\dagger(\hat{z})\mathfrak{a}_\nu(\hat{z}) \right\rangle_\Psi &= G\left\langle \mathfrak{a}_\nu^\dagger(0)\mathfrak{a}_\nu(0) \right\rangle_\Psi + \frac{G-1}{2} \\ &\quad + \left\langle \left|\mathscr{F}_{\nu,C}\right|^2 \right\rangle + \left\langle \left|\mathscr{F}_{\nu,Q}\right|^2 \right\rangle_\Psi,\end{aligned} \tag{C.60}$$

where absolute square of the quantum operator $\mathcal{F}_Q$ obeys the inequality (C.45), while the classical part is an ensemble average of

$$\left|\mathscr{F}_{\nu,C}\right|^2 = \frac{e^{\sqrt{3}\hat{z}}}{9qN_e}\sum_{j,\ell}^{N_e} e^{-i\nu(\theta_j^c - \theta_\ell^c)} = \frac{G}{qN_e}\sum_{j,\ell}^{N_e} e^{-i\nu(\theta_j^c - \theta_\ell^c)}. \tag{C.61}$$

The first line in Equation (C.60) contains the coherent amplification and noise term due to vacuum fluctuations, while the second line has contributions from classical and quantum amplifier noise, respectively. To find the power spectral density from the number of photons per mode (C.60), recall that each mode has an energy $\hbar\omega \approx \hbar\omega_1$ and that the quantized mode spacing is $d\omega = 2\pi/T$. Hence

$$\frac{dP}{d\omega} = \frac{\hbar\omega_1 \langle \mathfrak{a}_\nu^\dagger(\hat{z})\mathfrak{a}_\nu(\hat{z})\rangle_\Psi / T}{2\pi/T} = \frac{\hbar\omega_1}{2\pi}\left\langle \mathfrak{a}_\nu^\dagger(\hat{z})\mathfrak{a}_\nu(\hat{z}) \right\rangle_\Psi, \tag{C.62}$$

and using $q^{-1} \equiv \rho\gamma_r mc^2/\hbar\omega_1$, we have

$$\frac{dP}{d\omega} = G\left\{ \left.\frac{dP}{d\omega}\right|_0 + \frac{\rho\gamma_r mc^2}{2\pi}\left[\left\langle \frac{1}{N_e}\sum_{j,\ell}^{N_e} e^{-i\nu(\theta_j^c - \theta_\ell^c)} \right\rangle + q\left(\frac{1}{2} + \left\langle \left|\mathscr{F}_{\nu,Q}\right|^2 \right\rangle_\Psi\right)\right]\right\} \tag{C.63}$$

$$\geq G\left\{\left.\frac{dP}{d\omega}\right|_0 + \frac{\rho\gamma_r mc^2}{2\pi}\left[\left\langle\frac{1}{N_e}\sum_{j,\ell}^{N_e} e^{-i\nu(\theta_j^c-\theta_\ell^c)}\right\rangle + q\right]\right\}. \tag{C.64}$$

If the classical distribution is randomly distributed in phase as we have assumed thus far, the classical (shot noise) contribution is $1/q \gg 1$ times larger than the quantum noise. If, on the other hand, the beam distribution is prepared in such a way that the classical sum vanishes (for example, by adjusting the phases to be equidistant), than the quantum noise will be the only seeding source for SASE. For X-ray FELs, q is typically a few tenths of one percent or smaller.

C.3 Madey's Theorem

We first derived Madey's theorem for the simple low-gain FEL in Section 3.3. While that calculation used entirely classical physics, the first derivation of Madey's theorem, like that of the FEL, was based on quantum mechanics. Here we include a simple proof of Madey's theorem in the quantum setting. To begin, we write out the net energy production as the difference between the rate of emission and that of absorption. We use Dirac's bra-ket notation to write $|n,p\rangle$ as the state with n photons and the electron momentum p, and label the momentum states after emission (absorption) as p_e (p_a). For a low-gain device where the initial quantum state depends only on the initial electron energy we have

$$\begin{aligned}\frac{dU}{d\omega} &\propto N_e\hbar\omega\left(\left|\langle n+1,p_e|(\mathfrak{a}\mathfrak{J}^\dagger + \mathfrak{a}^\dagger\mathfrak{J})|n,p\rangle\right|^2\right.\\ &\qquad\left.-\left|\langle n-1,p_a|(\mathfrak{a}\mathfrak{J}^\dagger + \mathfrak{a}^\dagger\mathfrak{J})|n,p\rangle\right|^2\right)\end{aligned} \tag{C.65}$$

$$= N_e\hbar\omega\left(\left|\langle n+1|\mathfrak{a}^\dagger|n\rangle\langle p_e|\mathfrak{J}|p\rangle\right|^2 - \left|\langle n-1|\mathfrak{a}|n\rangle\langle p_a|\mathfrak{J}^\dagger|p\rangle\right|^2\right). \tag{C.66}$$

Here, the matrix element for the interaction Hamiltonian involves the radiation operators $\mathfrak{a}$ and $\mathfrak{a}^\dagger$ and current operators $\mathfrak{J}$ and $\mathfrak{J}^\dagger$; in the 1D model from the previous section $\mathfrak{J} = \mathscr{B}$.

To determine the matrix elements, we recall that the creation and annihilation operators for the EM field act on a state with n photons via $\mathfrak{a}^\dagger|n\rangle \propto |n+1\rangle$ and $\mathfrak{a}|n\rangle \propto |n-1\rangle$. From these and the commutation rule $[\mathfrak{a}, \mathfrak{a}^\dagger] = 1$ it follows that $\mathfrak{a}^\dagger\mathfrak{a}$ is the photon number operator with $\mathfrak{a}^\dagger\mathfrak{a}|n\rangle = n|n\rangle$, and furthermore that

$$\langle n+1|\mathfrak{a}^\dagger|n\rangle = \sqrt{n+1} \qquad \langle n-1|\mathfrak{a}|n\rangle = \sqrt{n}. \tag{C.67}$$

Inserting this into (C.66), we write the energy per unit frequency as

$$\frac{dU}{d\omega} = N_e\hbar\omega\left[(n+1)\Gamma_e - n\Gamma_a\right] = N_e\hbar\omega\left[n(\Gamma_e - \Gamma_a) + \Gamma_e\right], \tag{C.68}$$

where the emission rate $\Gamma_e \propto |\langle p_e|\mathfrak{J}|p\rangle|^2$ and the absorption rate $\Gamma_a \propto |\langle p_a|\mathfrak{J}^\dagger|p\rangle|^2$. The second term in (C.68) corresponds to the usual rate for spontaneous emission in the undulator; the first scales with the number of photons n and gives the contribution due to stimulated emission associated with FEL gain.

The emission rates for the gain depend on the electron energy of the final state, $\Gamma_e = \Gamma(\gamma_e)$ and $\Gamma_a = \Gamma(\gamma_a)$, so that Equation (C.68) can be simplified by considering the relationship between the energies γ_e and γ_a. Hence, we write out the two-vector momentum of the highly relativistic electron moving in the wiggler field and that of the photon as

$$p = mc\left(\gamma, \gamma - \frac{1+K^2/2}{2\gamma}\right), \qquad \hbar\omega/c = \hbar(k,k). \tag{C.69}$$

The electron and radiation field interact via the exchange of a virtual undulator photon with zero frequency (its static device) and wavevector magnitude k_u; for emission, the two-vector undulator momentum is

$$q_e = \hbar(0, -k_u), \tag{C.70}$$

and imposing energy-momentum conservation $p + q_e = p_e + k$ yields

$$mc^2\gamma = mc^2\gamma_e + \hbar\omega \tag{C.71}$$

$$mc\left(\gamma - \frac{1+K^2/2}{2\gamma}\right) - \hbar k_u = mc\left(\gamma_e - \frac{1+K^2/2}{2\gamma_e}\right) + \hbar\omega/c \tag{C.72}$$

for the emission of one photon. Solving this system of equations, we find that the emission frequency

$$\omega_e = \frac{2\gamma^2 ck_u}{(1+K^2/2) + 2\gamma\hbar k_u/mc} \approx \omega_r\left(1 - \frac{\hbar\omega_r}{\gamma mc^2}\right), \tag{C.73}$$

where $\omega_r \equiv 2\gamma^2 ck_u/(1+K^2/2)$ satisfies the classical FEL resonance condition, and we have assumed that the photon energy is much less than the electron energy, $\hbar\omega_r \ll \gamma mc^2$. Note the leading order quantum correction to the classical result, which arises due to the electron recoil during emission, scales as the ratio of the photon energy to e-beam energy.

The absorption process changes the sign of the undulator momentum, so that repeating momentum conservation with $p + q_a + k = p_a$ leads to the same equations but with $k_u \to -k_u$ and $\omega \to -\omega$. Hence, we find that

$$\omega_a = \frac{2\gamma^2 ck_u}{(1+K^2/2) - 2\gamma\hbar k_u/mc} \approx \omega_r\left(1 + \frac{\hbar\omega_r}{\gamma mc^2}\right). \tag{C.74}$$

Comparing (C.73) and (C.74), we see that $\omega_a(\gamma) \approx \omega_e(\gamma + \hbar\omega/mc^2)$, which in turn implies that $\Gamma_a(\gamma) = \Gamma_e(\gamma + \hbar\omega/mc^2)$. We can derive this more formally by considering the relationship of the probability amplitude for spontaneous emission to its matrix element:

$$\Gamma_e = \Gamma(\gamma, \omega, q) \propto |\langle p_e|\mathcal{J}|p\rangle|^2. \tag{C.75}$$

On the other hand, the matrix element for absorption is

$$\left|\langle p_a|\mathcal{J}^\dagger|p\rangle\right|^2 = \left|\langle p|\mathcal{J}|p_a\rangle^*\right|^2 = |\langle p|\mathcal{J}|p_a\rangle|^2 \propto \Gamma(\gamma + \hbar\omega/mc^2, \omega, q). \tag{C.76}$$

We now use (C.75) and (C.76) to write out the difference in the rates of emission and absorption as

$$\Gamma_e - \Gamma_a = \Gamma(\gamma mc^2) - \Gamma(\gamma mc^2 + \hbar\omega) \approx -\frac{\hbar\omega}{mc^2}\frac{d\Gamma}{d\gamma}, \tag{C.77}$$

which in turn implies that emitted energy per unit frequency is

$$\frac{dU}{d\omega} = N_e \hbar\omega \left(-\frac{n\hbar\omega}{mc^2}\frac{d\Gamma}{d\gamma} + \Gamma_e \right). \tag{C.78}$$

Equation (C.78) makes apparent the physical content of Madey's theorem: the change in energy (and gain) is proportional to the derivative of the spontaneous emission spectrum with respect to the electron energy. To reproduce the classical formula, we identify the energy change with that due to stimulated emission,

$$\langle \Delta\gamma \rangle = \frac{\Delta\omega}{N_e mc^2}\frac{dU_{\rm stim}}{d\omega} = -\frac{2\pi}{T}\frac{\hbar\omega}{N_e mc^2}\frac{n\hbar\omega}{mc^2}\frac{d\Gamma}{d\gamma}, \tag{C.79}$$

where for the single mode external field $\Delta\omega = 2\pi/T$. Now, replace the photon energy $n\hbar\omega$ and the emission rate with the previously introduced classical EM energy and 1D spectrum

$$n\hbar\omega = \frac{\epsilon_0}{2}\,|E_0|^2\,(2\pi\sigma_x^2 cT) \tag{C.80}$$

$$\Gamma = \frac{T}{\hbar\omega}\Delta\phi_{1D}\,\frac{dP}{d\omega d\boldsymbol{\phi}}\bigg|_{\boldsymbol{\phi}=0} = \frac{T}{\hbar\omega}\frac{\lambda^2}{2\pi\sigma_x^2}\frac{dP}{d\omega d\boldsymbol{\phi}}\bigg|_{\boldsymbol{\phi}=0}. \tag{C.81}$$

When we insert (C.80)–(C.81) into (C.79), the classical formula equation (3.41) falls right out:

$$\langle \Delta\gamma \rangle = -\frac{\pi\lambda^2 cT}{N_e}\frac{\epsilon_0\,|E_0|^2}{(mc^2)^2}\frac{\partial}{\partial\gamma}\frac{dP}{d\omega d\boldsymbol{\phi}}\bigg|_{\boldsymbol{\phi}=0}. \tag{C.82}$$

References

[1] J. M. J. Madey, "Stimulated emission of bremsstrahlung in a periodic magnetic field," *J. Appl. Phys.*, vol. 42, p. 1906, 1971.

[2] R. Bonifacio and F. Casagrande, "Instability threshold, quantum initiation and photon statistics in high-gain free electron lasers," *Nucl. Instrum. Methods Phys. Res., Sect. A*, vol. 237, p. 168, 1985.

[3] —, "Instabilities and quantum initiation in the free-electron laser," *Opt. Commun.*, vol. 50, p. 251, 1984.

[4] C. Schroeder, C. Pelligrini, and P. Chen, "Quantum effects in high-gain free-electron lasers," *Phys. Rev. E*, vol. 64, p. 056502, 2001.

[5] R. Bonifacio, N. Piovella, and G. Robb, "Quantum theory of SASE FEL," *Nucl. Instrum. Methods Phys. Res., Sect. A*, vol. 543, p. 645, 2005.

[6] R. Bonifacio, N. Piovella, G. Robb, and A. Schiavi, "Quantum regime of free electron lasers starting from noise," *Phys. Rev. ST Accel. Beams*, vol. 9, p. 090701, 2006.

[7] C. M. Caves, "Quantum limits on noise in linear amplifiers," *Phys. Rev. D*, vol. 26, p. 1817, 1982.

Appendix D Transverse Gradient Undulators

FEL gain requires a resonant interaction between the electrons and the radiation field. We have seen that this occurs when the resonance condition

$$\lambda_1 = \lambda_u \frac{1 + K^2/2}{2\gamma^2} \tag{D.1}$$

is approximately satisfied for most of the particles in the electron beam. For this reason, traditional FELs require e-beams with very small energy spreads ($\sigma_\eta \ll 1/N_u$ for low-gain devices while $\sigma_\eta \ll \rho$ for high-gain FELs). In Ref. [1], Smith and collaborators proposed a scheme for reducing the energy spread requirement by designing the undulator field to vary transversely (see Figure D.1). If one also sorts the incoming electrons such that those with higher energy see a stronger field than those with lower energy, one could in principle satisfy the resonance condition (D.1) for all electrons, regardless of the initial energy spread. While transverse gradient undulators were originally proposed for low-gain FELs, use of these devices has received renewed interest in high-gain applications including their potential to increase the FEL gain of large energy spread beams such as those generated from laser plasma accelerators [2], and for enhancing harmonic generation efficiency in a laser-seeded FEL [3].

To show how a transverse gradient undulator (TGU) might work, we consider the TGU in Figure D.1 whose undulator gap decreases along x. In this case the resulting B-field (and, hence, K parameter) is an increasing function of x, so that in order to satisfy (D.1) everywhere one would ideally position the electrons so that their energy is a monotonically increasing function of x. Specifically, for each electron indexed by j we would like

$$\lambda_1 = \lambda_u \frac{1 + K^2(\bar{x}_j)/2}{2\gamma_j^2}, \tag{D.2}$$

where now the resonance condition depends on both the energy γ_j and the average particle position $\bar{x}_j$ due to the varying magnetic field strength. We try to satisfy (D.2) for all electrons by introducing some dispersion upstream of the undulator that correlates the electron's energy and position according to

$$\bar{x}_j = D\eta_j + x_{\beta_j}. \tag{D.3}$$

The linear transformation (D.3) implies that x_β is the initial particle position before the dispersion section; in what follows we consider x_β to be the dynamical variable of

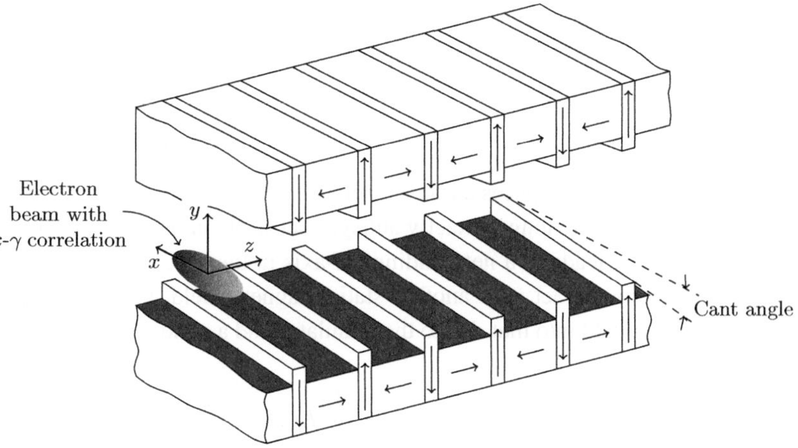

Figure D.1 Schematic of a transverse gradient undulator. Here, the field strength K increases along x, so that the energy spread requirement can be mitigated if the particle γ is an increasing function of x such that (D.3) is always satisfied.

interest and assume that $|k_u x_\beta| \ll 1$. We can now use the definition (D.3) to eliminate the energy in the resonance condition (D.2) to find that

$$\lambda_1 = \lambda_u \frac{1 + K^2(\bar{x}_j)/2}{2\gamma_r^2\left[1 + (\bar{x}_j - x_{\beta j})/D\right]^2} \approx \lambda_u \frac{1 + K^2(\bar{x}_j)/2}{2\gamma_r^2(1 + \bar{x}_j/D)^2}. \tag{D.4}$$

Thus, the TGU field profile required for a given dispersion D is

$$\frac{1 + K^2(\bar{x})/2}{(1 + \bar{x}/D)^2} = 1 + K_0^2/2, \tag{D.5}$$

where K_0 is the nominal undulator strength in the absence of any transverse variation, and we have $k_u = k_1(1 + K_0^2/2)/2\gamma_r^2$.

Assuming that the undulator field obeys (D.5) and that $\eta = (\bar{x} - x_\beta)/D$, the ponderomotive phase equation becomes

$$\frac{d\theta}{dz} = k_u - k_1 \frac{1 + K^2(\bar{x})/2}{2\gamma_r^2(1 + \eta)^2} \approx k_u - k_1 \frac{1 + K^2(\bar{x})/2}{2\gamma_r^2(1 + \bar{x}/D)^2}\left(1 + 2x_\beta/D\right)$$
$$= -\frac{2k_u}{D} x_\beta, \tag{D.6}$$

and we find that a transverse gradient effectively replaces the energy deviation η in the phase equation with the quantity $-x_\beta/D$. For this reason, we now consider the motion along x using the Lorentz force equation

$$\frac{d}{dt}\left(\gamma m \frac{dx}{dt}\right) = -e\left[B_z \frac{dy}{dt} - B_y \frac{dz}{dt}\right]. \tag{D.7}$$

To simplify the analysis, we assume that $|d\boldsymbol{x}/dt| \ll 1$, which permits us to approximate $dz/dt \approx c$ for highly relativistic electrons and to drop the force $\sim B_z(dy/dt)$ assuming that $B_z = 0$ along the axis. If we additionally assume that $d\gamma/dt \ll \gamma$ and switch the dependent variable to z, the Lorentz force law (D.7) simplifies to

$$\frac{d^2x}{dz^2} \approx \frac{eB_y}{mc\gamma}. \tag{D.8}$$

Since the TGU imposes a transversely varying undulator field, the oscillating motion of an electron is no longer exactly antisymmetric about one half-period of the motion. This asymmetry leads to a net bending off-axis in the absence of any correction, which we wish to counteract with another static magnetic field. Hence, we decompose the total magnetic field into that due to the undulator and a z-independent externally applied field as

$$B_y \approx -\frac{mck_u}{e}K(x)\sin(k_u z) + B_s(x). \tag{D.9}$$

To reduce (D.8) to an equation for x_β, we first separate the transverse position into its wiggle motion due to the undulator and the slowly evolving oscillation center as

$$x = x_w + \bar{x} \equiv \frac{K(\bar{x})}{k_u\gamma}\sin(k_u z) + \bar{x}. \tag{D.10}$$

We insert the magnetic field (D.9) and the coordinate (D.10) into the force equation (D.8), expand assuming $|x_w| \sim K/\gamma \ll 1$, and average over an undulator period to derive the following averaged equation of motion

$$\frac{d^2\bar{x}}{dz^2} = -\frac{1}{2\gamma^2}K(\bar{x})\frac{\partial K}{\partial \bar{x}} + \frac{e}{mc\gamma}B_s(\bar{x}). \tag{D.11}$$

We now introduce the betatron coordinate using (D.3) and $\gamma = \gamma_r(1+\eta) = \gamma_r[1+(\bar{x} - x_\beta)/D]$ to find that

$$\begin{aligned}\frac{d^2\bar{x}}{dz^2} &= -\frac{K(\bar{x})(\partial K/\partial \bar{x})}{2\gamma_r^2[1+(\bar{x}-x_\beta)/D]^2} + \frac{eB_s(\bar{x})}{mc\gamma_r[1+(\bar{x}-x_\beta)/D]} \\ &\approx -\frac{K(\bar{x})(\partial K/\partial \bar{x})}{2\gamma_r^2(1+\bar{x}/D)^2}\left(1+\frac{2x_\beta}{D+\bar{x}}\right) + \frac{eB_s(\bar{x})}{mc\gamma_r(1+\bar{x}/D)}\left(1+\frac{x_\beta}{D+\bar{x}}\right).\end{aligned}$$

The constant force can be canceled if we choose

$$B_s(\bar{x}) = \frac{mc}{e}\frac{K(\bar{x})(\partial K/\partial \bar{x})}{2\gamma_r(1+\bar{x}/D)}. \tag{D.12}$$

Additionally, we use the TGU condition (D.5) to derive that $K(\bar{x})(\partial K/\partial \bar{x}) = (2 + K_0^2)(D+\bar{x})/D^2$, so that the transverse betatron equation of motion is

$$\begin{aligned}x_\beta'' = -\frac{2+K_0^2}{2D^2\gamma_r^2(1+\bar{x}/D)^2}x_\beta - D\eta'' &\approx -\frac{2+K_0^2}{2D^2\gamma^2}x_\beta - D\eta'' \\ &\equiv -\frac{k_\beta^2}{(1+\eta)^2}x_\beta - D\eta''.\end{aligned} \tag{D.13}$$

Here, we have defined the nominal betatron frequency

$$k_\beta \equiv \frac{\sqrt{1+K_0^2/2}}{D\gamma_r}. \tag{D.14}$$

In the absence of the FEL interaction $\eta' = 0$ and the betatron motion is purely oscillatory with a frequency that depends on energy. On the other hand, energy exchange with the field acts as a source for transverse oscillations.

As a concrete example, we consider the simplest transverse gradient undulator field that satisfies the vacuum Maxwell equations [4]:

$$\begin{aligned} B = &-B_0(1+\alpha x)\cosh(k_u y)\sin(k_u z)\hat{y} \\ &- B_0(1+\alpha x)\sinh(k_u y)\cos(k_u z)\hat{z} - B_0\frac{\alpha}{k_u}\sinh(k_u y)\sin(k_u z)\hat{x}. \end{aligned} \tag{D.15}$$

This magnetic field cannot exactly satisfy the TGU phase condition (D.5) in general, but rather only works in the limit of large $K_0^2 \gg 2$ or small transverse displacement, $|\bar{x}(\partial K/\partial \bar{x})| = |\alpha \bar{x}| \ll 1$. Typically one or both of these conditions apply, in which case (D.5) reduces to

$$\alpha D = \frac{2+K_0^2}{K_0^2}. \tag{D.16}$$

Additionally, from (D.12) we derive that the lowest-order x-dependence of the externally applied B_y should be

$$B_s(x) \approx \frac{mc}{e}\frac{\alpha K_0^2}{2\gamma_r}\left(1+\frac{2\alpha x}{2+K_0^2}\right). \tag{D.17}$$

D.1 Low-Gain Analysis

In this section we assume that FEL gain is small, and begin with a low-gain analysis that largely follows the work by Kroll *et al.* [5], albeit with more modern notation consistent with the rest of this book. We then try and connect the work of Ref. [5] to our discussion on low-gain FELs using some arguments taken from Ref. [6].

For a low-gain device where the change in energy is small, we can approximate the transverse motion as being described by

$$x_\beta'' + \tilde{k}_\beta^2 x_\eta = -D\eta'', \tag{D.18}$$

with $\tilde{k}_\beta = k_\beta/[1+\eta(0)]$ depending on the initial energy of the particle. The solution of the driven oscillator equation (D.18) is

$$x_\beta(z) = \begin{cases} x_{\beta+}^H(z) - \dfrac{D}{k_\beta}\displaystyle\int_0^z dz' \sin\!\left[\tilde{k}_\beta(z-z')\right]\eta''(z') & \text{if } z \geq 0 \\ x_\beta^H(z) & \text{if } z < 0. \end{cases} \tag{D.19}$$

Here, $x_\beta^H(z)$ is the homogeneous solution of the betatron motion and the index '+' is introduced to distinguish the region in the undulator where $z \geq 0$ from the force-free motion where $z < 0$. Note that since the total transverse coordinate $\bar{x}$ and its derivative $\bar{x}'$ is everywhere continuous while η' is not (outside of the undulator $\eta'(z<0)=0$ but $\eta'(z=0) \neq 0$ in general), the function $x'_\beta(z)$ is discontinuous at $z=0$. From this it follows that

$$x_{\beta+}^H(z) = x_\beta^H(z) - \frac{D\eta'(0)}{\tilde{k}_\beta}\sin\left(\tilde{k}_\beta z\right). \tag{D.20}$$

We integrate (D.19) by parts as follows:

$$\begin{aligned} x_\beta(z) &= x_{\beta+}^H(z) + \frac{D\eta'(0)}{\tilde{k}_\beta}\sin\left(\tilde{k}_\beta z\right) - D\int_0^z dz'\ \cos\left[\tilde{k}_\beta(z-z')\right]\eta'(z') \\ &= x_\beta^H(z) - D\int_0^z dz'\ \cos\left[\tilde{k}_\beta(z-z')\right]\eta'(z'). \end{aligned} \tag{D.21}$$

The equation of motion for the ponderomotive phase can be found by inserting the solution (D.21) into (D.6); we find that

$$\frac{d\theta}{dz} = -\frac{2k_u}{D}x_\beta^H(z) + 2k_u\int_0^z dz'\ \cos\left[\tilde{k}_\beta(z-z')\right]\eta'(z'). \tag{D.22}$$

Finally, we consider the FEL-induced energy change by a plane-wave laser field $E \sim e^{ik_1(1+\Delta\nu)(z-ct)} = e^{i(1+\Delta\nu)\theta - i\Delta\nu k_u z} \approx e^{i(\theta-\Delta\nu k_u z)}$,

$$\frac{d\eta}{dz} = -\frac{\epsilon}{2k_e L_u^2}\sin(\theta - \Delta\nu k_u z), \tag{D.23}$$

where $\epsilon \equiv eE_0[\mathrm{JJ}]N_u L_u/(2\gamma_r^2 mc^2)$ is the dimensionless field strength. Equations (D.22)–(D.23) are the "pendulum equations" appropriate for the transverse gradient undulator FEL. The detuning term on the far right of (D.22) vanishes at $z=0$ by design of the TGU, while it becomes nonzero as z increases because of the induced betatron motion.

To study the low-gain behavior we assume that $\epsilon \ll 1$ and solve the pendulum equations (D.22)–(D.23) to order ϵ^2 in a manner very similar to that of Section 3.3. Specifically, the solution can be found by expanding the dynamical variables as $\theta(z) = \theta_0 + \epsilon\theta_1 + \epsilon^2\theta_2$ and $\eta(z) = \eta_0 + \epsilon\eta_1 + \epsilon^2\eta_2$ and solving the resulting equations at each order of ϵ. Rather than repeat the steps, we merely write down the equation of the second-order energy loss after averaging over the uniformly distributed initial phase $\theta(z=0)$:

$$\begin{aligned} \left\langle \epsilon^2\frac{d\eta_2}{dz}\right\rangle &= -\frac{\epsilon^2}{4k_u L_u^4}\int_0^z dz'\int_0^{z'} dz''\ \Big\langle \cos\left[\tilde{k}_\beta(z'-z'')\right] \\ &\quad \times \sin[k_u\Delta\nu(z''-z) - \Theta_\beta(z'') + \Theta_\beta(z)]\Big\rangle, \end{aligned} \tag{D.24}$$

where we have defined

$$\Theta_\beta(z) \equiv \frac{2k_u}{D}\int_0^z dz'\, x_\beta^H(z'). \tag{D.25}$$

The change in electron energy is obtained by integrating (D.24) over the undulator length. However, this integral is significantly complicated due to the presence of the betatron terms Θ_β. For now, we consider the case $\Theta_\beta(z) = 0$ and later mention under what conditions this approximation applies.

Introducing the scaled frequency difference from resonance $x = \pi N_u \Delta\nu$, the energy change in the limit $\Theta_\beta \to 0$ is

$$\langle \Delta\eta(L_u)\rangle = \frac{e^2 E_0^2 [\mathrm{JJ}]^2}{4\gamma_r^4 (mc^2)^2}\frac{k_u L_u^3}{4} \times \left\langle \frac{\sin^2\!\left(x + \tilde{k}_\beta L_u/2\right)}{\tilde{k}_\beta L_u\left(x + \tilde{k}_\beta L_u/2\right)^2} - \frac{\sin^2\!\left(x - \tilde{k}_\beta L_u/2\right)}{\tilde{k}_\beta L_u\left(x - \tilde{k}_\beta L_u/2\right)^2}\right\rangle. \tag{D.26}$$

The energy loss now depends both on the scaled frequency difference from resonance $x = -\pi N_u \Delta\nu$ and the betatron phase advance $\tilde{k}_\beta L_u$. This is the same expression as that given by Kroll, and it reproduces the usual gain formula in terms of Madey's theorem in the limit $\tilde{k}_\beta L_u \to 0$.

Low Gain for "Strong" TGU Gradient Focusing

Here, we assume that the focusing provided by the gradient field is strong enough that an electron undergoes many betatron periods. In this limit $k_\beta L_u \gg 1$ and the gain characteristics are quite different than that discussed in Section 3.3.1. For example, the gain is maximized when the normalized detuning $x = \tilde{k}_\beta L_u/2$, and the resulting energy change is

$$k_\beta L_u/2 \gg 1: \ \min_x \langle \Delta\eta(L_u)\rangle = -\frac{e^2 E_0^2 [JJ]^2}{4\gamma_r^4 (mc^2)^2}\frac{k_u L_u^2}{4\tilde{k}_\beta}; \tag{D.27}$$

the ratio of TGU to "normal" FEL gain is therefore

$$\frac{G_{\mathrm{TGU}}}{G_{\mathrm{normal}}} \approx \frac{2}{k_\beta L_u} \ll 1. \tag{D.28}$$

Defining the energy acceptance by $\Delta\left(\tilde{k}_\beta L_u/2\right) \approx (k_\beta L_u/2)\Delta\eta = 1$, we find that the TGU energy acceptance is

$$(\Delta\eta)_{\mathrm{TGU}} = \frac{2}{k_\beta L_u}. \tag{D.29}$$

Similar considerations for the normal FEL yield $(\Delta\eta)_{\mathrm{normal}} = 1/k_u L_u$, so that the increase in energy acceptance due to the TGU is

$$R \equiv \frac{(\Delta\eta)_{\mathrm{TGU}}}{(\Delta\eta)_{\mathrm{normal}}} = \frac{2k_u}{k_\beta}. \tag{D.30}$$

However, in conjunction with the expanded energy acceptance (D.30), the TGU also requires a smaller emittance. This can be seen by considering the size of the neglected

$\Theta_\beta(z)$ terms due to the induced betatron motion. For these to be small so that the energy loss formula (D.24) applies, we require

$$\Theta_\beta(z) = \frac{2k_u}{D}\int_0^z dz'\, x_\beta^H(z') \sim \frac{2k_u}{D}\frac{\sigma_x}{k_\beta} < 1 \tag{D.31}$$

if $k_\beta L_u \gg 1$. Eliminating D in favor of k_β with (D.14) and using $\sigma_x = \sqrt{\varepsilon_x/k_\beta}$, we can rewrite (D.31) as

$$\varepsilon_x < \frac{\lambda_1}{4\pi}\frac{k_\beta}{k_u} = \frac{\lambda_1}{4\pi}\frac{2}{R}. \tag{D.32}$$

The price of increasing the energy acceptance by a factor R is a more stringent emittance requirement, where the emittance must now be less than $2/R$ times the minimum radiation emittance.

The emittance requirement (D.32) is typically very difficult to achieve at X-ray wavelengths. Fortunately, the limit $k_\beta L_u \gg 1$ does not usually apply to low-gain devices appropriate for X-ray generation.

Low Gain for "Weak" Gradients

The focusing provided by the TGU is a factor $\sim (1/k_u)(\partial K/\partial x)$ smaller than the natural undulator focusing derived in Section 5.2.1. Since at high energy and large K one can typically neglect the natural focusing to first order, in these cases one also has $k_\beta L_u \ll 1$. When k_β is small the TGU physics is much more comparable to that of a "normal" FEL; first, if $k_\beta L_u \ll 1$ the energy loss formula (D.24) is maximized when $x = \pi N_u \Delta\nu \approx 1.3$, in which case the energy loss is the same as in the usual low-gain FEL:

$$k_\beta L_u/2 \ll 1: \ \min_x \langle \Delta\eta(L_u)\rangle \approx -0.54\frac{e^2E_0^2[JJ]^2}{4\gamma_r^4(mc^2)^2}\frac{k_u L_u^3}{4}. \tag{D.33}$$

The energy loss above applies provided the betatron term $\Theta_\beta < 1$, which from (D.25) leads to

$$\frac{2}{D}\left[\sigma_x(0) + \frac{L_u}{2}\sigma_{x'}(0)\right] < \frac{1}{k_u L_u}. \tag{D.34}$$

This is the same condition that one would obtain by requiring the TGU phase advance (D.22) due to transverse motion be small, and replaces the usual condition $\sigma_\eta < 1/k_u L_u$. The price of this different requirement is an increase of the e-beam size along x, $\sigma_x \to (\sigma_x^2 + D^2\sigma_\eta^2)^{1/2}$, and a resulting decrease in gain. To quantify this effect, assume that the initial e-beam has an energy spread that is a factor R too large, i.e., the incoming beam has $\sigma_\eta = R/k_u L_u$. We arrange the subsequent dispersion and TGU such that $\sigma_x \sim D/k_u L_u$, which in turn implies that in the FEL the new beam size along x increases to $\sigma_x(1 + R^2)^{1/2}$. This then leads to a reduction in gain by

$$\frac{G_{\text{TGU}}}{G_{\text{normal}}} = \frac{1}{\sqrt{1+R^2}} \approx \frac{1}{R}. \tag{D.35}$$

Hence, using a TGU with $k_\beta L_u \ll 1$ to increase the energy acceptance by a factor of R decreases the FEL gain by an amount $1/R$. Note that the comparison here is to the gain

G_{normal} in the *zero* energy spread limit. Since one would only employ a TGU when the energy spread effect is large, $\sigma_\eta N_u \gg 1$, the gain G_{TGU} can be much larger than that with no TGU. To see this more clearly, we consider the effects of energy spread on the gain using the 1D gain formula equation (4.29). We generalize this formula to include different beam sizes in x and y, and set $\sigma_{x,y} \to \Sigma_{x,y}$ as implied by the 3D Equation (5.4.1) [6]:

$$G = -4\pi^2 \frac{I}{I_A} \frac{K^2[\text{JJ}]^2}{(1+K^2/2)^2} \frac{\gamma\lambda_1 N_u^3}{2\pi\,\Sigma_x\Sigma_y} \times \int_{-1/2}^{1/2} dzds\,(z-s)\sin[x_0(z-s)]e^{-2[2\pi N_u(z-s)\sigma_\eta]^2}. \tag{D.36}$$

If there is no TGU, G is largest when the electron beam size matches the radiation field size with Rayleigh range $Z_R \approx L_u/2\pi$:

$$\Sigma_x = \Sigma_y \approx \sqrt{2}\sigma_r = \sqrt{\frac{\lambda_1 Z_R}{2\pi}} \approx \frac{\sqrt{\lambda_1 L_u}}{2\pi}. \tag{D.37}$$

In that case the gain becomes

$$G = -QN_u^2 \int_{-1/2}^{1/2} dzds\,(z-s)\sin[x_0(z-s)]e^{-2[2\pi N_u(z-s)\sigma_\eta]^2}, \tag{D.38}$$

where Q is a constant that is independent of N_u. If the energy spread is negligible, then the integration gives the usual gain function $g(x_0)/2$, implying that $G \propto N_u^2$ when $(2\pi\sigma_\eta)^2 \ll 1/N_u^2$. In the opposite limit of a large energy spread, the integrand is negligible unless $|z-s| \lesssim 1/N_u\sigma_\eta$, so that the integral scales as $1/N_u^2\sigma_\eta^2$ and G becomes independent of N_u. A reasonably accurate fitting formula that matches these two limits is

$$G_{\text{FEL}} \approx \frac{Q}{0.27} \frac{N_u^2}{1+(5.46N_u\sigma_\eta)^2}. \tag{D.39}$$

On the other hand, we have seen that for a TGU-enabled FEL the energy spread is effectively replaced by the ratio of the beam size to the dispersion; this is reflected in the energy spread-like condition (D.34) when $\sigma_x' \approx 2\pi\sigma_x/L_u$. Taking also into account larger beam size, we can understand the TGU's gain properties by making the replacements $\sigma_\eta \to \sigma_x/D$ and $\Sigma_x \to (\sigma_x^2 + \sigma_r^2 + D^2\sigma_\eta^2)^{1/2} \approx D\sigma_\eta$ in the gain formula (D.36). Applying similar reasoning as that which led to (D.38) results in the TGU gain expression

$$G_{\text{TGU}} \approx \frac{Q}{0.27} \frac{\sqrt{2}N_u^2/\sigma_\eta^2}{D/\sigma_x + (5.46N_u)^2\sigma_x/D}. \tag{D.40}$$

For large energy spreads $N_u^2\sigma_\eta^2 \gg 1$ the ratio of the TGU gain to that of a standard FEL is given by

$$\frac{G_{\text{TGU}}}{G_{\text{FEL}}} \approx \frac{D\sigma_\eta}{\sigma_x} \frac{\sqrt{2}}{1+[D/(5.46N_u\sigma_x)]^2}. \tag{D.41}$$

We see that the TGU gain first increases with increasing dispersion D as the larger x-y correlation mitigates the variation of the FEL resonance condition. Once the energy

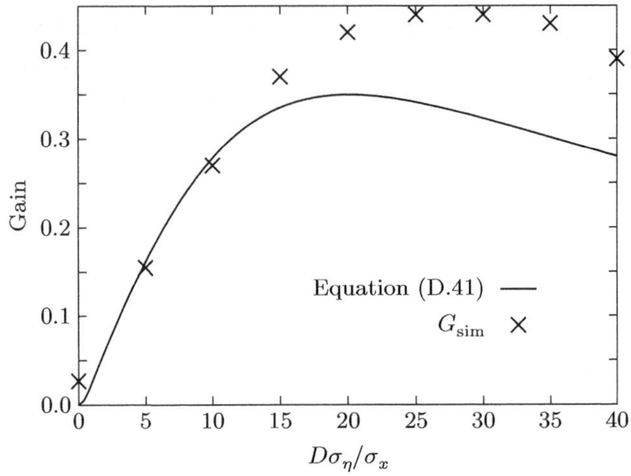

Figure D.2 TGU gain as a function of the dispersion for a normalized energy spread $2\pi N_u\sigma_\eta \approx 22$. The line shows the prediction (D.40), while the points are from FEL simulations including a gradient undulator.

spread is effectively cancelled, however, increasing the dispersion and e-beam size further reduces the FEL coupling along with the gain. This simple reasoning implies that a TGU can increase the gain in a long undulator by the amount $D\sigma_\eta/\sigma_x \gg 1$. We show an example that compares the predictions of Equation (D.40) to that of full FEL simulations in Figure D.2.

D.2 High-Gain Analysis

In the low-gain section we have shown that the betatron motion induced by the transverse gradient can significantly alter the TGU FEL dynamics. However, when $k_\beta L_u \ll 1$ the gain physics is quite similar to that of a traditional FEL whose phase variation is given by σ_x/D rather than σ_η, and whose transverse beam size is larger in x due to the dispersion. We will use these observations to derive some simple scalings of a high-gain TGU device when $k_\beta L_u \ll 1$.

We recall that the FEL Pierce parameter $\rho^3 \sim 1/(\sigma_x\sigma_y)$. Thus, increasing the beam size in a TGU leads to the effective FEL parameter

$$\rho_T = \frac{\rho}{(\sigma_T/\sigma_x)^{1/3}} \qquad \sigma_T = \sqrt{\sigma_x^2 + D^2\sigma_\eta^2}. \tag{D.42}$$

Furthermore, we can incorporate the phase advance σ_x/D as an effective energy spread

$$\sigma_\eta^{\rm eff} = \frac{\sigma_x}{D} = \frac{K_0^2\alpha\sigma_x}{2+K_0^2}, \tag{D.43}$$

where we have used the explicit TGU field (D.16). Recalling that the increase in gain length due to energy spread is an approximately quadratic function of σ_η/ρ that we wrote in (4.62), Equation (D.43) implies that the gain length of a TGU is well-described by [2]

$$L_G = \frac{\lambda_u}{4\pi\sqrt{3}\rho_T}\left[1+\left(\frac{\sigma_\eta^{\text{eff}}}{\rho_T}\right)^2\right] = \frac{\lambda_u}{4\pi\sqrt{3}\rho_T}\left[1+\frac{K_0^4\alpha^2\sigma_x^2}{\rho_T^2(2+K_0^2)^2}\right]$$
$$= \frac{\lambda_u}{4\pi\sqrt{3}\rho_T}\left[1+\frac{K_0^4\alpha^2\varepsilon_x L_u}{2\rho_T^2(2+K_0^2)^2}\right], \tag{D.44}$$

where we have approximated $\sigma_x^2 \approx \varepsilon_x L_u/2$ as is appropriate for an FEL with negligible natural focusing. This equation can also be approximated as

$$L_g^T \approx \frac{\lambda_u}{4\pi\sqrt{3}\rho}\left[\left(\frac{\sigma_T}{\sigma_x}\right)^{1/3} + \frac{\sigma_\eta^2}{\rho^2}\left(\frac{\sigma_T}{\sigma_x}\right)^{-1}\right]. \tag{D.45}$$

Let us suppose that we can optimize the dispersion to minimize the gain length while Equation (D.16) is satisfied. From Equation (D.45), we have the optimal dispersion and gain length

$$D \approx 2.28\frac{\sigma_x\sigma_\eta^{1/2}}{\rho^{3/2}}, \qquad L_g^T \approx 1.75\frac{\lambda_u}{4\pi\sqrt{3}\rho}\left(\frac{\sigma_\eta}{\rho}\right)^{1/2}. \tag{D.46}$$

Hence, with the appropriate parameters the TGU gain length increases as the square root of the initial energy spread. If we compare this to Equation (4.62) which predicts a quadratic dependence of the gain length on energy spread for a nominal undulator, we see that the effects of a large energy spread can be greatly reduced in a TGU.

A self-consistent theoretical analysis of a TGU-based, high-gain FEL which takes into account three-dimensional (3D) effects including beam size variations along the undulator has been presented in Ref. [4]. The calculated gain length compares favorably with simulations and also confirms the simple 1D theory for the optimum dispersion. Figure D.3 shows such a comparison using 1-GeV, soft X-ray FEL parameters used

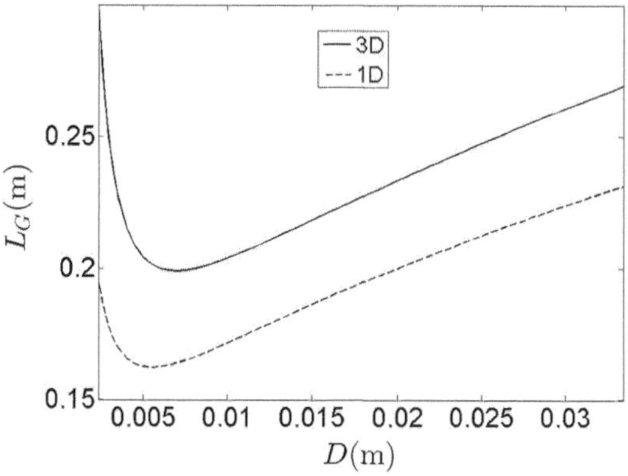

Figure D.3 Frequency-optimized gain length as a function of dispersion for parameters relevant to future laser plasma based accelerators (see Ref. [4] for more details). The data shown were derived using a 3D theory (solid) and the 1D formula of Equation (D.46) (dashed).

in Ref. [2]. Since the beam cross-section is highly asymmetric in a TGU, higher-order transverse modes can develop for such an FEL. Their effects on transverse coherence are analyzed in Ref. [7]. The degradation to the transverse coherence can be mitigated by choosing a dispersion function somewhat smaller than the optimal dispersion given in Equation (D.46).

References

[1] T. I. Smith, L. R. Elias, J. M. J. Madey, and D. A. G. Deacon, "Reducing the sensitivity of a free electron laser to electron energy," *J. Appl. Phys.*, vol. 50, p. 4580, 1979.

[2] Z. Huang, Y. Ding, and C. B. Schroeder, "Compact X-ray free-electron laser from a laser-plasma accelerator using a transverse-gradient undulator," *Phys. Rev. Lett.*, vol. 109, p. 204801, 2012.

[3] H. Deng and C. Feng, "Using off-resonance laser modulation for beam-energy-spread cooling in generation of short-wavelength radiation," *Phys. Rev. Lett.*, vol. 111, p. 084801, 2013.

[4] P. Baxevanis, Y. Ding, Z. Huang, and R. Ruth, "3D theory of a high-gain free-electron laser based on a transverse gradient undulator," *Phys. Rev. ST Accel. Beams*, vol. 17, p. 020701, 2014.

[5] N. M. Kroll, P. L. Morton, M. N. Rosenbluth, J. N. Eckstein, and J. M. J. Madey, "Theory of the transverse gradient wiggler," *IEEE J. Quantum Electron.*, vol. 17, p. 1496, 1981.

[6] R. R. Lindberg, K.-J. Kim, Y. Cai, Y. Ding, and Z. Huang, "Transverse gradient undulators for a storage ring X-ray FEL oscillator," in *Proceedings of FEL 2013*, p. 740, 2013.

[7] P. Baxevanis, Z. Huang, R. Ruth, and C. B. Schroeder, "Eigenmode analysis of a high-gain free-electron laser based on a transverse gradient undulator," *Phys. Rev. ST Accel. Beams*, vol. 18, p. 010701, 2015.

Further Reading

There are several other excellent books dealing with the topics discussed in this book. Books on synchrotron radiation include:

D. Attwood, *Soft X-rays and Extreme Ultraviolet Radiation*, Cambridge University Press, 1999. This book is mainly for X-ray optics but contains an introductory level discussion of synchrotron radiation and free-electron lasers.

H. Wiedemann, *Synchrotron Radiation*, Springer Verlag, 2002.

A. Hoffman, *The Physics of Synchrotron Radiation*, Cambridge University Press, 2003. A thorough treatment of angular distribution of synchrotron radiation.

H. Onuki and P. Ellaume, *Undulators, Wigglers, and Their Applications*, Taylor and Francis, 2003

Books on free-electron lasers include:

T. C. Marshall, *Free-Electron Lasers*, MacMillan, New York, 1985.

C. Brau, *Free-Electron Lasers*, Academic Press Inc., 1990

P. Luchini and H. Motz, *Undulators and Free-Electron Lasers*, Oxford University Press, 1990. Gives discussion of both the undulator radiation and low-gain as well as the high-gain FEL physics, emphasizing the fundamental principles as well as discussion of accelerators.

H. P. Freund and T. M. Antonsen, *Principles of Free-Electron Lasers*, Chapman and Hall, 1994. Contains detailed discussion of space-charge-dominant regime.

E. L. Saldin, E. A. Schneidmiller and M. V. Yurkov, *Physics of Free-Electron Lasers*, Springer, 2000. Gives a comprehensive treatment of high-gain FEL theory including 3D effects, except for the effect of the betatron oscillation.

P. Schmüser, M. Dohlus, J. Rossbach, and C. Behrens, *Free-Electron Lasers in the UV and X-Ray Regime*, Springer, 2014. Gives introductory-level discussion on high-gain FEL physics.

Review articles:

S. Krinsky, M. L. Perlman, and R. E. Watson, "Characteristics of Synchrotron Radiation and of Its Sources," in *Handbook of Synchrotron Radiation*, E. E. Koch, Ed., vol. 1. North Holland Publishing Co., pp. 65–171, 1983.

K.-J. Kim, "Characteristics of Synchrotron Radiation," in *Proc. US Particle Accelerator School*, M. Month and M. Dienes, Eds. AIP Conference Proceedings no. 184, pp. 565–632, 1989.

J. B. Murphy and C. Pellegrini, "Introduction to Physics of the Free-Electron Laser," in *Laser Handbook*, W. B. Colson, C. Pellegrini, and A. Renieri, Eds., vol. 6. Elsevier Science Publishers B.V., pp. 9–69, 1990.

W. B. Colson, "Classical Free-Electron Laser Theory," in *Laser Handbook*, W. B. Colson, C. Pellegrini, and A. Renieri, Eds., vol. 6. Elsevier Science Publishers B.V., pp. 115–194, 1990.

Z. Huang and K.-J. Kim, "Review of X-ray free-electron laser theory," *Phys. Rev. ST-AB*, vol. 10, p. 034801, 2007.

C. Pellegrini, A. Marinelli, and S. Reiche, "The physics of X-ray free-electron lasers," *Rev. Mod. Phys.* vol. 88, p. 015006, 2016.

Index